Saturn
Automotive
Repair
Manual

by Mark Ryan
and John H Haynes
Member of the Guild of Motoring Writers

Models covered:
All Saturn models
1991 through 1999

(12D11 - 87010)
(2083)

ABCDE
FGHIJ
KLM

Haynes Publishing Group
Sparkford Nr Yeovil
Somerset BA22 7JJ England

Haynes North America, Inc
861 Lawrence Drive
Newbury Park
California 91320 USA

Acknowledgements

We are grateful to Saturn of Thousand Oaks, who provided one of the vehicles used in certain photographs. Wiring diagrams were provided by Mitchell International. Technical writers who contributed to this project include Rob Maddox, Mike Stubblefield, Larry Warren and Tim Imhoff.

A book in the Haynes Automotive Repair Manual Series

Printed in the U.S.A.

ISBN 1 56392 333 5

Library of Congress Catalog Card Number 99-60418

Contents

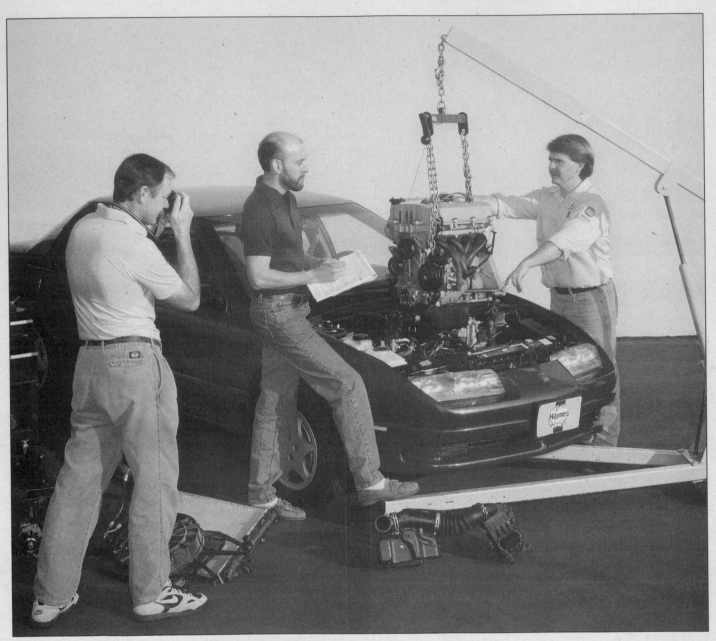

Haynes photographer, author and mechanic with 1991 Saturn

About this manual

Its purpose

The purpose of this manual is to help you get the best value from your vehicle. It can do so in several ways. It can help you decide what work must be done, even if you choose to have it done by a dealer service department or a repair shop; it provides information and procedures for routine maintenance and servicing; and it offers diagnostic and repair procedures to follow when trouble occurs.

We hope you use the manual to tackle the work yourself. For many simpler jobs, doing it yourself may be quicker than arranging an appointment to get the vehicle into a shop and making the trips to leave it and pick it up. More importantly, a lot of money can be saved by avoiding the expense the shop must pass on to you to cover its labor and overhead costs. An added benefit is the sense of satisfaction and accomplishment that you feel after doing the job yourself.

Using the manual

The manual is divided into Chapters. Each Chapter is divided into numbered Sections, which are headed in bold type between horizontal lines. Each Section consists of consecutively numbered paragraphs.

At the beginning of each numbered Section you will be referred to any illustrations which apply to the procedures in that Section. The reference numbers used in illustration captions pinpoint the pertinent Section and the Step within that Section. That is, illustration 3.2 means the illustration refers to Section 3 and Step (or paragraph) 2 within that Section.

Procedures, once described in the text, are not normally repeated. When it's necessary to refer to another Chapter, the reference will be given as Chapter and Section number. Cross references given without use of the word "Chapter" apply to Sections and/or paragraphs in the same Chapter. For example, "see Section 8" means in the same Chapter.

References to the left or right side of the vehicle assume you are sitting in the driver's seat, facing forward.

Even though we have prepared this manual with extreme care, neither the publisher nor the author can accept responsibility for any errors in, or omissions from, the information given.

NOTE

A **Note** provides information necessary to properly complete a procedure or information which will make the procedure easier to understand.

CAUTION

A **Caution** provides a special procedure or special steps which must be taken while completing the procedure where the Caution is found. Not heeding a Caution can result in damage to the assembly being worked on.

WARNING

A **Warning** provides a special procedure or special steps which must be taken while completing the procedure where the Warning is found. Not heeding a Warning can result in personal injury.

Introduction to the Saturn

Saturn models are available in four-door sedan and station wagon as well as two-door coupe body styles.

The transversely mounted Single Overhead Cam (SOHC) and Dual Overhead Cam (DOHC) four-cylinder engines are equipped with electronic fuel injection. The 1994 and earlier SOHC engine uses Throttle Body Injection (TBI), while the 1995 and later SOHC and all DOHC engines use the Multi Port Fuel Injection (MPFI) system.

The engine drives the front wheels through either a five-speed manual or a four-speed automatic transaxle via independent driveaxles.

Independent suspension, featuring coil spring/strut damper units, is used on all four wheels. The power-assisted rack-and-pinion steering unit is mounted behind the engine.

The brakes are disc at the front and drums or discs at the rear, with power assist standard.

Vehicle identification numbers

Modifications are a continuing and unpublicized process in vehicle manufacturing. Since spare parts lists and manuals are compiled on a numerical basis, the individual vehicle numbers are necessary to correctly identify the component required.

Vehicle Identification Number (VIN)

This very important identification number is stamped on a plate attached to the dashboard inside the windshield on the driver's side of the vehicle **(see illustration)**. The VIN also appears on the Vehicle Certificate of Title and Registration. It contains information such as where and when the vehicle was manufactured, the model year and the body style.

VIN engine and model year codes

Two particularly important pieces of information found in the VIN are the engine code and the model year code. Counting from the left, the engine code letter designation is the 8th digit and the model year code letter designation is the 10th digit.

On the models covered by this manual the engine codes are:

7 - 1.9L DOHC four-cylinder with Multi Port Fuel Injection (MPFI)
8 - 1.9L SOHC four-cylinder with Multi Port Fuel Injection (MPFI)
9 - 1.9L SOHC four-cylinder with Throttle Body Injection (TBI)

On the models covered by this manual the model year codes are:

M - 1991	**S** - 1995
N - 1992	**T** - 1996
P - 1993	**V** - 1997
R - 1994	**W** - 1998

Vehicle Certification Label

The Vehicle Certification Label is attached to the end of the driver's side door. Information on this label includes the name of the manufacturer, the month and year of production, the Gross Vehicle Weight Rating (GVWR), the Gross Axle Weight Rating (GAWR) and the certification statement.

Engine number

The engine number is stamped onto a machined pad at the rear edge of the engine block, where the engine mates with the transaxle.

Transaxle number

The transaxle number is located on the bellhousing, between the two upper transaxle-to-engine mounting bolts.

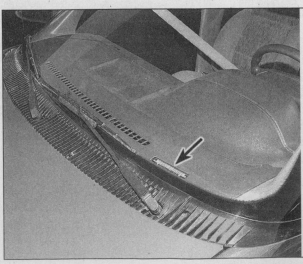

The Vehicle Identification Number (VIN) is important for identifying the vehicle and engine type - it is on the front of the dash, visible from outside the vehicle, looking through the windshield on the driver's side. The chart below shows the information conveyed by the VIN. The eighth digit identifies the engine and the tenth digit denotes model year - see the text for further explanation

Buying parts

Replacement parts are available from many sources, which generally fall into one of two categories - authorized dealer parts departments and independent retail auto parts stores. Our advice concerning these parts is as follows:

Retail auto parts stores: Good auto parts stores will stock frequently needed components which wear out relatively fast, such as clutch components, exhaust systems, brake parts, tune-up parts, etc. These stores often supply new or reconditioned parts on an exchange basis, which can save a considerable amount of money. Discount auto parts stores are often very good places to buy materials and parts needed for general vehicle maintenance such as oil, grease, filters, spark plugs, belts, touch-up paint, bulbs, etc. They also usually sell tools and general accessories, have convenient hours, charge lower prices and can often be found not far from home.

Authorized dealer parts department: This is the best source for parts which are unique to the vehicle and not generally available elsewhere (such as major engine parts, transmission parts, trim pieces, etc.).

Warranty information: If the vehicle is still covered under warranty, be sure that any replacement parts purchased - regardless of the source - do not invalidate the warranty!

To be sure of obtaining the correct parts, have engine and chassis numbers available and, if possible, take the old parts along for positive identification.

Maintenance techniques, tools and working facilities

Maintenance techniques

There are a number of techniques involved in maintenance and repair that will be referred to throughout this manual. Application of these techniques will enable the home mechanic to be more efficient, better organized and capable of performing the various tasks properly, which will ensure that the repair job is thorough and complete.

Fasteners

Fasteners are nuts, bolts, studs and screws used to hold two or more parts together. There are a few things to keep in mind when working with fasteners. Almost all of them use a locking device of some type, either a lockwasher, locknut, locking tab or thread adhesive. All threaded fasteners should be clean and straight, with undamaged threads and undamaged corners on the hex head where the wrench fits. Develop the habit of replacing all damaged nuts and bolts with new ones. Special locknuts with nylon or fiber inserts can only be used once. If they are removed, they lose their locking ability and must be replaced with new ones.

Rusted nuts and bolts should be treated with a penetrating fluid to ease removal and prevent breakage. Some mechanics use turpentine in a spout-type oil can, which works quite well. After applying the rust penetrant, let it work for a few minutes before trying to loosen the nut or bolt. Badly rusted fasteners may have to be chiseled or sawed off or removed with a special nut breaker, available at tool stores.

If a bolt or stud breaks off in an assembly, it can be drilled and removed with a special tool commonly available for this purpose. Most automotive machine shops can perform this task, as well as other repair procedures, such as the repair of threaded holes that have been stripped out.

Flat washers and lockwashers, when removed from an assembly, should always be replaced exactly as removed. Replace any damaged washers with new ones. Never use a lockwasher on any soft metal surface (such as aluminum), thin sheet metal or plastic.

Fastener sizes

For a number of reasons, automobile manufacturers are making wider and wider use of metric fasteners. Therefore, it is important to be able to tell the difference between standard (sometimes called U.S. or SAE) and metric hardware, since they cannot be interchanged.

All bolts, whether standard or metric, are sized according to diameter, thread pitch and length. For example, a standard 1/2 - 13 x 1 bolt is 1/2 inch in diameter, has 13 threads per inch and is 1 inch long. An M12 - 1.75 x 25 metric bolt is 12 mm in diameter, has a thread pitch of 1.75 mm (the distance between threads) and is 25 mm long. The two bolts are nearly identical, and easily confused, but they are not interchangeable.

In addition to the differences in diameter, thread pitch and length, metric and standard bolts can also be distinguished by examining the bolt heads. To begin with, the distance across the flats on a standard bolt head is measured in inches, while the same dimension on a metric bolt is sized in millimeters (the same is true for nuts). As a result, a standard wrench should not be used on a metric bolt and a metric wrench should not be used on a standard bolt. Also, most standard bolts have slashes radiating out from the center of the head to denote the grade or strength of the bolt, which is an indication of the amount of torque that can be applied to it. The greater the number of slashes, the greater the strength of the bolt. Grades 0 through 5 are commonly used on automobiles. Metric bolts have a property class (grade) number, rather than a slash, molded into their heads to indicate bolt strength. In this case, the higher the number, the stronger the bolt. Property class numbers 8.8, 9.8 and 10.9 are commonly used on automobiles.

Strength markings can also be used to distinguish standard hex nuts from metric hex nuts. Many standard nuts have dots stamped into one side, while metric nuts are marked with a number. The greater the number of dots, or the higher the number, the greater the strength of the nut.

Metric studs are also marked on their ends according to property class (grade). Larger studs are numbered (the same as metric bolts), while smaller studs carry a geometric code to denote grade.

It should be noted that many fasteners, especially Grades 0 through 2, have no distinguishing marks on them. When such is the case, the only way to determine whether it is standard or metric is to measure the thread pitch or compare it to a known fastener of the same size.

Standard fasteners are often referred to as SAE, as opposed to metric. However, it should be noted that SAE technically refers to a non-metric fine thread fastener only. Coarse thread non-metric fasteners are referred to as USS sizes.

Grade 1 or 2   Grade 5   Grade 8

Bolt strength marking (standard/SAE/USS; bottom - metric)

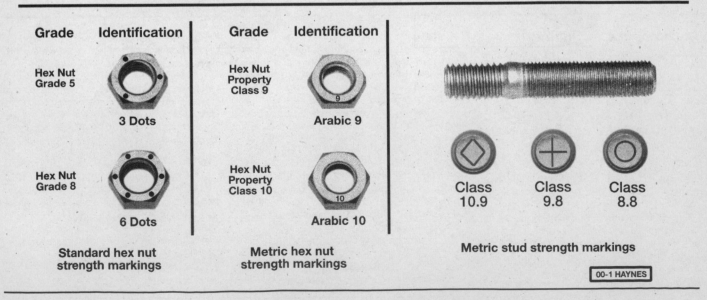

Grade	Identification
Hex Nut Grade 5	3 Dots
Hex Nut Grade 8	6 Dots

Standard hex nut strength markings

Grade	Identification
Hex Nut Property Class 9	Arabic 9
Hex Nut Property Class 10	Arabic 10

Metric hex nut strength markings

Class 10.9   Class 9.8   Class 8.8

Metric stud strength markings

Since fasteners of the same size (both standard and metric) may have different strength ratings, be sure to reinstall any bolts, studs or nuts removed from your vehicle in their original locations. Also, when replacing a fastener with a new one, make sure that the new one has a strength rating equal to or greater than the original.

Tightening sequences and procedures

Most threaded fasteners should be tightened to a specific torque value (torque is the twisting force applied to a threaded com-ponent such as a nut or bolt). Overtightening the fastener can weaken it and cause it to break, while undertightening can cause it to eventually come loose. Bolts, screws and studs, depending on the material they are made of and their thread diameters, have specific torque values, many of which are noted in the Specifications at the beginning of each Chapter. Be sure to follow the torque recommendations closely. For fasteners not assigned a specific torque, a general torque value chart is presented here as a guide. These torque values are for dry (unlubricated) fasteners threaded into steel or cast iron (not aluminum). As was previously mentioned, the size and grade of a fastener determine the amount of torque that can safely be applied to it. The figures listed here are approximate for Grade 2 and Grade 3 fasteners. Higher grades can tolerate higher torque values.

Fasteners laid out in a pattern, such as cylinder head bolts, oil pan bolts, differential cover bolts, etc., must be loosened or tight-ened in sequence to avoid warping the com-ponent. This sequence will normally be shown in the appropriate Chapter. If a spe-cific pattern is not given, the following proce-dures can be used to prevent warping.

Metric thread sizes	Ft-lbs	Nm
M-6	6 to 9	9 to 12
M-8	14 to 21	19 to 28
M-10	28 to 40	38 to 54
M-12	50 to 71	68 to 96
M-14	80 to 140	109 to 154

Pipe thread sizes		
1/8	5 to 8	7 to 10
1/4	12 to 18	17 to 24
3/8	22 to 33	30 to 44
1/2	25 to 35	34 to 47

U.S. thread sizes		
1/4 - 20	6 to 9	9 to 12
5/16 - 18	12 to 18	17 to 24
5/16 - 24	14 to 20	19 to 27
3/8 - 16	22 to 32	30 to 43
3/8 - 24	27 to 38	37 to 51
7/16 - 14	40 to 55	55 to 74
7/16 - 20	40 to 60	55 to 81
1/2 - 13	55 to 80	75 to 108

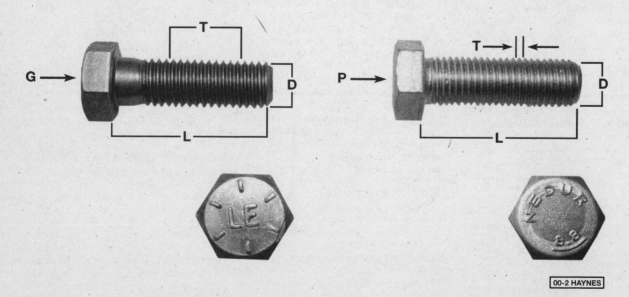

Standard (SAE and USS) bolt dimensions/grade marks

- G   Grade marks (bolt strength)
- L   Length (in inches)
- T   Thread pitch (number of threads per inch)
- D   Nominal diameter (in inches)

Metric bolt dimensions/grade marks

- P   Property class (bolt strength)
- L   Length (in millimeters)
- T   Thread pitch (distance between threads in millimeters)
- D   Diameter

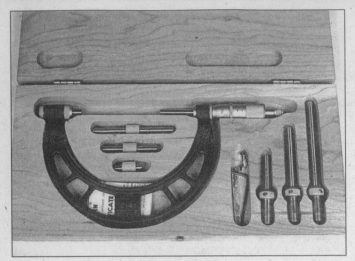

Micrometer set

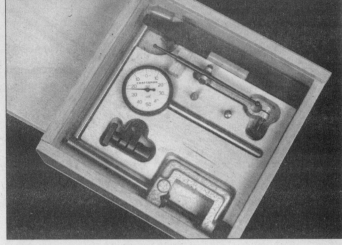

Dial indicator set

Initially, the bolts or nuts should be assembled finger-tight only. Next, they should be tightened one full turn each, in a criss-cross or diagonal pattern. After each one has been tightened one full turn, return to the first one and tighten them all one-half turn, following the same pattern. Finally, tighten each of them one-quarter turn at a time until each fastener has been tightened to the proper torque. To loosen and remove the fasteners, the procedure would be reversed.

Component disassembly

Component disassembly should be done with care and purpose to help ensure that the parts go back together properly. Always keep track of the sequence in which parts are removed. Make note of special characteristics or marks on parts that can be installed more than one way, such as a grooved thrust washer on a shaft. It is a good idea to lay the disassembled parts out on a clean surface in the order that they were removed. It may also be helpful to make sketches or take instant photos of components before removal.

When removing fasteners from a component, keep track of their locations. Sometimes threading a bolt back in a part, or putting the washers and nut back on a stud, can prevent mix-ups later. If nuts and bolts cannot be returned to their original locations, they should be kept in a compartmented box or a series of small boxes. A cupcake or muffin tin is ideal for this purpose, since each cavity can hold the bolts and nuts from a particular area (i.e. oil pan bolts, valve cover bolts, engine mount bolts, etc.). A pan of this type is especially helpful when working on assemblies with very small parts, such as the carburetor, alternator, valve train or interior dash and trim pieces. The cavities can be marked with paint or tape to identify the contents.

Whenever wiring looms, harnesses or connectors are separated, it is a good idea to identify the two halves with numbered pieces of masking tape so they can be easily reconnected.

Gasket sealing surfaces

Throughout any vehicle, gaskets are used to seal the mating surfaces between two parts and keep lubricants, fluids, vacuum or pressure contained in an assembly.

Many times these gaskets are coated with a liquid or paste-type gasket sealing compound before assembly. Age, heat and pressure can sometimes cause the two parts to stick together so tightly that they are very difficult to separate. Often, the assembly can be loosened by striking it with a soft-face hammer near the mating surfaces. A regular hammer can be used if a block of wood is placed between the hammer and the part. Do not hammer on cast parts or parts that could be easily damaged. With any particularly stubborn part, always recheck to make sure that every fastener has been removed.

Avoid using a screwdriver or bar to pry apart an assembly, as they can easily mar the gasket sealing surfaces of the parts, which must remain smooth. If prying is absolutely necessary, use an old broom handle, but keep in mind that extra clean up will be necessary if the wood splinters.

After the parts are separated, the old gasket must be carefully scraped off and the gasket surfaces cleaned. Stubborn gasket material can be soaked with rust penetrant or treated with a special chemical to soften it so it can be easily scraped off. A scraper can be fashioned from a piece of copper tubing by flattening and sharpening one end. Copper is recommended because it is usually softer than the surfaces to be scraped, which reduces the chance of gouging the part. Some gaskets can be removed with a wire brush, but regardless of the method used, the mating surfaces must be left clean and smooth. If for some reason the gasket surface is gouged, then a gasket sealer thick enough to fill scratches will have to be used during reassembly of the components. For most applications, a non-drying (or semi-drying) gasket sealer should be used.

Hose removal tips

Warning: *If the vehicle is equipped with air conditioning, do not disconnect any of the A/C hoses without first having the system depressurized by a dealer service department or a service station.*

Hose removal precautions closely parallel gasket removal precautions. Avoid scratching or gouging the surface that the hose mates against or the connection may leak. This is especially true for radiator hoses. Because of various chemical reactions, the rubber in hoses can bond itself to the metal spigot that the hose fits over. To remove a hose, first loosen the hose clamps that secure it to the spigot. Then, with slip-joint pliers, grab the hose at the clamp and rotate it around the spigot. Work it back and forth until it is completely free, then pull it off. Silicone or other lubricants will ease removal if they can be applied between the hose and the outside of the spigot. Apply the same lubricant to the inside of the hose and the outside of the spigot to simplify installation.

As a last resort (and if the hose is to be replaced with a new one anyway), the rubber can be slit with a knife and the hose peeled from the spigot. If this must be done, be careful that the metal connection is not damaged.

If a hose clamp is broken or damaged, do not reuse it. Wire-type clamps usually weaken with age, so it is a good idea to replace them with screw-type clamps whenever a hose is removed.

Tools

A selection of good tools is a basic requirement for anyone who plans to maintain and repair his or her own vehicle. For the owner who has few tools, the initial investment might seem high, but when compared to the spiraling costs of professional auto maintenance and repair, it is a wise one.

To help the owner decide which tools are needed to perform the tasks detailed in this manual, the following tool lists are offered: *Maintenance and minor repair,*

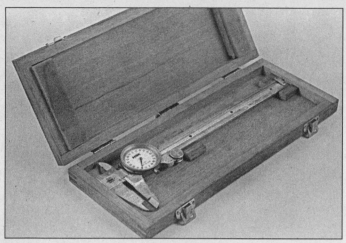

Dial caliper

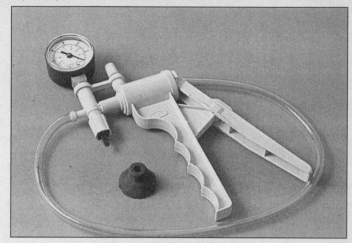

Hand-operated vacuum pump

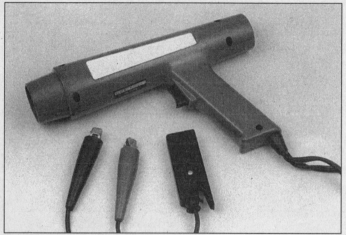

Timing light

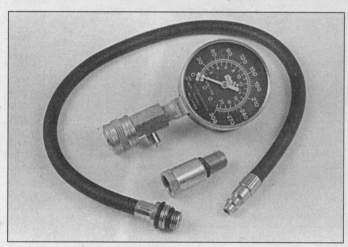

Compression gauge with spark plug hole adapter

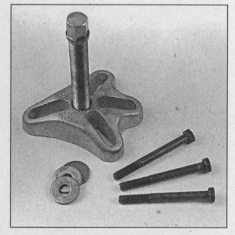

Damper/steering wheel puller

General purpose puller

Hydraulic lifter removal tool

Repair/overhaul and *Special.*

The newcomer to practical mechanics should start off with the *maintenance and minor repair* tool kit, which is adequate for the simpler jobs performed on a vehicle. Then, as confidence and experience grow, the owner can tackle more difficult tasks, buying additional tools as they are needed.

Eventually the basic kit will be expanded into the *repair and overhaul* tool set. Over a period of time, the experienced do-it-yourselfer will assemble a tool set complete enough for most repair and overhaul procedures and will add tools from the special category when it is felt that the expense is justified by the frequency of use.

Maintenance and minor repair tool kit

The tools in this list should be considered the minimum required for performance of routine maintenance, servicing and minor repair work. We recommend the purchase of combination wrenches (box-end and open-

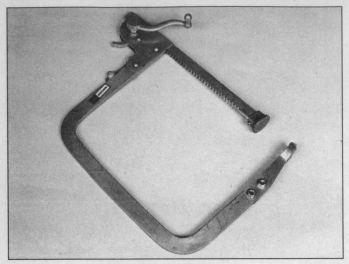

Valve spring compressor

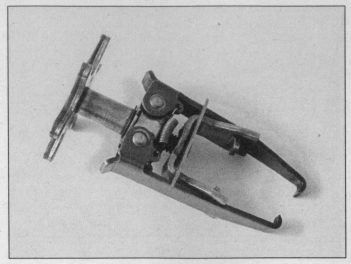

Valve spring compressor

Ridge reamer

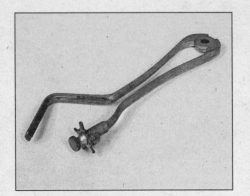

Piston ring groove cleaning tool

Ring removal/installation tool

end combined in one wrench). While more expensive than open end wrenches, they offer the advantages of both types of wrench.

> Combination wrench set (1/4-inch to
>   1 inch or 6 mm to 19 mm)
> Adjustable wrench, 8 inch
> Spark plug wrench with rubber insert
> Spark plug gap adjusting tool
> Feeler gauge set
> Brake bleeder wrench
> Standard screwdriver (5/16-inch x
>   6 inch)
> Phillips screwdriver (No. 2 x 6 inch)
> Combination pliers - 6 inch
> Hacksaw and assortment of blades
> Tire pressure gauge
> Grease gun
> Oil can
> Fine emery cloth
> Wire brush
> Battery post and cable cleaning tool
> Oil filter wrench
> Funnel (medium size)
> Safety goggles
> Jackstands (2)
> Drain pan

Note: If basic tune-ups are going to be part of routine maintenance, it will be necessary to purchase a good quality stroboscopic timing

light and combination tachometer/dwell meter. Although they are included in the list of special tools, it is mentioned here because they are absolutely necessary for tuning most vehicles properly.

Repair and overhaul tool set

These tools are essential for anyone who plans to perform major repairs and are in addition to those in the maintenance and minor repair tool kit. Included is a comprehensive set of sockets which, though expensive, are invaluable because of their versatility, especially when various extensions and drives are available. We recommend the 1/2-inch drive over the 3/8-inch drive. Although the larger drive is bulky and more expensive, it has the capacity of accepting a very wide range of large sockets. Ideally, however, the mechanic should have a 3/8-inch drive set and a 1/2-inch drive set.

> Socket set(s)
> Reversible ratchet
> Extension - 10 inch
> Universal joint
> Torque wrench (same size drive as
>   sockets)
> Ball peen hammer - 8 ounce
> Soft-face hammer (plastic/rubber)

Ring compressor

> Standard screwdriver (1/4-inch x 6 inch)
> Standard screwdriver (stubby -
>   5/16-inch)
> Phillips screwdriver (No. 3 x 8 inch)
> Phillips screwdriver (stubby - No. 2)
> Pliers - vise grip
> Pliers - lineman's
> Pliers - needle nose
> Pliers - snap-ring (internal and external)
> Cold chisel - 1/2-inch

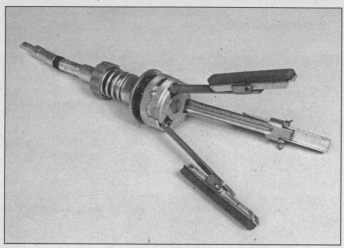

Cylinder hone

Brake hold-down spring tool

Scribe
Scraper (made from flattened copper
    tubing)
Centerpunch
Pin punches (1/16, 1/8, 3/16-inch)
Steel rule/straightedge -- 12 inch
Allen wrench set (1/8 to 3/8-inch or
    4 mm to 10 mm)
A selection of files

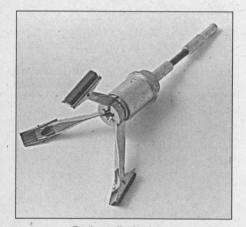

Brake cylinder hone

Wire brush (large)
Jackstands (second set)
Jack (scissor or hydraulic type)

Note: Another tool which is often useful is an electric drill with a chuck capacity of 3/8-inch and a set of good quality drill bits.

Special tools

The tools in this list include those which are not used regularly, are expensive to buy, or which need to be used in accordance with their manufacturer's instructions. Unless these tools will be used frequently, it is not very economical to purchase many of them. A consideration would be to split the cost and use between yourself and a friend or friends. In addition, most of these tools can be obtained from a tool rental shop on a temporary basis.

This list primarily contains only those tools and instruments widely available to the public, and not those special tools produced by the vehicle manufacturer for distribution to dealer service departments. Occasionally, references to the manufacturer's special tools are included in the text of this manual. Generally, an alternative method of doing the job without the special tool is offered. How-

ever, sometimes there is no alternative to their use. Where this is the case, and the tool cannot be purchased or borrowed, the work should be turned over to the dealer service department or an automotive repair shop.

Valve spring compressor
Piston ring groove cleaning tool
Piston ring compressor
Piston ring installation tool
Cylinder compression gauge
Cylinder ridge reamer
Cylinder surfacing hone
Cylinder bore gauge
Micrometers and/or dial calipers
Hydraulic lifter removal tool
Balljoint separator
Universal-type puller
Impact screwdriver
Dial indicator set
Stroboscopic timing light (inductive
    pick-up)
Hand operated vacuum/pressure pump
Tachometer/dwell meter
Universal electrical multimeter
Cable hoist
Brake spring removal and installation
    tools
Floor jack

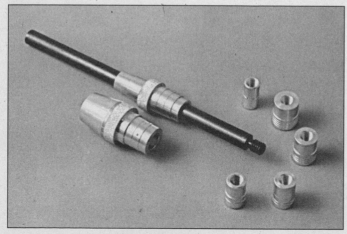

Clutch plate alignment tool

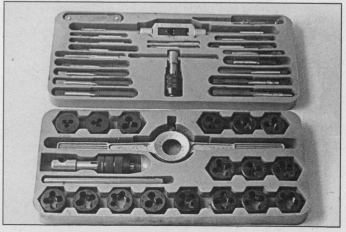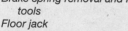

Tap and die set

Buying tools

For the do-it-yourselfer who is just starting to get involved in vehicle maintenance and repair, there are a number of options available when purchasing tools. If maintenance and minor repair is the extent of the work to be done, the purchase of individual tools is satisfactory. If, on the other hand, extensive work is planned, it would be a good idea to purchase a modest tool set from one of the large retail chain stores. A set can usually be bought at a substantial savings over the individual tool prices, and they often come with a tool box. As additional tools are needed, add-on sets, individual tools and a larger tool box can be purchased to expand the tool selection. Building a tool set gradually allows the cost of the tools to be spread over a longer period of time and gives the mechanic the freedom to choose only those tools that will actually be used.

Tool stores will often be the only source of some of the special tools that are needed, but regardless of where tools are bought, try to avoid cheap ones, especially when buying screwdrivers and sockets, because they won't last very long. The expense involved in replacing cheap tools will eventually be greater than the initial cost of quality tools.

Care and maintenance of tools

Good tools are expensive, so it makes sense to treat them with respect. Keep them clean and in usable condition and store them properly when not in use. Always wipe off any dirt, grease or metal chips before putting them away. Never leave tools lying around in the work area. Upon completion of a job, always check closely under the hood for tools that may have been left there so they won't get lost during a test drive.

Some tools, such as screwdrivers, pliers, wrenches and sockets, can be hung on a panel mounted on the garage or workshop wall, while others should be kept in a tool box or tray. Measuring instruments, gauges, meters, etc. must be carefully stored where they cannot be damaged by weather or impact from other tools.

When tools are used with care and stored properly, they will last a very long time. Even with the best of care, though, tools will wear out if used frequently. When a tool is damaged or worn out, replace it. Subsequent jobs will be safer and more enjoyable if you do.

How to repair damaged threads

Sometimes, the internal threads of a nut or bolt hole can become stripped, usually from overtightening. Stripping threads is an all-too-common occurrence, especially when working with aluminum parts, because aluminum is so soft that it easily strips out.

Usually, external or internal threads are only partially stripped. After they've been cleaned up with a tap or die, they'll still work. Sometimes, however, threads are badly damaged. When this happens, you've got three choices:

1) *Drill and tap the hole to the next suitable oversize and install a larger diameter bolt, screw or stud.*

2) *Drill and tap the hole to accept a threaded plug, then drill and tap the plug to the original screw size. You can also buy a plug already threaded to the original size. Then you simply drill a hole to the specified size, then run the threaded plug into the hole with a bolt and jam nut. Once the plug is fully seated, remove the jam nut and bolt.*

3) *The third method uses a patented thread repair kit like Heli-Coil or Slimsert. These easy-to-use kits are designed to repair damaged threads in straight-through holes and blind holes. Both are available as kits which can handle a variety of sizes and thread patterns. Drill the hole, then tap it with the special included tap. Install the Heli-Coil and the hole is back to its original diameter and thread pitch.*

Regardless of which method you use, be sure to proceed calmly and carefully. A little impatience or carelessness during one of these relatively simple procedures can ruin your whole day's work and cost you a bundle if you wreck an expensive part.

Working facilities

Not to be overlooked when discussing tools is the workshop. If anything more than routine maintenance is to be carried out, some sort of suitable work area is essential.

It is understood, and appreciated, that many home mechanics do not have a good workshop or garage available, and end up removing an engine or doing major repairs outside. It is recommended, however, that the overhaul or repair be completed under the cover of a roof.

A clean, flat workbench or table of comfortable working height is an absolute necessity. The workbench should be equipped with a vise that has a jaw opening of at least four inches.

As mentioned previously, some clean, dry storage space is also required for tools, as well as the lubricants, fluids, cleaning solvents, etc. which soon become necessary.

Sometimes waste oil and fluids, drained from the engine or cooling system during normal maintenance or repairs, present a disposal problem. To avoid pouring them on the ground or into a sewage system, pour the used fluids into large containers, seal them with caps and take them to an authorized disposal site or recycling center. Plastic jugs, such as old antifreeze containers, are ideal for this purpose.

Always keep a supply of old newspapers and clean rags available. Old towels are excellent for mopping up spills. Many mechanics use rolls of paper towels for most work because they are readily available and disposable. To help keep the area under the vehicle clean, a large cardboard box can be cut open and flattened to protect the garage or shop floor.

Whenever working over a painted surface, such as when leaning over a fender to service something under the hood, always cover it with an old blanket or bedspread to protect the finish. Vinyl covered pads, made especially for this purpose, are available at auto parts stores.

Booster battery (jump) starting

Observe the following precautions when using a booster battery to start a vehicle:

a) *Before connecting the booster battery, make sure the ignition switch is in the Off position.*

b) *Ensure that all electrical equipment (lights, heater, wipers etc.) are switched off.*

c) *Make sure that the booster battery is the same voltage as the discharged battery in the vehicle.*

d) *If the battery is being jump started from the battery in another vehicle, the two vehicles MUST NOT TOUCH each other.*

e) *Make sure the transaxle is in Neutral (manual transaxle) or Park (automatic transaxle).*

f) *Wear eye protection when jump starting a vehicle.*

Connect one jumper lead between the positive (+) terminals of the two batteries. Connect the other jumper lead first to the negative (-) terminal of the booster battery, then to a good engine ground on the vehicle to be started **(see illustration)**. Attach the lead at least 18 inches from the battery, if possible. Make sure that the jumper leads will not contact the fan, drivebelt of other moving parts of the engine.

Start the engine using the booster battery and allow the engine idle speed to stabilize. Disconnect the jumper leads in the reverse order of connection.

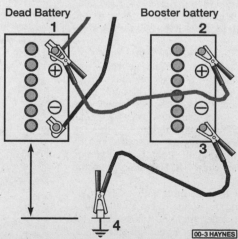

Make the booster battery cable connections in the numerical order shown (note that the negative cable of the booster battery is NOT attached to the negative terminal of the dead battery)

Jacking and towing

Jacking

Warning: *The jack supplied with the vehicle should only be used for changing a tire or placing jackstands under the frame. Never work under the vehicle or start the engine while this jack is being used as the only means of support.*

The vehicle should be on level ground. Place the shift lever in Park, if you have an automatic, or Reverse if you have a manual transaxle. Block the wheel diagonally opposite the wheel being changed. Set the parking brake.

Remove the spare tire and jack from stowage. Remove the wheel cover and trim ring (if so equipped) with the tapered end of the lug nut wrench by inserting and twisting the handle and then prying against the back of the wheel cover. Loosen, but do not remove, the lug nuts (one-half turn is sufficient).

Place the scissors-type jack under the side of the vehicle and adjust the jack height until it engages with the vertical rocker panel flange nearest the wheel to be changed Position the jack as close to the wheel opening as possible, but make sure the entire jack head is making contact with the flange.

Turn the jack handle clockwise until the tire clears the ground. Remove the lug nuts and pull the wheel off. Replace it with the spare.

Install the lug nuts with the beveled edges facing in. Tighten them snugly. Don't attempt to tighten them completely until the vehicle is lowered or it could slip off the jack. Turn the jack handle counterclockwise to lower the vehicle. Remove the jack and tighten the lug nuts in a criss-cross pattern.

Install the cover (and trim ring, if used)

and be sure it's snapped into place all the way around.

Stow the tire, jack and wrench. Unblock the wheels.

Towing

As a general rule, the vehicle should be towed with the front (drive) wheels off the ground. If they can't be raised, place them on a dolly. The ignition key must be in the OFF position, since the steering lock mechanism isn't strong enough to hold the front wheels straight while towing.

Vehicles equipped with an automatic transaxle can be towed from the front only with all four wheels on the ground, provided that speeds don't exceed 30 mph and the distance is not over 50 miles. Before towing, check the transaxle fluid level (see Chapter 1). If the level is below the HOT line on the

dipstick, add fluid or use a towing dolly. Release the parking brake, put the transaxle in Neutral and place the ignition key in the OFF position. **Caution:** *Never tow a vehicle with an automatic transaxle from the rear with the front wheels on the ground.*

When towing a vehicle equipped with a manual transaxle with all four wheels on the ground, be sure to place the shift lever in neutral and release the parking brake.

Equipment specifically designed for towing should be used. It should be attached to the main structural members of the vehicle, not the bumpers or brackets.

Safety is a major consideration when towing and all applicable state and local laws must be obeyed. A safety chain system must be used at all times. Remember that power steering and power brakes will not work with the engine off.

Vehicle jacking points

Automotive chemicals and lubricants

A number of automotive chemicals and lubricants are available for use during vehicle maintenance and repair. They include a wide variety of products ranging from cleaning solvents and degreasers to lubricants and protective sprays for rubber, plastic and vinyl.

Cleaners

Carburetor cleaner and choke cleaner is a strong solvent for gum, varnish and carbon. Most carburetor cleaners leave a dry-type lubricant film which will not harden or gum up. Because of this film it is not recommended for use on electrical components.

Brake system cleaner is used to remove grease and brake fluid from the brake system, where clean surfaces are absolutely necessary. It leaves no residue and often eliminates brake squeal caused by contaminants.

Electrical cleaner removes oxidation, corrosion and carbon deposits from electrical contacts, restoring full current flow. It can also be used to clean spark plugs, carburetor jets, voltage regulators and other parts where an oil-free surface is desired.

Demoisturants remove water and moisture from electrical components such as alternators, voltage regulators, electrical connectors and fuse blocks. They are non-conductive, non-corrosive and non-flammable.

Degreasers are heavy-duty solvents used to remove grease from the outside of the engine and from chassis components. They can be sprayed or brushed on and, depending on the type, are rinsed off either with water or solvent.

Lubricants

Motor oil is the lubricant formulated for use in engines. It normally contains a wide variety of additives to prevent corrosion and reduce foaming and wear. Motor oil comes in various weights (viscosity ratings) from 0 to 50. The recommended weight of the oil depends on the season, temperature and the demands on the engine. Light oil is used in cold climates and under light load conditions. Heavy oil is used in hot climates and where high loads are encountered. Multi-viscosity oils are designed to have characteristics of both light and heavy oils and are available in a number of weights from 5W-20 to 20W-50.

Gear oil is designed to be used in differentials, manual transmissions and other areas where high-temperature lubrication is required.

Chassis and wheel bearing grease is a heavy grease used where increased loads and friction are encountered, such as for wheel bearings, balljoints, tie-rod ends and universal joints.

High-temperature wheel bearing grease is designed to withstand the extreme temperatures encountered by wheel bearings

in disc brake equipped vehicles. It usually contains molybdenum disulfide (moly), which is a dry-type lubricant.

White grease is a heavy grease for metal-to-metal applications where water is a problem. White grease stays soft under both low and high temperatures (usually from -100 to +190-degrees F), and will not wash off or dilute in the presence of water.

Assembly lube is a special extreme pressure lubricant, usually containing moly, used to lubricate high-load parts (such as main and rod bearings and cam lobes) for initial start-up of a new engine. The assembly lube lubricates the parts without being squeezed out or washed away until the engine oiling system begins to function.

Silicone lubricants are used to protect rubber, plastic, vinyl and nylon parts.

Graphite lubricants are used where oils cannot be used due to contamination problems, such as in locks. The dry graphite will lubricate metal parts while remaining uncontaminated by dirt, water, oil or acids. It is electrically conductive and will not foul electrical contacts in locks such as the ignition switch.

Moly penetrants loosen and lubricate frozen, rusted and corroded fasteners and prevent future rusting or freezing.

Heat-sink grease is a special electrically non-conductive grease that is used for mounting electronic ignition modules where it is essential that heat is transferred away from the module.

Sealants

RTV sealant is one of the most widely used gasket compounds. Made from silicone, RTV is air curing, it seals, bonds, waterproofs, fills surface irregularities, remains flexible, doesn't shrink, is relatively easy to remove, and is used as a supplementary sealer with almost all low and medium temperature gaskets.

Anaerobic sealant is much like RTV in that it can be used either to seal gaskets or to form gaskets by itself. It remains flexible, is solvent resistant and fills surface imperfections. The difference between an anaerobic sealant and an RTV-type sealant is in the curing. RTV cures when exposed to air, while an anaerobic sealant cures only in the absence of air. This means that an anaerobic sealant cures only after the assembly of parts, sealing them together.

Thread and pipe sealant is used for sealing hydraulic and pneumatic fittings and vacuum lines. It is usually made from a Teflon compound, and comes in a spray, a paint-on liquid and as a wrap-around tape.

Chemicals

Anti-seize compound prevents seizing, galling, cold welding, rust and corrosion in

fasteners. High-temperature anti-seize, usually made with copper and graphite lubricants, is used for exhaust system and exhaust manifold bolts.

Anaerobic locking compounds are used to keep fasteners from vibrating or working loose and cure only after installation, in the absence of air. Medium strength locking compound is used for small nuts, bolts and screws that may be removed later. High-strength locking compound is for large nuts, bolts and studs which aren't removed on a regular basis.

Oil additives range from viscosity index improvers to chemical treatments that claim to reduce internal engine friction. It should be noted that most oil manufacturers caution against using additives with their oils.

Gas additives perform several functions, depending on their chemical makeup. They usually contain solvents that help dissolve gum and varnish that build up on carburetor, fuel injection and intake parts. They also serve to break down carbon deposits that form on the inside surfaces of the combustion chambers. Some additives contain upper cylinder lubricants for valves and piston rings, and others contain chemicals to remove condensation from the gas tank.

Miscellaneous

Brake fluid is specially formulated hydraulic fluid that can withstand the heat and pressure encountered in brake systems. Care must be taken so this fluid does not come in contact with painted surfaces or plastics. An opened container should always be resealed to prevent contamination by water or dirt.

Weatherstrip adhesive is used to bond weatherstripping around doors, windows and trunk lids. It is sometimes used to attach trim pieces.

Undercoating is a petroleum-based, tar-like substance that is designed to protect metal surfaces on the underside of the vehicle from corrosion. It also acts as a sound-deadening agent by insulating the bottom of the vehicle.

Waxes and polishes are used to help protect painted and plated surfaces from the weather. Different types of paint may require the use of different types of wax and polish. Some polishes utilize a chemical or abrasive cleaner to help remove the top layer of oxidized (dull) paint on older vehicles. In recent years many non-wax polishes that contain a wide variety of chemicals such as polymers and silicones have been introduced. These non-wax polishes are usually easier to apply and last longer than conventional waxes and polishes.

Conversion factors

Length (distance)
Inches (in)	X 25.4	= Millimetres (mm)	X 0.0394	= Inches (in)
Feet (ft)	X 0.305	= Metres (m)	X 3.281	= Feet (ft)
Miles	X 1.609	= Kilometres (km)	X 0.621	= Miles

Volume (capacity)
Cubic inches (cu in; in³)	X 16.387	= Cubic centimetres (cc; cm³)	X 0.061	= Cubic inches (cu in; in³)
Imperial pints (Imp pt)	X 0.568	= Litres (l)	X 1.76	= Imperial pints (Imp pt)
Imperial quarts (Imp qt)	X 1.137	= Litres (l)	X 0.88	= Imperial quarts (Imp qt)
Imperial quarts (Imp qt)	X 1.201	= US quarts (US qt)	X 0.833	= Imperial quarts (Imp qt)
US quarts (US qt)	X 0.946	= Litres (l)	X 1.057	= US quarts (US qt)
Imperial gallons (Imp gal)	X 4.546	= Litres (l)	X 0.22	= Imperial gallons (Imp gal)
Imperial gallons (Imp gal)	X 1.201	= US gallons (US gal)	X 0.833	= Imperial gallons (Imp gal)
US gallons (US gal)	X 3.785	= Litres (l)	X 0.264	= US gallons (US gal)

Mass (weight)
Ounces (oz)	X 28.35	= Grams (g)	X 0.035	= Ounces (oz)
Pounds (lb)	X 0.454	= Kilograms (kg)	X 2.205	= Pounds (lb)

Force
Ounces-force (ozf; oz)	X 0.278	= Newtons (N)	X 3.6	= Ounces-force (ozf; oz)
Pounds-force (lbf; lb)	X 4.448	= Newtons (N)	X 0.225	= Pounds-force (lbf; lb)
Newtons (N)	X 0.1	= Kilograms-force (kgf; kg)	X 9.81	= Newtons (N)

Pressure
Pounds-force per square inch (psi; lbf/in²; lb/in²)	X 0.070	= Kilograms-force per square centimetre (kgf/cm²; kg/cm²)	X 14.223	= Pounds-force per square inch (psi; lbf/in²; lb/in²)
Pounds-force per square inch (psi; lbf/in²; lb/in²)	X 0.068	= Atmospheres (atm)	X 14.696	= Pounds-force per square inch (psi; lbf/in²; lb/in²)
Pounds-force per square inch (psi; lbf/in²; lb/in²)	X 0.069	= Bars	X 14.5	= Pounds-force per square inch (psi; lbf/in²; lb/in²)
Pounds-force per square inch (psi; lbf/in²; lb/in²)	X 6.895	= Kilopascals (kPa)	X 0.145	= Pounds-force per square inch (psi; lbf/in²; lb/in²)
Kilopascals (kPa)	X 0.01	= Kilograms-force per square centimetre (kgf/cm²; kg/cm²)	X 98.1	= Kilopascals (kPa)

Torque (moment of force)
Pounds-force inches (lbf in; lb in)	X 1.152	= Kilograms-force centimetre (kgf cm; kg cm)	X 0.868	= Pounds-force inches (lbf in; lb in)
Pounds-force inches (lbf in; lb in)	X 0.113	= Newton metres (Nm)	X 8.85	= Pounds-force inches (lbf in; lb in)
Pounds-force inches (lbf in; lb in)	X 0.083	= Pounds-force feet (lbf ft; lb ft)	X 12	= Pounds-force inches (lbf in; lb in)
Pounds-force feet (lbf ft; lb ft)	X 0.138	= Kilograms-force metres (kgf m; kg m)	X 7.233	= Pounds-force feet (lbf ft; lb ft)
Pounds-force feet (lbf ft; lb ft)	X 1.356	= Newton metres (Nm)	X 0.738	= Pounds-force feet (lbf ft; lb ft)
Newton metres (Nm)	X 0.102	= Kilograms-force metres (kgf m; kg m)	X 9.804	= Newton metres (Nm)

Vacuum
Inches mercury (in. Hg)	X 3.377	= Kilopascals (kPa)	X 0.2961	= Inches mercury
Inches mercury (in. Hg)	X 25.4	= Millimeters mercury (mm Hg)	X 0.0394	= Inches mercury

Power
Horsepower (hp)	X 745.7	= Watts (W)	X 0.0013	= Horsepower (hp)

Velocity (speed)
Miles per hour (miles/hr; mph)	X 1.609	= Kilometres per hour (km/hr; kph)	X 0.621	= Miles per hour (miles/hr; mph)

Fuel consumption*
Miles per gallon, Imperial (mpg)	X 0.354	= Kilometres per litre (km/l)	X 2.825	= Miles per gallon, Imperial (mpg)
Miles per gallon, US (mpg)	X 0.425	= Kilometres per litre (km/l)	X 2.352	= Miles per gallon, US (mpg)

Temperature
Degrees Fahrenheit = (°C x 1.8) + 32

Degrees Celsius (Degrees Centigrade; °C) = (°F - 32) x 0.56

*It is common practice to convert from miles per gallon (mpg) to litres/100 kilometres (l/100km),
where mpg (Imperial) x l/100 km = 282 and mpg (US) x l/100 km = 235

Safety first!

Regardless of how enthusiastic you may be about getting on with the job at hand, take the time to ensure that your safety is not jeopardized. A moment's lack of attention can result in an accident, as can failure to observe certain simple safety precautions. The possibility of an accident will always exist, and the following points should not be considered a comprehensive list of all dangers. Rather, they are intended to make you aware of the risks and to encourage a safety conscious approach to all work you carry out on your vehicle.

Essential DOs and DON'Ts

DON'T rely on a jack when working under the vehicle. Always use approved jackstands to support the weight of the vehicle and place them under the recommended lift or support points.

DON'T attempt to loosen extremely tight fasteners (i.e. wheel lug nuts) while the vehicle is on a jack - it may fall.

DON'T start the engine without first making sure that the transmission is in Neutral (or Park where applicable) and the parking brake is set.

DON'T remove the radiator cap from a hot cooling system - let it cool or cover it with a cloth and release the pressure gradually.

DON'T attempt to drain the engine oil until you are sure it has cooled to the point that it will not burn you.

DON'T touch any part of the engine or exhaust system until it has cooled sufficiently to avoid burns.

DON'T siphon toxic liquids such as gasoline, antifreeze and brake fluid by mouth, or allow them to remain on your skin.

DON'T inhale brake lining dust - it is potentially hazardous (see *Asbestos* below).

DON'T allow spilled oil or grease to remain on the floor - wipe it up before someone slips on it.

DON'T use loose fitting wrenches or other tools which may slip and cause injury.

DON'T push on wrenches when loosening or tightening nuts or bolts. Always try to pull the wrench toward you. If the situation calls for pushing the wrench away, push with an open hand to avoid scraped knuckles if the wrench should slip.

DON'T attempt to lift a heavy component alone - get someone to help you.

DON'T rush or take unsafe shortcuts to finish a job.

DON'T allow children or animals in or around the vehicle while you are working on it.

DO wear eye protection when using power tools such as a drill, sander, bench grinder, etc. and when working under a vehicle.

DO keep loose clothing and long hair well out of the way of moving parts.

DO make sure that any hoist used has a safe working load rating adequate for the job.

DO get someone to check on you periodically when working alone on a vehicle.

DO carry out work in a logical sequence and make sure that everything is correctly assembled and tightened.

DO keep chemicals and fluids tightly capped and out of the reach of children and pets.

DO remember that your vehicle's safety affects that of yourself and others. If in doubt on any point, get professional advice.

Asbestos

Certain friction, insulating, sealing, and other products - such as brake linings, brake bands, clutch linings, torque converters, gaskets, etc. - may contain asbestos. Extreme care must be taken to avoid inhalation of dust from such products, since it is hazardous to health. If in doubt, assume that they do contain asbestos.

Fire

Remember at all times that gasoline is highly flammable. Never smoke or have any kind of open flame around when working on a vehicle. But the risk does not end there. A spark caused by an electrical short circuit, by two metal surfaces contacting each other, or even by static electricity built up in your body under certain conditions, can ignite gasoline vapors, which in a confined space are highly explosive. Do not, under any circumstances, use gasoline for cleaning parts. Use an approved safety solvent.

Always disconnect the battery ground (-) cable at the battery before working on any part of the fuel system or electrical system. Never risk spilling fuel on a hot engine or exhaust component. It is strongly recommended that a fire extinguisher suitable for use on fuel and electrical fires be kept handy in the garage or workshop at all times. Never try to extinguish a fuel or electrical fire with water.

Fumes

Certain fumes are highly toxic and can quickly cause unconsciousness and even death if inhaled to any extent. Gasoline vapor falls into this category, as do the vapors from some cleaning solvents. Any draining or pouring of such volatile fluids should be done in a well ventilated area.

When using cleaning fluids and solvents, read the instructions on the container carefully. Never use materials from unmarked containers.

Never run the engine in an enclosed space, such as a garage. Exhaust fumes contain carbon monoxide, which is extremely poisonous. If you need to run the engine, always do so in the open air, or at least have the rear of the vehicle outside the work area.

If you are fortunate enough to have the use of an inspection pit, never drain or pour gasoline and never run the engine while the vehicle is over the pit. The fumes, being heavier than air, will concentrate in the pit with possibly lethal results.

The battery

Never create a spark or allow a bare light bulb near a battery. They normally give off a certain amount of hydrogen gas, which is highly explosive.

Always disconnect the battery ground (-) cable at the battery before working on the fuel or electrical systems.

If possible, loosen the filler caps or cover when charging the battery from an external source (this does not apply to sealed or maintenance-free batteries). Do not charge at an excessive rate or the battery may burst.

Take care when adding water to a non maintenance-free battery and when carrying a battery. The electrolyte, even when diluted, is very corrosive and should not be allowed to contact clothing or skin.

Always wear eye protection when cleaning the battery to prevent the caustic deposits from entering your eyes.

Household current

When using an electric power tool, inspection light, etc., which operates on household current, always make sure that the tool is correctly connected to its plug and that, where necessary, it is properly grounded. Do not use such items in damp conditions and, again, do not create a spark or apply excessive heat in the vicinity of fuel or fuel vapor.

Secondary ignition system voltage

A severe electric shock can result from touching certain parts of the ignition system (such as the spark plug wires) when the engine is running or being cranked, particularly if components are damp or the insulation is defective. In the case of an electronic ignition system, the secondary system voltage is much higher and could prove fatal.

Troubleshooting

Contents

This section provides an easy reference guide to the more common problems which may occur during the operation of your vehicle. These problems and their possible causes are grouped under headings denoting various components or systems, such as Engine, Cooling system, etc. They also refer you to the chapter and/or section which deals with the problem.

Remember that successful troubleshooting is not a mysterious black art practiced only by professional mechanics. It is simply the result of the right knowledge combined with an intelligent, systematic approach to the problem. Always work by a process of elimination, starting with the simplest solution and working through to the most complex - and never overlook the obvious. Anyone can run the gas tank dry or leave the lights on overnight, so don't assume that you are exempt from such oversights.

Finally, always establish a clear idea of why a problem has occurred and take steps to ensure that it doesn't happen again. If the electrical system fails because of a poor connection, check the other connections in the system to make sure that they don't fail as well. If a particular fuse continues to blow, find out why - don't just replace one fuse after another. Remember, failure of a small component can often be indicative of potential failure or incorrect functioning of a more important component or system.

Engine

1 Engine will not rotate when attempting to start

1 Battery terminal connections loose or corroded (Chapter 1).
2 Battery discharged or faulty (Chapter 1).
3 Automatic transmission not completely engaged in Park (Chapter 7B) or clutch not completely depressed (Chapter 8).
4 Broken, loose or disconnected wiring in the starting circuit (Chapters 5 and 12).
5 Starter motor pinion jammed in flywheel ring gear (Chapter 5).
6 Starter solenoid faulty (Chapter 5).
7 Starter motor faulty (Chapter 5).
8 Ignition switch faulty (Chapter 12).
9 Starter pinion or flywheel teeth worn or broken (Chapter 5).

2 Engine rotates but will not start

1 Fuel tank empty.
2 Battery discharged (engine rotates slowly) (Chapter 5).
3 Battery terminal connections loose or corroded (Chapter 1).
4 Leaking fuel injector(s), faulty fuel pump, pressure regulator, etc. (Chapter 4).
5 Fuel not reaching fuel rail or TBI unit (Chapter 4).

6 Ignition components damp or damaged (Chapter 5).
7 Worn, faulty or incorrectly gapped spark plugs (Chapter 1).
8 Broken, loose or disconnected wiring in the starting circuit (Chapter 5).
9 Broken, loose or disconnected wires at the ignition coil or faulty coil (Chapter 5).
10 Broken timing chain.

3 Engine hard to start when cold

1 Battery discharged or low (Chapter 1).
2 Malfunctioning fuel system (Chapter 4).
3 Injector(s) leaking (Chapter 4).

4 Engine hard to start when hot

1 Air filter clogged (Chapter 1).
2 Fuel not reaching the fuel injection system (Chapter 4).
3 Corroded battery connections, especially ground (Chapter 1).

5 Starter motor noisy or excessively rough in engagement

1 Pinion or flywheel gear teeth worn or broken (Chapter 5).
2 Starter motor mounting bolts loose or missing (Chapter 5).

6 Engine starts but stops immediately

1 Loose or faulty electrical connections at module, coil or alternator (Chapter 5).
2 Insufficient fuel reaching the fuel injection system (Chapters 1 and 4).
3 Vacuum leak at the gasket between the intake manifold and throttle body (Chapters 1 and 4).

7 Oil puddle under engine

1 Oil pan gasket and/or oil pan drain bolt washer leaking (Chapter 2).
2 Oil pressure sending unit leaking (Chapter 2).
3 Valve covers leaking (Chapter 2).
4 Engine oil seals leaking (Chapter 2).

8 Engine lopes while idling or idles erratically

1 Vacuum leakage (Chapters 2 and 4).
2 Leaking EGR valve (Chapter 6).
3 Air filter clogged (Chapter 1).
4 Fuel pump not delivering sufficient fuel to the fuel injection system (Chapter 4).
5 Leaking head gasket (Chapter 2).

6 Timing chain and/or sprockets worn (Chapter 2).
7 Camshaft lobes worn (Chapter 2).

9 Engine misses at idle speed

1 Spark plugs worn or not gapped properly (Chapter 1).
2 Faulty spark plug wires (Chapter 1).
3 Vacuum leaks (Chapter 1).
4 Uneven or low compression (Chapter 2).

10 Engine misses throughout driving speed range

1 Fuel filter clogged and/or impurities in the fuel system (Chapter 1).
2 Low fuel output at the injector(s) (Chapter 4).
3 Faulty or incorrectly gapped spark plugs (Chapter 1).
4 Defective coil or module (Chapter 5).
5 Leaking spark plug wires (Chapters 1 or 5).
6 Faulty emission system components (Chapter 6).
7 Low or uneven cylinder compression pressures (Chapter 2).
8 Weak or faulty ignition system (Chapter 5).
9 Vacuum leak in fuel injection system, intake manifold or vacuum hoses (Chapter 4).

11 Engine stumbles on acceleration

1 Spark plugs fouled (Chapter 1).
2 Fuel injection system faulty (Chapter 4).
3 Fuel filter clogged (Chapters 1 and 4).
4 Intake manifold air leak (Chapters 2 and 4).

12 Engine surges while holding accelerator steady

1 Intake air leak (Chapter 4).
2 Fuel pump faulty (Chapter 4).
3 Loose fuel injector wire harness connectors (Chapter 4).
4 Defective PCM or information sensor (Chapter 6).

13 Engine stalls

1 Fuel filter clogged and/or water and impurities in the fuel system (Chapters 1 and 4).
2 Faulty emissions system components (Chapter 6).
3 Faulty or incorrectly gapped spark plugs (Chapter 1).
4 Faulty spark plug wires (Chapter 1).
5 Vacuum leak in the fuel injection system,

intake manifold or vacuum hoses (Chapters 2 and 4).

14 Engine lacks power

1 Faulty or incorrectly gapped spark plugs (Chapter 1).
2 Fuel injection system malfunction (Chapter 4).
3 Faulty coil or spark plug wire (Chapter 5).
4 Brakes binding (Chapter 9).
5 Automatic transaxle fluid level incorrect (Chapter 1).
6 Clutch slipping (Chapter 8).
7 Fuel filter clogged and/or impurities in the fuel system (Chapters 1 and 4).
8 Emission control system not functioning properly (Chapter 6).
9 Low or uneven cylinder compression pressures (Chapter 2).
10 Obstructed exhaust system (Chapter 4).

15 Engine backfires

1 Emission control system not functioning properly (Chapter 6).
2 Faulty secondary ignition system (cracked spark plug insulator, faulty plug wires or coil) (Chapters 1 and 5).
3 Fuel injection system faulty (Chapter 4).
4 Vacuum leak at fuel injector(s), intake manifold, air control valve or vacuum hoses (Chapters 2 and 4).

16 Pinging or knocking engine sounds during acceleration or uphill

1 Incorrect grade of fuel.
2 Fuel injection system faulty (Chapter 4).
3 Improper or damaged spark plugs or wires (Chapter 1).
4 EGR valve not functioning (Chapter 6).
5 Vacuum leak (Chapters 2 and 4).

17 Engine runs with oil pressure light on

1 Low oil level (Chapter 1).
2 Short in wiring circuit (Chapter 12).
3 Faulty oil pressure sender (Chapter 2).
4 Worn engine bearings and/or oil pump (Chapter 2).

18 Engine diesels (continues to run) after switching off

1 Idle speed too high (Chapter 5)
2 Excessive engine operating temperature (Chapter 3).

Engine electrical system

19 Battery will not hold a charge

1 Alternator drivebelt defective or not adjusted properly (Chapter 1).
2 Battery electrolyte level low (Chapter 1).
3 Battery terminals loose or corroded (Chapter 1).
4 Alternator not charging properly (Chapter 5).
5 Loose, broken or faulty wiring in the charging circuit (Chapter 5).
6 Short in vehicle wiring (Chapter 12).
7 Internally defective battery (Chapters 1 and 5).

20 Alternator light fails to go out

1 Faulty alternator or charging circuit (Chapter 5).
2 Alternator drivebelt defective or out of adjustment (Chapter 1).
3 Alternator voltage regulator inoperative (Chapter 5).

21 Alternator light fails to come on when key is turned on

1 Bulb burned out (Chapter 12).
2 Fault in the printed circuit, dash wiring or bulb holder (Chapter 12).

Fuel system

22 Excessive fuel consumption

1 Dirty or clogged air filter element (Chapter 1).
2 Emissions system not functioning properly (Chapter 6).
3 Fuel injection system malfunctioning (Chapter 4).
4 Low tire pressure or incorrect tire size (Chapter 1).

23 Fuel leakage and/or fuel odor

1 Leaking fuel feed or return line (Chapters 1 and 4).
2 Tank overfilled.
3 Evaporative canister filter clogged (Chapters 1 and 6).
4 Fuel injection system malfunctioning (Chapter 4).

Cooling system

24 Overheating

1 Insufficient coolant in system (Chapter 1).

2 Water pump drivebelt defective or out of adjustment (Chapter 1).
3 Radiator core blocked or grille restricted (Chapter 3).
4 Thermostat faulty (Chapter 3).
5 Electric coolant fan blades broken or cracked (Chapter 3).
6 Coolant cap not maintaining proper pressure (Chapter 3).

25 Overcooling

1 Faulty thermostat (Chapter 3).
2 Inaccurate temperature gauge sending unit (Chapter 3).

26 External coolant leakage

1 Deteriorated/damaged hoses; loose clamps (Chapters 1 and 3).
2 Water pump defective (Chapter 3).
3 Leakage from radiator core or coolant expansion tank (Chapter 3).
4 Engine drain or water jacket core plugs leaking (Chapter 2B).

27 Internal coolant leakage

1 Leaking cylinder head gasket (Chapter 2).
2 Cracked cylinder bore or cylinder head (Chapter 2).

28 Coolant loss

1 Coolant boiling away because of overheating (Chapter 3).
2 Internal or external leakage (Chapter 3).
3 Faulty coolant cap (Chapter 3).

29 Poor coolant circulation

1 Inoperative water pump (Chapter 3).
2 Restriction in cooling system (Chapters 1 and 3).
3 Water pump drivebelt defective/out of adjustment (Chapter 1).
4 Thermostat sticking (Chapter 3).

Clutch

30 Pedal travels to floor - no pressure or very little resistance

1 Damaged pedal mechanism (Chapter 8).
2 Leak in hydraulic system (Chapter 8).
3 Broken release bearing or fork (Chapter 8).

31 Unable to select gears

1 Faulty transaxle (Chapter 7).
2 Faulty clutch disc (Chapter 8).
3 Release lever and bearing not assembled properly (Chapter 8).
4 Faulty pressure plate (Chapter 8).
5 Pressure plate-to-flywheel bolts loose (Chapter 8).

32 Clutch slips (engine speed increases with no increase in vehicle speed)

1 Clutch plate worn (Chapter 8).
2 Clutch plate is oil soaked by leaking rear main seal (Chapter 8).
3 Clutch plate not seated.
4 Warped pressure plate or flywheel (Chapter 8).
5 Weak diaphragm spring (Chapter 8).
6 Clutch plate overheated. Allow to cool.

33 Grabbing (chattering) as clutch is engaged

1 Oil on clutch plate lining, burned or glazed facings (Chapter 8).
2 Worn or loose engine or transaxle mounts (Chapters 2 and 7).
3 Worn splines on clutch plate hub (Chapter 8).
4 Warped pressure plate or flywheel (Chapter 8).
5 Burned or smeared resin on flywheel or pressure plate (Chapter 8).

34 Transaxle rattling (clicking)

1 Release lever loose (Chapter 8).
2 Clutch plate damper spring failure (Chapter 8).

35 Noise in clutch area

1 Fork shaft improperly installed (Chapter 8).
2 Faulty bearing (Chapter 8).

36 Clutch pedal stays on floor

2 Leak in hydraulic system (Chapter 8).
3 Broken release bearing or fork (Chapter 8).

37 High pedal effort

1 Clutch pedal binding (Chapter 8).
2 Pressure plate faulty (Chapter 8).

Manual transaxle

38 Knocking noise at low speeds

1 Worn driveaxle constant velocity (CV) joints (Chapter 8).
2 Worn side gear shaft counterbore in differential case (Chapter 7A).*

39 Noise most pronounced when turning

Differential gear noise (Chapter 7A).*

40 Clunk on acceleration or deceleration

1 Loose engine or transaxle mounts (Chapters 2 and 7A).
2 Worn differential pinion shaft in case.*
3 Worn side gear shaft counterbore in differential case (Chapter 7A).*
4 Worn or damaged driveaxle inboard CV joints (Chapter 8).

41 Clicking noise in turns

Worn or damaged outboard CV joint (Chapter 8).

42 Vibration

1 Rough wheel bearing (Chapter 10).
2 Damaged driveaxle (Chapter 8).
3 Out of round tires (Chapter 1).
4 Tire out of balance (Chapters 1 and 10).
5 Worn CV joint (Chapter 8).

43 Noisy in neutral with engine running

1 Damaged input gear bearing (Chapter 7A).*
2 Damaged clutch release bearing (Chapter 8).

44 Noisy in one particular gear

1 Damaged or worn constant mesh gears (Chapter 7A).*
2 Damaged or worn synchronizers (Chapter 7A).*
3 Bent reverse fork (Chapter 7A).*
4 Damaged fourth speed gear or output gear (Chapter 7A).*
5 Worn or damaged reverse idler gear or idler bushing (Chapter 7A).*

45 Noisy in all gears

1 Insufficient lubricant (Chapter 7A).
2 Damaged or worn bearings (Chapter 7A).*
3 Worn or damaged input gear shaft and/or output gear shaft (Chapter 7A).*

46 Slips out of gear

1 Worn or improperly adjusted linkage (Chapter 7A).
2 Transaxle loose on engine (Chapter 7A).
3 Shift linkage does not work freely, binds (Chapter 7A).
4 Input gear bearing retainer broken or loose (Chapter 7A).*
5 Dirt between clutch cover and engine housing (Chapter 7A).
6 Worn shift fork (Chapter 7A).*

47 Leaks lubricant

1 Driveaxle seals worn (Chapter 7).
2 Excessive amount of lubricant in transaxle (Chapters 1 and 7A).
3 Loose or broken input gear shaft bearing retainer (Chapter 7A).*
4 Input gear bearing retainer O-ring and/or lip seal damaged (Chapter 7A).*

48 Locked in gear

Lock pin or interlock pin missing (Chapter 7A).*

Although the corrective action necessary to remedy the symptoms described is beyond the scope of the home mechanic, the above information should be helpful in isolating the cause of the condition so that the owner can communicate clearly with a professional mechanic.

Automatic transaxle

Note: *Due to the complexity of the automatic transaxle, it is difficult for the home mechanic to properly diagnose and service this component. For problems other than the following, the vehicle should be taken to a dealer or transmission shop.*

49 Fluid leakage

1 Automatic transmission fluid is a deep red color. Fluid leaks should not be confused with engine oil, which can easily be blown onto the transaxle by air flow.
2 To pinpoint a leak, first remove all built-up dirt and grime from the transaxle housing with degreasing agents and/or steam clean-

ing. Then drive the vehicle at low speeds so air flow will not blow the leak far from its source. Raise the vehicle and determine where the leak is coming from. Common areas of leakage are:

a) *Filter (Chapter 1)*
b) *Dipstick tube (Chapters 1 and 7)*
c) *Transaxle oil lines (Chapter 7)*
d) *Speed sensor (Chapter 7)*

50 Transaxle fluid brown or has a burned smell

Transaxle fluid burned (Chapter 1).

51 General shift mechanism problems

1 Chapter 7, Part B, deals with checking and adjusting the shift linkage on automatic transaxles. Common problems which may be attributed to poorly adjusted linkage are:

a) *Engine starting in gears other than Park or Neutral.*
b) *Indicator on shifter pointing to a gear other than the one actually being used.*
c) *Vehicle moves when in Park.*

2 Refer to Chapter 7B for the shift linkage adjustment procedure.
Throttle valve cable out of adjustment (Chapter 7B).

52 Engine will start in gears other than Park or Neutral

Neutral start switch malfunctioning (Chapter 7B).

53 Transaxle slips, shifts roughly, is noisy or has no drive in forward or reverse gears

There are many probable causes for the above problems, but the home mechanic should be concerned with only one possibility - fluid level. Before taking the vehicle to a repair shop, check the level and condition of the fluid as described in Chapter 1. Correct the fluid level as necessary or change the fluid and filter if needed. If the problem persists, have a professional diagnose the cause.

Driveaxles

54 Clicking noise in turns

Worn or damaged outboard CV joint (Chapter 8).

55 Shudder or vibration during acceleration

1 Excessive toe-in (Chapter 10).
2 Incorrect spring heights (Chapter 10).
3 Worn or damaged inboard or outboard CV joints (Chapter 8).
4 Sticking inboard CV joint assembly (Chapter 8).

56 Vibration at highway speeds

1 Out of balance front wheels and/or tires (Chapters 1 and 10).
2 Out of round front tires (Chapters 1 and 10).
3 Worn CV joint(s) (Chapter 8).

Brakes

Note: *Before assuming that a brake problem exists, make sure that:*

a) *The tires are in good condition and properly inflated (Chapter 1).*
b) *The front end alignment is correct (Chapter 10).*
c) *The vehicle is not loaded with weight in an unequal manner.*

57 Vehicle pulls to one side during braking

1 Incorrect tire pressures (Chapter 1).
2 Front end out of line (have the front end aligned).
3 Front, or rear, tires not matched to one another.
4 Restricted brake lines or hoses (Chapter 9).
5 Malfunctioning drum brake or caliper assembly (Chapter 9).
6 Loose suspension parts (Chapter 10).
7 Loose calipers (Chapter 9).
8 Excessive wear of brake shoe or pad material (or disc/drum) on one side.

58 Noise (high-pitched squeal when the brakes are applied)

Front disc brake pads worn out. The noise comes from the wear sensor rubbing against the disc (does not apply to all vehicles). Replace pads with new ones immediately (Chapter 9).

59 Brake roughness or chatter (pedal pulsates)

1 Excessive lateral runout (Chapter 9).
2 Uneven pad wear (Chapter 9).
3 Defective disc (Chapter 9).

60 Excessive brake pedal effort required to stop vehicle

1 Malfunctioning power brake booster (Chapter 9).
2 Partial system failure (Chapter 9).
3 Excessively worn pads or shoes (Chapter 9).
4 Piston in caliper or wheel cylinder stuck or sluggish (Chapter 9).
5 Brake pads or shoes contaminated with oil or grease (Chapter 9).
6 New pads or shoes installed and not yet seated. It will take a while for the new material to seat against the disc or drum.

61 Excessive brake pedal travel

1 Partial brake system failure (Chapter 9).
2 Insufficient fluid in master cylinder (Chapters 1 and 9).
3 Air trapped in system (Chapters 1 and 9).

62 Dragging brakes

1 Incorrect adjustment of brake light switch (Chapter 9).
2 Master cylinder pistons not returning correctly (Chapter 9).
3 Restricted brakes lines or hoses (Chapters 1 and 9).
4 Incorrect parking brake adjustment (Chapter 9).

63 Grabbing or uneven braking action

1 Malfunction of proportioning valve (Chapter 9).
2 Malfunction of power brake booster unit (Chapter 9).
3 Binding brake pedal mechanism (Chapter 9).

64 Brake pedal feels spongy when depressed

1 Air in hydraulic lines (Chapter 9).
2 Master cylinder mounting bolts loose (Chapter 9).
3 Master cylinder defective (Chapter 9).

65 Brake pedal travels to the floor with little resistance

1 Little or no fluid in the master cylinder reservoir (Chapter 9).
2 Loose, damaged or disconnected brake lines (Chapter 9).

66 Parking brake does not hold

Parking brake linkage improperly adjusted (Chapters 1 and 9).

Suspension and steering systems

Note: *Before attempting to diagnose the suspension and steering systems, perform the following preliminary checks:*
 a) *Tires for wrong pressure and uneven wear.*
 b) *Steering universal joints from the column to the rack-and-pinion for loose connectors or wear.*
 c) *Front and rear suspension and the rack-and-pinion assembly for loose or damaged parts.*
 d) *Out-of-round or out-of-balance tires, bent rims and loose and/or rough wheel bearings.*

67 Vehicle pulls to one side

1 Mismatched or uneven tires (Chapter 10).
2 Broken or sagging springs (Chapter 10).
3 Wheel alignment (Chapter 10).
4 Front brake dragging (Chapter 9).

68 Abnormal or excessive tire wear

1 Wheel alignment (Chapter 10).
2 Sagging or broken springs (Chapter 10).
3 Tire out of balance (Chapter 10).
4 Worn strut damper (Chapter 10).
5 Overloaded vehicle.
6 Tires not rotated regularly.

69 Wheel makes a thumping noise

1 Blister or bump on tire (Chapter 10).
2 Improper strut damper action (Chapter 10).

70 Shimmy, shake or vibration

1 Tire or wheel out-of-balance or out-of-round (Chapter 10).
2 Loose or worn front hub or wheel bearings (Chapters 1, 8 and 10).
3 Worn tie-rod ends (Chapter 10).
4 Worn lower balljoints (Chapters 1 and 10).
5 Excessive wheel runout (Chapter 10).
6 Blister or bump on tire (Chapter 10).

71 Hard steering

1 Lack of lubrication at balljoints, tie-rod ends and rack-and-pinion assembly (Chapter 10).
2 Front wheel alignment (Chapter 10).
3 Low tire pressure(s) (Chapters 1 and 10).

72 Poor returnability of steering to center

1 Lack of lubrication at balljoints and tie-rod ends (Chapter 10).
2 Binding in balljoints (Chapter 10).
3 Binding in steering column (Chapter 10).
4 Lack of lubricant in steering gear assembly (Chapter 10).
5 Front wheel alignment (Chapter 10).

73 Abnormal noise at the front end

1 Lack of lubrication at balljoints and tie-rod ends (Chapters 1 and 10).
2 Damaged strut mounting (Chapter 10).
3 Worn control arm bushings or tie-rod ends (Chapter 10).
4 Loose stabilizer bar (Chapter 10).
5 Loose wheel nuts (Chapters 1 and 10).
6 Loose suspension bolts (Chapter 10)

74 Wander or poor steering stability

1 Mismatched or uneven tires (Chapter 10).
2 Lack of lubrication at balljoints and tie-rod ends (Chapters 1 and 10).
3 Worn strut assemblies (Chapter 10).
4 Loose stabilizer bar (Chapter 10).
5 Broken or sagging springs (Chapter 10).
6 Wheels out of alignment (Chapter 10).

75 Erratic steering when braking

1 Front hub bearings worn (Chapter 10).
2 Broken or sagging springs (Chapter 10).
3 Leaking wheel cylinder or caliper (Chapter 10).
4 Warped discs or drums (Chapter 10).

76 Excessive pitching and/or rolling around corners or during braking

1 Loose stabilizer bar (Chapter 10).
2 Worn strut dampers or mountings (Chapter 10).
3 Broken or sagging springs (Chapter 10).
4 Overloaded vehicle.

77 Suspension bottoms

1 Overloaded vehicle.
2 Worn strut dampers (Chapter 10).
3 Incorrect, broken or sagging springs (Chapter 10).

78 Cupped tires

1 Front wheel or rear wheel alignment (Chapter 10).
2 Worn strut dampers (Chapter 10).
3 Wheel bearings worn (Chapter 10).
4 Excessive tire or wheel runout (Chapter 10).
5 Worn balljoints (Chapter 10).

79 Excessive tire wear on outside edge

1 Inflation pressures incorrect (Chapter 1).
2 Excessive speed in turns.
3 Front end alignment incorrect (excessive toe-in). Have professionally aligned.
4 Suspension arm bent or twisted (Chapter 10).

80 Excessive tire wear on inside edge

1 Inflation pressures incorrect (Chapter 1).
2 Front end alignment incorrect (toe-out). Have professionally aligned.
3 Loose or damaged steering or suspension components (Chapter 10).

81 Tire tread worn in one place

1 Tires out of balance.
2 Damaged or buckled wheel. Inspect and replace if necessary.
3 Defective tire (Chapter 1).

82 Excessive play or looseness in steering system

1 Front hub bearing(s) worn (Chapter 10).
2 Tie-rod end loose (Chapter 10).
3 Steering gear loose (Chapter 10).
4 Worn or loose steering intermediate shaft (Chapter 10).

83 Rattling or clicking noise in steering gear

1 Steering gear loose (Chapter 10).
2 Steering gear defective.

Troubleshooting

Notes

Chapter 1
Tune-up and routine maintenance

Contents

Specifications

Recommended lubricants and fluids

Note: *Listed here are manufacturer recommendations at the time this manual was written. Manufacturers occasionally upgrade their fluid and lubricant specifications, so check with your local auto parts store for current recommendations.*

Engine oil type .... API grade SG or SG/CD multigrade and fuel efficient oil
Viscosity.. See accompanying chart

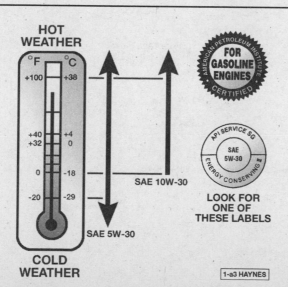

Engine oil viscosity chart - for best fuel economy and cold starting, select the lowest SAE viscosity grade for the expected temperature range

1-a3 HAYNES

Recommended lubricants and fluids

Fuel .... Unleaded gasoline, 87 octane or higher
Automatic transaxle fluid type .... DEXRON IIE automatic transmission fluid
Manual transaxle lubricant type .... DEXRON IIE automatic transmission fluid
Brake fluid type .... DOT 3 brake fluid
Clutch fluid type .... DOT 3 brake fluid
Engine coolant type .... 50/50 mixture of non-phosphate ethylene glycol
base antifreeze and water
Power steering system fluid .... GM power steering fluid part no. 10552884 or equivalent

Capacities*

Engine oil (including filter) .... 4 qts
Coolant .... 7 qts
Transaxle
    Automatic (including filter) .... 3.5 qts
    Manual .... 2.6 qts

All capacities approximate. Add as necessary to bring appropriate level.

Cylinder locations and coil-pack terminal diagram

Ignition system

Spark plug type and gap
1994 and earlier
    SOHC
        Type .... AC FR4LS or equivalent
        Gap .... 0.040 inch
    DOHC
        Type .... AC FR3LS or equivalent
        Gap .... 0.040 inch
1995
    Type .... AC FR4LSJ or equivalent
    Gap .... 0.040 inch
1996
    SOHC
        Type .... AC 41-608 or equivalent
        Gap .... 0.060 inch
    DOHC
        Type .... AC 41-607 or equivalent
        Gap .... 0.060 inch
1997 through 1999
    SOHC
        Type .... AC 41-628 or equivalent
        Gap .... 0.040 inch
    DOHC
        Type .... AC 41-627 or equivalent
        Gap .... 0.040 inch
Spark plug wire resistance .... 12,000 ohms
Engine firing order 1-3-4-2

Cooling system

Thermostat rating
    Starts to open .... 190-degrees F
    Fully open .... 212-degrees F

Brakes

Disc brake pad lining thickness (minimum) .... 1/16 inch
Drum brake shoe lining thickness (minimum) .... 1/16 inch
Parking brake adjustment .... 4 to 6 clicks

Suspension and steering

Steering wheel freeplay limit .... 1-3/16 inch
Balljoint allowable movement 0-inch

Torque specifications

                                                                       Ft-lbs (unless otherwise indicated)
Transaxle drain plug 20
Spark plugs 20
Wheel lug nuts 100

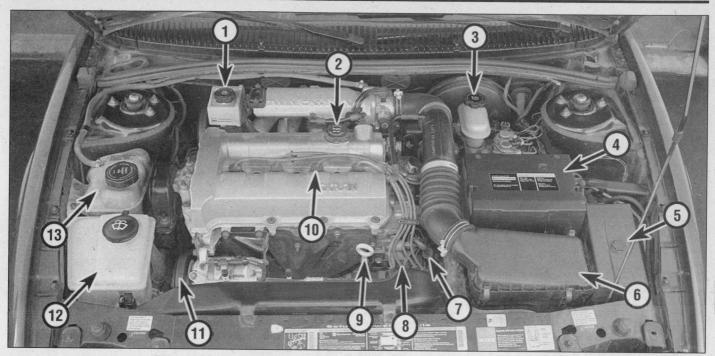

Engine compartment components (1993 twin cam model shown, others similar)

1 Power steering fluid reservoir	6 Air filter	11 Engine drivebelt
2 Engine oil filler cap	7 Automatic transaxle dipstick	12 Windshield washer fluid reservoir
3 Brake fluid reservoir	8 Ignition coil and wires	13 Engine coolant expansion tank
4 Battery	9 Engine oil dipstick	
5 Fuse box	10 Spark plug and wires	

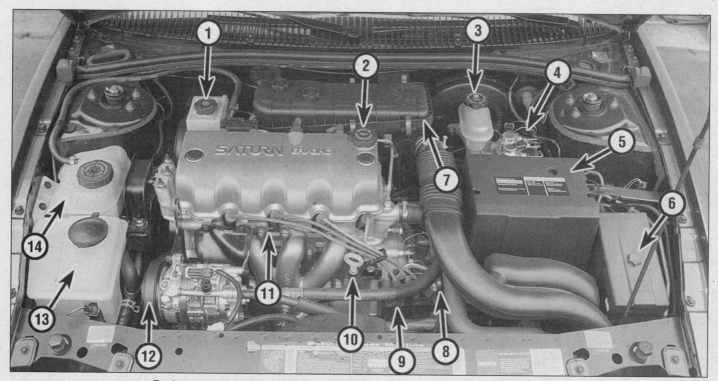

Engine compartment components (1994 single cam model shown, others similar)

1 Power steering fluid reservoir	6 Fuse box	11 Spark plug and wires
2 Engine oil filler cap	7 Air filter	12 Engine drivebelt
3 Brake fluid reservoir	8 Automatic transaxle dipstick	13 Windshield washer fluid reservoir
4 Clutch fluid reservoir	9 Ignition coil and wires	14 Engine coolant expansion tank
5 Battery	10 Engine oil dipstick	

1

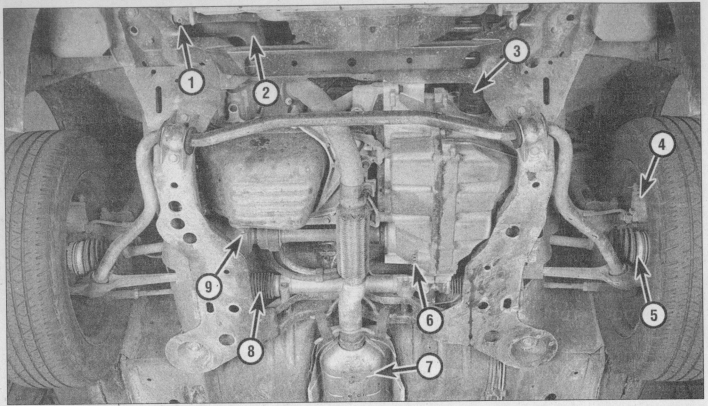

Engine compartment underside components (1993 model shown, others similar)

1	Radiator drain location	4	Disc brake caliper	7	Exhaust system
2	Lower radiator hose	5	Driveaxle CV joint boot	8	Steering gear boot
3	Automatic transaxle spin-off type filter	6	Transaxle drain plug	9	Engine oil drain plug

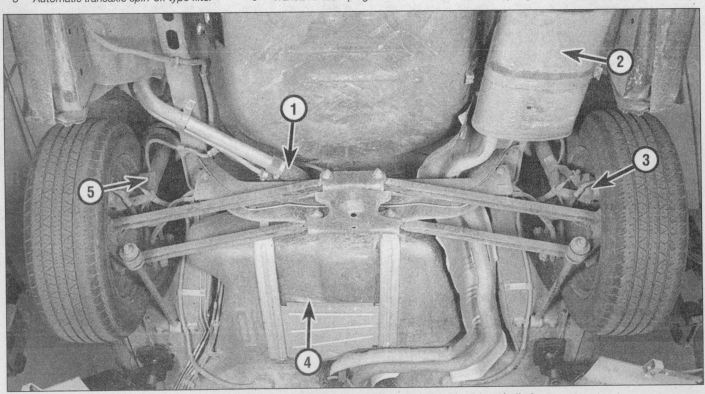

Typical rear underside components (1993 model shown, others similar)

1	Fuel tank inlet hose	3	Disc brake caliper	5	Rear suspension strut
2	Muffler	4	Fuel tank		

2 Saturn Maintenance schedule

The maintenance intervals in this manual are provided with the assumption that you, not the dealer, will be doing the work. These are the minimum maintenance intervals recommended by the factory for Saturns that are driven daily. If you wish to keep your vehicle in peak condition at all times, you may wish to perform some of these procedures even more often. Because frequent maintenance enhances the efficiency, performance and resale value of your car, we encourage you to do so. If you drive in dusty areas, tow a trailer, idle or drive at low speeds for extended periods or drive for short distances (less than four miles) in below freezing temperatures, shorter intervals are also recommended.

When your vehicle is new, it should be serviced by a factory authorized dealer service department to protect the factory warranty. In many cases, the initial maintenance check is done at no cost to the owner.

Every 250 miles or weekly, whichever comes first

Check the engine oil level (Section 4)
Check the engine coolant level (Section 4)
Check the windshield washer fluid level (Section 4)
Check the brake fluid level (Section 4)
Check the tires and tire pressures (Section 5)

Every 3000 miles or 3 months, whichever comes first

All items listed above, plus:
Check the power steering fluid level (Section 6)
Check the automatic transaxle fluid level (Section 7)
Change the engine oil and oil filter (Section 8)

Every 6000 miles or 6 months, whichever comes first

All items listed above, plus:
Inspect and replace if necessary the windshield wiper blades (Section 9)
Check and service the battery (Section 10)
Check the engine drivebelt (Section 11)
Inspect and replace if necessary all underhood hoses (Section 12)
Check the cooling system (Section 13)
Rotate the tires (Section 14)
Lubricate the automatic seat belt track (Section 15)

Every 15,000 miles or 12 months, whichever comes first

All items listed above, plus:
Inspect the brake system (Section 16)*
Replace the air filter (Section 17)
Inspect the fuel system (Section 18)

Check the manual transaxle lubricant level (Section 19)*
Inspect the suspension and steering components (Section 20)*
Check the driveaxle boots (Section 21)

Every 30,000 miles or 24 months, whichever comes first

All items listed above, plus:
Replace the fuel filter (Section 22)
Check and, if necessary, replace the spark plugs (Section 23)
Inspect and replace if necessary the spark plug and coil wires (Section 24)
Service the cooling system (drain, flush and refill) (Section 25)
Inspect the exhaust system (Section 26)
Change the automatic transaxle fluid and filter (Section 27) **
Change the manual transaxle lubricant (Section 28)
Check the Exhaust Gas Recirculation (EGR) valve (1991 models only) (see Chapter 6)

* This item is affected by "severe" operating conditions as described below. If your vehicle is operated under "severe" conditions, perform all maintenance indicated with an asterisk (*) at 3000 mile/3 month intervals. Severe conditions are indicated if you mainly operate your vehicle under one or more of the following conditions:

Operating in dusty areas
Towing a trailer
Idling for extended periods and/or low speed operation
Operating when outside temperatures remain below freezing and when most trips are less than 4 miles

** If operated under one or more of the following conditions, change the automatic transaxle fluid and differential lubricant every 15,000 miles:

In heavy city traffic where the outside temperature regularly reaches 90-degrees F (32-degrees C) or higher
In hilly or mountainous terrain
Frequent trailer pulling

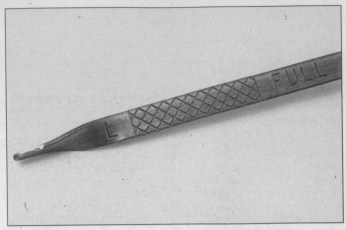

4.4 The oil level should be at or near the upper hatched area on the dipstick - if it isn't, add enough oil to bring the level to near the upper mark (it takes one quart of oil to raise the level from the lower to upper mark)

4.2 The engine oil dipstick (arrow) is located on the front side of the engine, behind the radiator

2   Introduction

This chapter is designed to help the home mechanic maintain the Saturn for peak performance, economy, safety and long life.

On the following pages is a master maintenance schedule, followed by sections dealing specifically with each item on the schedule. Visual checks, adjustments, component replacement and other helpful items are included. Refer to the accompanying illustrations of the engine compartment and the underside of the vehicle for the location of various components.

Servicing your Saturn in accordance with the mileage/time maintenance schedule and the following Sections will provide it with a planned maintenance program that should result in a long and reliable service life. This is a comprehensive plan, so maintaining some items but not others at the specified service intervals will not produce the same results.

As you service your Saturn, you will discover that many of the procedures can - and should - be grouped together because of the nature of the particular procedure you're performing or because of the close proximity of two otherwise unrelated components to one another.

For example, if the vehicle is raised for any reason, you should inspect the exhaust, suspension, steering and fuel systems while you're under the vehicle. When you're rotating the tires, it makes good sense to check the brakes and wheel bearings since the wheels are already removed.

Finally, let's suppose you have to borrow or rent a torque wrench. Even if you only need to tighten the spark plugs, you might as well check the torque of as many critical fasteners as time allows.

The first step of this maintenance program is to prepare yourself before the actual work begins. Read through all sections pertinent to the procedures you're planning to do,

then make a list of and gather together all the parts and tools you will need to do the job. If it looks as if you might run into problems during a particular segment of some procedure, seek advice from your local parts man or dealer service department.

3   Tune-up general information

The term tune-up is used in this manual to represent a combination of individual operations rather than one specific procedure.

If, from the time the vehicle is new, the routine maintenance schedule is followed closely and frequent checks are made of fluid levels and high wear items, as suggested throughout this manual, the engine will be kept in relatively good running condition and the need for additional work will be minimized.

More likely than not, however, there will be times when the engine is running poorly due to lack of regular maintenance. This is even more likely if a used vehicle, which has not received regular and frequent maintenance checks, is purchased. In such cases, an engine tune-up will be needed outside of the regular routine maintenance intervals.

The first step in any tune-up or engine diagnosis to help correct a poor running engine would be a cylinder compression check. A check of the engine compression (Chapter 2 Part B) will give valuable information regarding the overall performance of many internal components and should be used as a basis for tune-up and repair procedures. If, for instance, a compression check indicates serious internal engine wear, a conventional tune-up will not help the running condition of the engine and would be a waste of time and money.

The following series of operations are those most often needed to bring a generally poor running engine back into a proper state of tune.

Minor tune-up

Clean, inspect and test the battery (Section 10)
Check all engine related fluids (Section 4)
Check the drivebelt (Section 11)
Replace the spark plugs (Section 23)
Inspect the spark plug and coil wires (Section 24)
Check the air filter (Section 17)
Check the cooling system (Section 13)
Check all underhood hoses (Section 12)

Major tune-up

All items listed under Minor tune-up, plus . . .
Check the charging system (Chapter 5)
Check the fuel system (Section 18)
Replace the air filter (Section 17)
Replace the spark plug and coil wires (Section 24)

4   Fluid level checks (every 250 miles or weekly)

1   Fluids are an essential part of the lubrication, cooling, brake, clutch and other systems. Because these fluids gradually become depleted and/or contaminated during normal operation of the vehicle, they must be periodically replenished. See *Recommended lubricants and fluids* and *capacities* at the beginning of this Chapter before adding fluid to any of the following components. **Note:** *The vehicle must be on level ground before fluid levels can be checked.*

Engine oil

Refer to illustrations 4.2, 4.4 and 4.6
2   The engine oil level is checked with a dipstick located at the front side of the engine **(see illustration)**. The dipstick extends through a metal tube from which it protrudes down into the engine oil pan.
3   The oil level should be checked before

4.6 The oil filler cap is located on the valve cover - always make sure the area around the opening is clean before unscrewing the cap to prevent dirt from contaminating the engine

4.8 The coolant expansion tank is located on the right side of the engine compartment - keep the level near the FULL COLD line

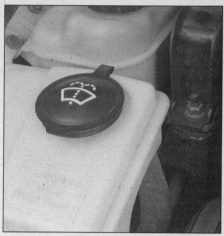

4.14 The windshield fluid reservoir is located in the right front corner of the engine compartment

1

the vehicle has been driven, or about 15 minutes after the engine has been shut off. If the oil is checked immediately after driving the vehicle, some of the oil will remain in the upper engine components, producing an inaccurate reading on the dipstick.

4　Pull the dipstick from the tube and wipe all the oil from the end with a clean rag or paper towel. Insert the clean dipstick all the way back into its metal tube and pull it out again. Observe the oil at the end of the dipstick. At its highest point, the level should be between the L and F marks **(see illustration)**.

5　It takes one quart of oil to raise the level from the L mark to the F mark on the dipstick. Do not allow the level to drop below the L mark or oil starvation may cause engine damage. Conversely, overfilling the engine (adding oil above the F mark) may cause oil fouled spark plugs, oil leaks or oil seal failures.

6　Remove the cap from the valve cover to add oil **(see illustration)**. Use a funnel to prevent spills. After adding the oil, install the filler cap hand tight. Start the engine and look carefully for any small leaks around the oil filter or drain plug. Stop the engine and check the oil level again after it has had sufficient time to drain from the upper block and cylinder head galleys.

7　Checking the oil level is an important preventive maintenance step. A continually dropping oil level indicates oil leakage through damaged seals, from loose connections, or past worn rings or valve guides. If the oil looks milky in color or has water droplets in it, a cylinder head gasket may be blown. The engine should be checked immediately. The condition of the oil should also be checked. Each time you check the oil level, slide your thumb and index finger up the dipstick before wiping off the oil. If you see small dirt or metal particles clinging to the dipstick, the oil should be changed (see Section 8).

Engine coolant

Refer to illustration 4.8

Warning: *Do not allow antifreeze to come in contact with your skin or painted surfaces of the vehicle. Flush contaminated areas immediately with plenty of water. Don't store new coolant or leave old coolant lying around where it's accessible to children or pets – they're attracted by its sweet smell. Ingestion of even a small amount of coolant can be fatal! Wipe up garage floor and drip pan spills immediately. Keep antifreeze containers covered and repair cooling system leaks as soon as they're noticed.*

8　All vehicles covered by this manual are equipped with a pressurized coolant recovery system. A white coolant expansion tank located on the right (passenger) side of the engine compartment is connected by a hose to the engine **(see illustration)**. As the engine heats up during operation, the expanding coolant fills the tank. As the engine cools, the coolant level will recede to the Full Cold line on the tank.

9　The coolant level should be checked regularly. The level will vary with the temperature of the engine. When the engine is cold, the coolant level should be at or slightly above the Full Cold rib on the tank. Once the engine has warmed up, the level should be at or near the base of the filler neck. If it isn't, allow the fluid in the tank to cool, then remove the cap (see Step 11) from the reservoir and add coolant to bring the level up to the Full Cold line. Use only phosphate-free, low-silicate ethylene/glycol type coolant and water in the mixture ratio recommended in this Chapter's Specifications or in your owner's manual. Do not use supplemental inhibitors or additives. If only a small amount of coolant is required to bring the system up to the proper level, water can be used. However, repeated additions of water will dilute the recommended antifreeze and water solution. In order to maintain the proper ratio of antifreeze and water, it is advisable to top up

the coolant level with the correct mixture. Refer to your owner's manual for the recommended ratio.

10　If the coolant level drops within a short time after replenishment, there may be a leak in the system. Inspect the radiator, hoses, coolant filler cap, drain plugs, air bleeder plugs and water pump. If no leak is evident, have the cap pressure tested by your dealer.

Warning: *Never remove the coolant expansion tank cap when the engine is running or has just been shut down, because the cooling system is hot. Escaping steam and scalding liquid could cause serious injury.*

11　If it is necessary to open the pressure cap, wait until the system has cooled completely, then wrap a thick cloth around the cap and turn it to the first stop. If any steam escapes, wait until the system has cooled further, then remove the cap.

12　When checking the coolant level, always note its condition. It should be relatively clear. If it is murky or rust colored, the system should be drained, flushed and refilled. Even if the coolant appears to be normal, the corrosion inhibitors wear out with use, so it must be replaced at the specified intervals.

13　Do not allow antifreeze to come in contact with your skin or painted surfaces of the vehicle. Flush contacted areas immediately with plenty of water.

Washer fluid

Refer to illustration 4.14

14　Fluid for the windshield washer system is stored in a plastic reservoir which is located at the right front corner of the engine compartment **(see illustration)**. In milder climates, plain water can be used to top up the reservoir, but the reservoir should be kept no more than two-thirds full to allow for expansion should the water freeze. In colder climates, the use of a specially designed windshield washer fluid, available at your dealer and any auto parts store, will help lower the freezing point of the fluid. Mix the solution with water in accordance with the manufacturer's direc-

4.17 The brake fluid level should be near the neck of the translucent plastic reservoir

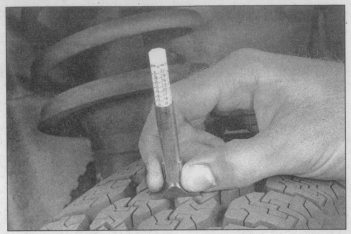

5.2 Use a tire tread depth indicator to monitor tire wear - they are available at auto parts stores and service stations and cost very little

tions on the container. Do not use regular antifreeze. It will damage the vehicle's paint.

Battery electrolyte

15   On models not equipped with a sealed battery, check the electrolyte level of all six battery cells. Remove the filler caps and check the level - it must be at or near the split ring. If the level is low, add distilled water. Install and securely retighten the cap. **Caution:** *Overfilling the cells may cause elec-*

trolyte to spill over during periods of heavy charging, causing corrosion or damage.

Brake and clutch fluid

Refer to illustration 4.17

16   The brake master cylinder is mounted on the front of the power booster unit in the engine compartment and the clutch master cylinder is located next to it.

17   To check the fluid level of the brake master cylinder reservoir, simply look at the

reservoir and make sure the fluid is at or near the base of the filler neck **(see illustration)**. To check the fluid level of the clutch master cylinder reservoir, remove the cap and note whether the fluid level is within 1/2-inch of the top. Make sure the rubber diaphragm is fully extended before installing the cap.

18   If the level is low for either reservoir, wipe the cap and the area around it with a clean rag to prevent contamination of the brake or clutch system before lifting the cover.

UNDERINFLATION

CUPPING

Cupping may be caused by:

- Underinflation and/or mechanical irregularities such as out-of-balance condition of wheel and/or tire, and bent or damaged wheel.
- Loose or worn steering tie-rod or steering idler arm.
- Loose, damaged or worn front suspension parts.

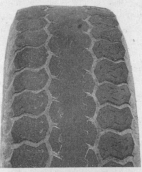

OVERINFLATION

INCORRECT TOE-IN OR EXTREME CAMBER

FEATHERING DUE TO MISALIGNMENT

5.3 This chart will help you determine the condition of the tires, the probable cause(s) of abnormal wear and the corrective action necessary

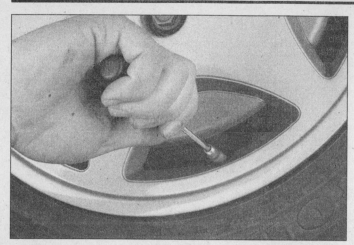

5.4a If a tire loses air on a steady basis, check the valve stem core first to make sure it's snug (special inexpensive wrenches are commonly available at auto parts stores)

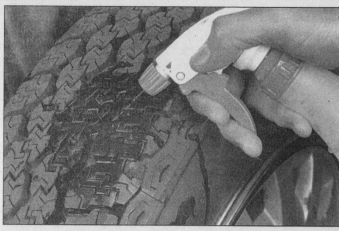

5.4b If the valve stem core is tight, raise the corner of the vehicle with the low tire and spray a soapy water solution onto the tread as the tire is turned slowly - leaks will cause small bubbles to appear

19   Add only the specified brake fluid to the brake or clutch reservoir (refer to *Recommended lubricants and fluids* at the front of this Chapter or to your owner's manual). Mixing different types of brake fluid can damage the system. Fill the brake master cylinder reservoir only to the dotted line - this brings the fluid to the correct level when you put the cover back on. **Warning:** *Use caution when filling either reservoir - brake fluid can harm your eyes and damage painted surfaces. Do not use brake fluid that has been opened for more than one year or has been left open. Brake fluid absorbs moisture from the air. Excess moisture can cause a dangerous loss of braking.*

20   While the reservoir cap is removed, inspect the master cylinder reservoir for contamination. If deposits, dirt particles or water droplets are present, the system should be drained and refilled (see Chapter 9).

21   After filling the reservoir to the proper level, make sure the lid is properly seated to prevent fluid leakage and/or system pressure loss.

22   The brake fluid in the master cylinder will drop slightly as the brake pads at each wheel wear down during normal operation. If the master cylinder requires repeated replenishing to keep it at the proper level, this is an indication of leakage in the brake system, which should be corrected immediately. Check all brake lines and connections, along with the wheel cylinders and booster (see Section 16 for more information).

23   If, upon checking the master cylinder fluid level, you discover one or both reservoirs empty or nearly empty, the brake system should be bled (see Chapter 9).

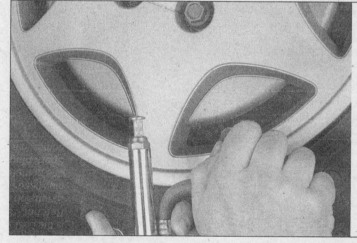

5.8 To extend the life of the tires, check the air pressure at least once a week with an accurate gauge (don't forget the spare!)

5   Tire and tire pressure checks (every 250 miles or weekly)

Refer to illustrations 5.2, 5.3, 5.4a, 5.4b and 5.8

1   Periodic inspection of the tires may spare you from the inconvenience of being stranded with a flat tire. It can also provide you with vital information regarding possible problems in the steering and suspension systems before major damage occurs.

2   Normal tread wear can be monitored with a simple, inexpensive device known as a tread depth indicator **(see illustration)**. When the tread depth reaches the specified minimum, replace the tire(s).

3   Note any abnormal tread wear **(see illustration)**. Tread pattern irregularities such as cupping, flat spots and more wear on one side than the other are indications of front end alignment and/or balance problems. If any of these conditions are noted, take the vehicle to a tire shop or service station to correct the problem.

4   Look closely for cuts, punctures and embedded nails or tacks. Sometimes a tire will hold its air pressure for a short time or leak down very slowly even after a nail has embedded itself into the tread. If a slow leak persists, check the valve stem core to make sure it is tight **(see illustration)**. Examine the tread for an object that may have embedded itself into the tire or for a "plug" that may have begun to leak (radial tire punctures are repaired with a plug that is installed in a puncture). If a puncture is suspected, it can be easily verified by spraying a solution of soapy water onto the puncture area **(see illustration)**. The soapy solution will bubble if there is a leak. Unless the puncture is inordinately large, a tire shop or gas station can usually repair the punctured tire.

5   Carefully inspect the inner sidewall of each tire for evidence of brake fluid leakage. If you see any, inspect the brakes immediately.

6   Correct tire air pressure adds miles to the life span of the tires, improves mileage and enhances overall ride quality. Tire pressure cannot be accurately estimated by looking at a tire, particularly if it is a radial. A tire pressure gauge is therefore essential. Keep an accurate gauge in the glove box. The pressure gauges fitted to the nozzles of air hoses at gas stations are often inaccurate.

7   Always check tire pressure when the tires are cold. "Cold," in this case, means the vehicle has not been driven over a mile in the three hours preceding a tire pressure check. A pressure rise of four to eight pounds is not uncommon once the tires are warm.

8   Unscrew the valve stem cap protruding from the wheel or hubcap and push the

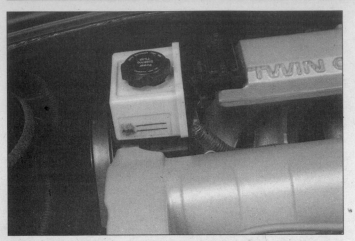

6.3 The power steering fluid reservoir is translucent so the fluid level can be checked without removing the cap

7.4a The automatic transaxle dipstick (arrow) is located next to the ignition coil

gauge firmly onto the valve **(see illustration)**. Note the reading on the gauge and compare this figure to the recommended tire pressure shown on the tire placard on the left door. Be sure to reinstall the valve cap to keep dirt and moisture out of the valve stem mechanism. Check all four tires and, if necessary, add enough air to bring them up to the recommended pressure levels.

9   Don't forget to keep the spare tire inflated to the specified pressure (consult your owner's manual). Note that the air pressure specified for the compact spare is significantly higher than the pressure of the regular tires.

7.4b Check the fluid with the transaxle at normal operating temperature - the level should be kept in the cross-hatched area (at the upper hole)

6   Power steering fluid level check (every 3000 miles or 3 months)

Refer to illustration 6.3

1   Unlike manual steering, the power steering system relies on fluid which may, over a period of time, require replenishing.

2   The fluid reservoir for the power steering pump is located on the inner fender panel near the front of the engine on the right (passenger) side of the engine compartment.

3   On these models the reservoir is translucent plastic and the fluid level can be checked visually **(see illustration)**.

4   The fluid level should be kept between the FULL and ADD marks on the reservoir.

5   If additional fluid is required, pour the specified type directly into the reservoir, using a funnel to prevent spills.

6   If the reservoir requires frequent fluid additions, all power steering hoses, hose connections, the power steering pump and the rack-and-pinion assembly should be carefully checked for leaks.

7   Automatic transaxle fluid level check (every 3000 miles or 3 months)

Refer to illustrations 7.4a and 7.4b

1   The level of the automatic transaxle fluid

should be carefully maintained. Low fluid level can lead to slipping or loss of drive, while overfilling can cause foaming, loss of fluid and transaxle damage.

2   The transaxle fluid level should only be checked when the transaxle is hot (at its normal operating temperature). If the vehicle has just been driven over 10 miles (15 miles in a frigid climate), and the fluid temperature is 160 to 175-degrees F, the transaxle is hot.

Caution: *If the vehicle has just been driven for a long time at high speed or in city traffic in hot weather, or if it has been pulling a trailer, an accurate fluid level reading cannot be obtained. Allow the fluid to cool down for about 30 minutes.*

3   Park the vehicle on level ground, set the parking brake and start the engine. While the engine is idling, depress the brake pedal and move the selector lever through all the gear ranges, beginning and ending in Park.

4   With the engine still idling, remove the dipstick from its tube, located at the left (drivers) side of the engine compartment, below the ignition coil **(see illustration)**. Check the level of the fluid on the dipstick **(see illustration)** and note its condition.

5   Wipe the fluid from the dipstick with a clean rag and reinsert it back into the filler tube until the cap seats.

6   Pull the dipstick out again and note the fluid level. The fluid level should be at the upper hole (in the middle of the cross-

hatched area) or near the FULL mark. If the level is low, add the specified automatic transmission fluid through the dipstick tube with a funnel.

7   Add just enough of the recommended fluid to fill the transaxle to the proper level. It takes about one pint to raise the level from the low mark to the high mark when the fluid is hot, so add the fluid a little at a time and keep checking the level until it is correct.

8   The condition of the fluid should also be checked along with the level. If the fluid at the end of the dipstick is black or a dark reddish brown color, or if it emits a burned smell, the fluid should be changed (see Section 27). If you are in doubt about the condition of the fluid, purchase some new fluid and compare the two for color and smell.

8   Engine oil and oil filter change (every 3000 miles or 3 months)

Refer to illustrations 8.2, 8.7, 8.13 and 8.15

1   Frequent oil changes are the best preventive maintenance the home mechanic can give the engine, because aging oil becomes diluted and contaminated, which leads to premature engine wear.

2   Make sure that you have all the necessary tools before you begin this procedure **(see illustration)**. You should also have plenty of rags or newspapers handy for mop-

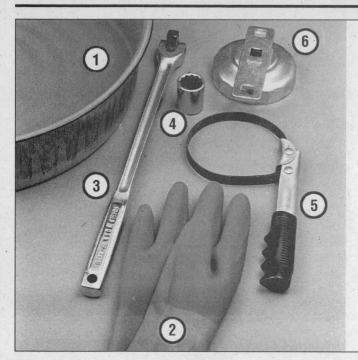

8.2 These tools are required when changing the engine oil and filter:

1   **Drain pan** - It should be fairly shallow in depth, but wide to prevent spills
2   **Rubber gloves** - When removing the drain plug and filter, you will get oil on your hands (the gloves will prevent burns)
3   **Breaker bar** - Sometimes the oil drain plug is tight, and a long breaker bar is needed to loosen it
4   **Socket** - To be used with the breaker bar or a ratchet (must be the correct size to fit the drain plug - six-point preferred)
5   **Filter wrench** - This is a metal band-type wrench, which requires clearance around the filter to be effective
6   **Filter wrench** - This type fits on the bottom of the filter and can be turned with a ratchet or breaker bar (different-size wrenches are available for different types of filters)

ping up any spills.

3   Access to the underside of the vehicle is greatly improved if the vehicle can be lifted on a hoist, driven onto ramps or supported by jackstands. **Warning:** *Do not work under a vehicle which is supported only by a bumper, hydraulic or scissors-type jack.*

4   If this is your first oil change, get under the vehicle and familiarize yourself with the location of the oil drain plug. The engine and exhaust components will be warm during the actual work, so try to anticipate any potential problems before the engine and accessories are hot.

5   Park the vehicle on a level spot. Start the engine and allow it to reach its normal operating temperature (the needle on the temperature gauge should be at least above the bottom mark). Warm oil and sludge will flow out more easily. Turn off the engine when it's warmed up. Remove the oil filler cap.

6   Raise the vehicle and support it on jackstands. **Warning:** *To avoid personal injury, never get beneath the vehicle when it is supported by only by a jack. The jack provided with your vehicle is designed solely for raising the vehicle to remove and replace the wheels. Always use jackstands to support the vehicle when it becomes necessary to place your body underneath the vehicle.*

7   Being careful not to touch the hot exhaust components, place the drain pan under the drain plug in the bottom of the pan and remove the plug **(see illustration)**. You may want to wear gloves while unscrewing the plug the final few turns if the engine is really hot.

8   Allow the old oil to drain into the pan. It may be necessary to move the pan farther under the engine as the oil flow slows to a trickle. Inspect the old oil for the presence of metal shavings and chips.

9   After all the oil has drained, wipe off the drain plug with a clean rag. Even minute metal particles clinging to the plug would immediately contaminate the new oil.

10   Clean the area around the drain plug opening, reinstall the plug and tighten it securely, but do not strip the threads.

11   Move the drain pan into position under the oil filter.

12   Remove all tools, rags, etc. from under the vehicle, being careful not to spill the oil in the drain pan, then lower the vehicle.

13   Loosen the oil filter **(see illustration)** by turning it counterclockwise with the filter wrench. Any standard filter wrench will work. Sometimes the oil filter is screwed on so tightly that it cannot be loosened. If this situation occurs, punch a metal bar or long screwdriver directly through the side of the canister and use it as a T-bar to turn the filter. Be prepared for oil to spurt out of the canister as it

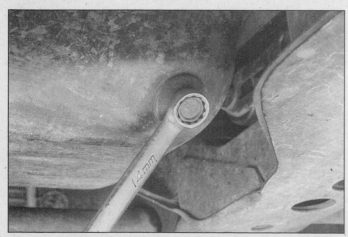

8.7 Use a proper size box-end wrench or socket to remove the oil drain plug without rounding it off

8.13 Since the oil filter (accessible from below) is on very tight, you'll need a special wrench for removal - DO NOT use the wrench to tighten the new filter

8.15 Lubricate the oil filter gasket with clean engine oil before installing the filter on the engine

9.6 Press the tab and push the wiper down out of the hook in the end of arm

is punctured. Once the filter is loose, use your hands to unscrew it from the block. Just as the filter is detached from the block, immediately tilt the open end up to prevent the oil inside the filter from spilling out. **Warning:** *The engine exhaust manifold may still be hot, so be careful.*

14   With a clean rag, wipe off the mounting surface on the block. If a residue of old oil is allowed to remain, it will smoke when the block is heated up. It will also prevent the new filter from seating properly. Also make sure that the none of the old gasket remains stuck to the mounting surface. It can be removed with a scraper if necessary.

15   Compare the old filter with the new one to make sure they are the same type. Smear some engine oil on the rubber gasket of the new filter and screw it into place **(see illustration)**. Because overtightening the filter will damage the gasket, do not use a filter wrench to tighten the filter. Tighten it by hand until the gasket contacts the seating surface. Then seat the filter by giving it an additional 3/4-turn.

16   Add new oil to the engine through the oil filler cap in the valve cover. Use a spout or funnel to prevent oil from spilling onto the top of the engine. Pour three quarts of fresh oil into the engine. Wait a few minutes to allow the oil to drain into the pan, then check the level on the oil dipstick (see Section 4 if necessary). If the oil level is at or near the F mark, install the filler cap hand tight, start the engine and allow the new oil to circulate.

17   Allow the engine to run for about a minute. While the engine is running, look under the vehicle and check for leaks at the oil pan drain plug and around the oil filter. If either is leaking, stop the engine and tighten the plug or filter slightly.

18   Wait a few minutes to allow the oil to trickle down into the pan, then recheck the level on the dipstick and, if necessary, add enough oil to bring the level to the F mark.

19   During the first few trips after an oil change, make it a point to check frequently for leaks and proper oil level.

20   The old oil drained from the engine cannot be reused in its present state and should be discarded. Oil reclamation centers, auto repair shops and gas stations will normally accept the oil, which can be refined and used again. After the oil has cooled, it can be drained into a suitable container (capped plastic jugs, topped bottles, milk cartons, etc.) for transport to one of these disposal sites.

9   Windshield wiper blade inspection and replacement (every 6000 miles or 6 months)

Refer to illustrations 9.6 and 9.7

1   The wiper and blade assembly should be inspected periodically for damage, loose components and cracked or worn blade elements.

2   Road film can build up on the wiper blades and affect their efficiency, so they should be washed regularly with a mild detergent solution.

3   The action of the wiping mechanism can loosen bolts, nuts and fasteners, so they should be checked and tightened, as necessary, at the same time the wiper blades are checked.

4   If the wiper blade elements are cracked, worn or warped, or no longer clean adequately, they should be replaced with new ones.

5   Lift the arm assembly away from the glass for clearance.

6   Depress the retaining tab and slide the blade assembly down, out of the hook in the end of the wiper arm **(see illustration)**.

7   Bend the end of the wiper element out of the way and use needle-nose pliers to pull the two support rods out of the element **(see illustration)**.

8   With the support rods removed, slide the element out of the blade assembly.

9   Slide the new element into place and insert the support rods.

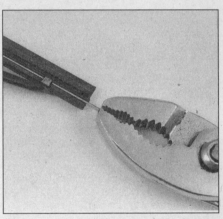

9.7 Use pliers to pull the support rods out of the wiper element

10   Battery check, maintenance and charging (every 6000 miles or 6 months)

Refer to illustrations 10.1 and 10.8

Warning: *Certain precautions must be followed when checking and servicing the battery. Hydrogen gas, which is highly flammable, is always present in the battery cells, so keep lighted tobacco and all other open flames and sparks away from the battery. The electrolyte inside the battery is actually dilute sulfuric acid, which will cause injury if splashed on your skin or in your eyes. It will also ruin clothes and painted surfaces. When removing the battery cables, always detach the negative cable first and hook it up last!*

1   A routine preventive maintenance program for the battery in your vehicle is the only way to ensure quick and reliable starts. But before performing any battery maintenance, make sure that you have the proper equipment necessary to work safely around the battery **(see illustration)**.

2   There are also several precautions that should be taken whenever battery maintenance is performed. Before servicing the battery, always turn the engine and all acces-

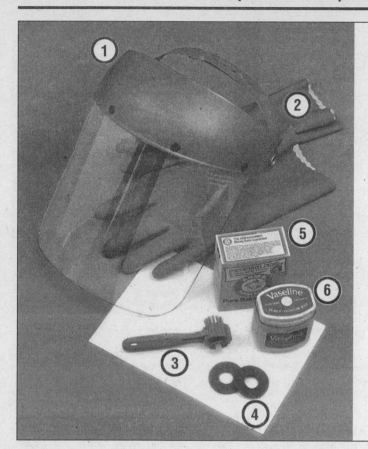

10.1 Tools and materials required for battery maintenance

1 **Face shield/safety goggles** - *When removing corrosion with a brush, the acidic particles can easily fly up into your eyes*
2 **Rubber gloves** - *Another safety item to consider when servicing the battery - remember that's acid inside the battery!*
3 **Battery terminal/cable cleaner** - *This wire brush cleaning tool will remove all traces of corrosion from the battery and cable*
4 **Treated felt washers** - *Placing one of these on each terminal, directly under the cable end, will help prevent corrosion (be sure to get the correct type for side-terminal batteries)*
5 **Baking soda** - *A solution of baking soda and water can be used to neutralize corrosion*
6 **Petroleum jelly** - *A layer of this on the battery terminal bolts will help prevent corrosion*

sories off and disconnect the cable from the negative terminal of the battery.

3 The battery produces hydrogen gas, which is both flammable and explosive. Never create a spark, smoke or light a match around the battery. Always charge the battery in a ventilated area.

4 Electrolyte contains poisonous and corrosive sulfuric acid. Do not allow it to get in your eyes, on your skin on your clothes. Never ingest it. Wear protective safety glasses when working near the battery. Keep children away from the battery.

5 Note the external condition of the battery. Look for any corroded or loose connections, cracks in the case or cover or loose hold-down clamp nut or bolt. Also check the entire length of each cable for cracks and frayed conductors.

6 If corrosion, which looks like white, fluffy deposits is evident, particularly around the terminals, the battery should be removed for cleaning. Remove the cable bolts, being careful to remove the ground (negative) cable first and detach the cables. Remove the clamp bolt and nut and lift the battery from the engine compartment.

7 Clean the cable terminals and cable connections with a battery brush or a terminal cleaner and a solution of warm water and baking soda. Wash the terminals and the top of the battery case with the same solution but make sure that the solution doesn't get into the battery. When cleaning the cables, terminals and battery top, wear safety goggles and rubber gloves to prevent any solution from

coming in contact with your eyes or hands. Wear old clothes too - even diluted, sulfuric acid splashed onto clothes will burn holes in them. If the terminals have been extensively corroded, clean them up with a terminal cleaner. Thoroughly wash all cleaned areas with plain water.

8 Make sure the battery tray is in good condition and the hold-down clamp bolt or nut is tight **(see illustration)**. If the battery is removed from the tray, make sure no parts remain in the bottom of the tray when the battery is reinstalled. When reinstalling the hold-down clamp bolt and nut, do not overtighten it. Be sure to reinstall the battery cover.

9 Information on removing and installing the battery can be found in Chapter 5. Information on jump starting can be found at the front of this manual. For more detailed battery checking procedures, refer to the *Haynes Automotive Electrical Manual*.

Cleaning

10 Corrosion on the hold-down components, battery case and surrounding areas can be removed with a solution of water and baking soda. Thoroughly rinse all cleaned areas with plain water.

11 Any metal parts of the vehicle damaged by corrosion should be covered with a zinc-based primer, then painted.

Charging

Warning: *When batteries are being charged, hydrogen gas, which is very explosive and*

10.8 Make sure the battery hold-down nut (arrow) and the bolt on the other side are tight

flammable, is produced. Do not smoke or allow open flames near a charging or a recently charged battery. Wear eye protection when near the battery during charging. Also, make sure the charger is unplugged before connecting or disconnecting the battery from the charger.

12 Slow-rate charging is the best way to restore a battery that's discharged to the point where it will not start the engine. It's also a good way to maintain the battery charge in a vehicle that's only driven a few miles between starts. Maintaining the battery charge is particularly important in the winter

when the battery must work harder to start the engine and electrical accessories that drain the battery are in greater use.

13   It's best to use a one or two-amp battery charger (sometimes called a "trickle" charger). They are the safest and put the least strain on the battery. They are also the least expensive. For a faster charge, you can use a higher amperage charger, but don't use one rated more than 1/10th the amp/hour rating of the battery. Rapid boost charges that claim to restore the power of the battery in one to two hours are hardest on the battery and can damage batteries not in good condition. This type of charging should only be used in emergency situations.

14   The average time necessary to charge a battery should be listed in the instructions that come with the charger. As a general rule, a trickle charger will charge a battery in 12 to 16 hours.

---

11   Drivebelt check and replacement (every 6000 miles or 6 months)

Refer to illustrations 11.2 and 11.4

Check

1   A single serpentine drivebelt is located at the front of the engine and plays an important role in the overall operation of the engine and its components. Due to its function and material make up, the belt is prone to wear and should be periodically inspected The single serpentine drivebelt drives the alternator, power steering pump and air conditioning compressor.

2   With the engine off, open the hood and locate the drivebelt at the right end (passenger side) of the engine. With a flashlight, check the belt for separation of the rubber on both sides of the core, separation of the ribs from the adhesive rubber, cracking or separation of the ribs, and torn or worn ribs or cracks in the inner ridges of the ribs **(see illustration)**. Also check for fraying and glazing, which gives the belt a shiny appearance.

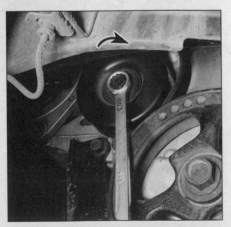

11.4 Use a 14 mm wrench to lift the tensioner in the direction shown to release the tension from the belt

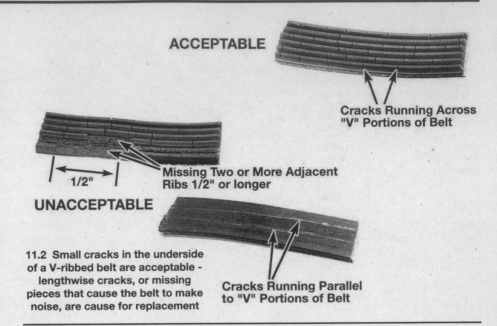

ACCEPTABLE

Cracks Running Across "V" Portions of Belt

1/2"

Missing Two or More Adjacent Ribs 1/2" or longer

UNACCEPTABLE

Cracks Running Parallel to "V" Portions of Belt

11.2 Small cracks in the underside of a V-ribbed belt are acceptable - lengthwise cracks, or missing pieces that cause the belt to make noise, are cause for replacement

Both sides of the belt should be inspected, which means you will have to twist the belt to check the underside. Use your fingers to feel the belt where you can't see it. If any of the above conditions are evident, replace the belt.

3   The tension of the belt is checked visually. Locate the belt tensioner at the front of the engine under the air conditioning compressor on the right (passenger)side, then find the tensioner operating marks located on the side of the tensioner. If the indicator mark is outside the operating range, the belt should be replaced.

Replacement

4   Use a 9/16-inch box-end wrench to rotate the tensioner clockwise and release the tension on the belt **(see illustration)**. Remove the belt and carefully release the tensioner. Route the new belt over the pulleys, again rotating the tensioner to allow the belt to be installed, then release the belt tensioner. Make sure the belt is properly positioned on each pulley.

---

12   Underhood hose check and replacement (every 6000 miles or 6 months)

Warning: *Replacement of air conditioning hoses must be left to a dealer service department or air conditioning shop that has the equipment to depressurize the system safely. Never remove air conditioning components or hoses until the system has been depressurized.*

General

1   High temperatures in the engine compartment can cause the deterioration of the rubber and plastic hoses used for engine,

accessory and emission systems operation. Periodic inspection should be made for cracks, loose clamps, material hardening and leaks.

2   Information specific to the cooling system hoses can be found in Section 13.

3   Some, but not all, hoses are secured to the fittings with clamps. Where clamps are used, check to be sure they haven't lost their tension, allowing the hose to leak. If clamps aren't used, make sure the hose has not expanded and/or hardened where it slips over the fitting, allowing it to leak.

Vacuum hoses

4   It's quite common for vacuum hoses, especially those in the emissions system, to be color coded or identified by colored stripes molded into them. Various systems require hoses with different wall thicknesses, collapse resistance and temperature resistance. When replacing hoses, be sure the new ones are made of the same material.

5   Often the only effective way to check a hose is to remove it completely from the vehicle. If more than one hose is removed, be sure to label the hoses and fittings to ensure correct installation.

6   When checking vacuum hoses, be sure to include any plastic T-fittings in the check. Inspect the fittings for cracks and the hose where it fits over the fitting for distortion, which could cause leakage.

7   A small piece of vacuum hose (1/4-inch inside diameter) can be used as a stethoscope to detect vacuum leaks. Hold one end of the hose to your ear and probe around vacuum hoses and fittings, listening for the "hissing" sound characteristic of a vacuum leak. **Warning:** *When probing with the vacuum hose stethoscope, be very careful not to come into contact with moving engine components such as the drivebelts, cooling fan, etc.*

Fuel hose

Warning: *There are certain precautions which must be taken when inspecting or servicing fuel system components. Work in a well ventilated area and do not allow open flames (cigarettes, appliance pilot lights, etc.) or bare light bulbs near the work area. Mop up any spills immediately and do not store fuel soaked rags where they could ignite.*

8   Check all rubber fuel lines for deterioration and chafing. Check especially for cracks in areas where the hose bends and just before fittings, such as where a hose attaches to the fuel filter.

9   High quality fuel line, usually identified by the word *Fluoroelastomer* printed on the hose, should be used for fuel line replacement. Never, under any circumstances, use vacuum line, clear plastic tubing or water hose for fuel lines.

10   Spring-type clamps are commonly used on fuel lines. These clamps often lose their tension over a period of time, and can be "sprung" during removal. Replace all spring-type clamps with screw clamps whenever a hose is replaced.

Metal lines

11   Sections of metal line are often used for fuel line between the fuel pump and fuel injection unit. Check carefully to be sure the line has not been bent or crimped and that cracks have not started in the line.

12   If a section of metal fuel line must be replaced, only seamless steel tubing should be used, since copper and aluminum tubing don't have the strength necessary to withstand normal engine vibration.

13   Check the metal brake lines where they enter the master cylinder and brake proportioning unit (if used) for cracks in the lines or loose fittings. Any sign of brake fluid leakage calls for an immediate thorough inspection of the brake system.

13   Cooling system check (every 6000 miles or 6 months)

Refer to illustration 13.4

1   Many major engine failures can be attributed to a faulty cooling system. If the vehicle is equipped with an automatic transaxle, the cooling system also cools the transaxle fluid and thus plays an important role in prolonging transaxle life.

2   The cooling system should be checked with the engine cold. Do this before the vehicle is driven for the day or after the engine has been shut off for at least three hours.

3   Remove the expansion tank cap by slowly unscrewing it. **Warning:** *If you hear a hissing sound (indicating the engine isn't cold and there is still pressure in the system), wait until it stops, then continue to unscrew the cap.* Thoroughly clean the cap, inside and out, with clean water. Also clean the opening on the tank. All traces of corrosion should be

Check for a chafed area that could fail prematurely.

Check for a soft area indicating the hose has deteriorated inside.

Overtightening the clamp on a hardened hose will damage the hose and cause a leak.

Check each hose for swelling and oil-soaked ends. Cracks and breaks can be located by squeezing the hose.

13.4 Hoses, like drivebelts, have a habit of failing at the worst possible time - to prevent the inconvenience of a blown radiator or heater hose, inspect them carefully as shown here

removed. The coolant inside the expansion tank should be relatively transparent. If it's rust colored, the system should be drained and refilled (see Section 25). If the coolant level isn't up to the top, add additional antifreeze/coolant mixture (see Section 4).

4   Carefully check the large upper and lower radiator hoses along with the smaller diameter heater hoses which run from the engine to the firewall. Inspect each hose along its entire length, replacing any hose which is cracked, swollen or shows signs of deterioration. Cracks may become more apparent if the hose is squeezed (**see illustration**). Regardless of condition, it's a good idea to replace hoses with new ones every two years.

5   Make sure that all hose connections are tight. A leak in the cooling system will usually show up as white or rust colored deposits on the areas adjoining the leak. If wire-type

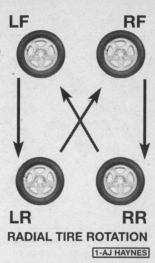

LF   RF

LR   RR

RADIAL TIRE ROTATION

1-AJ HAYNES

14.2 The recommended tire rotation pattern for these vehicles

clamps are used at the ends of the hoses, it may be a good idea to replace them with more secure screw-type clamps.

6   Use compressed air or a soft brush to remove bugs, leaves, etc. from the front of the radiator or air conditioning condenser. Be careful not to damage the delicate cooling fins or cut yourself on them.

7   Every other inspection, or at the first indication of cooling system problems, have the cap and system pressure tested. If you don't have a pressure tester, most gas stations and repair shops will do this for a minimal charge.

14   Tire rotation (every 6000 miles or 6 months)

Refer to illustration 14.2

1   The tires should be rotated at the specified intervals and whenever uneven wear is noticed. Since the vehicle will be raised and the tires removed anyway, check the brakes (see Section 16) at this time.

2   Radial tires must be rotated in a specific pattern (**see illustration**).

3   Refer to the information in *Jacking and towing* at the front of this manual for the proper procedures to follow when raising the vehicle and changing a tire. If the brakes are to be checked, do not apply the parking brake as stated. Make sure the tires are blocked to prevent the vehicle from rolling.

4   Preferably, the entire vehicle should be raised at the same time. This can be done on a hoist or by jacking up each corner and then lowering the vehicle onto jackstands placed under the frame rails. Always use four jackstands and make sure the vehicle is firmly supported.

5   After rotation, check and adjust the tire pressures as necessary and be sure to check the lug nut tightness.

6   For further information on the wheels and tires, refer to Chapter 10.

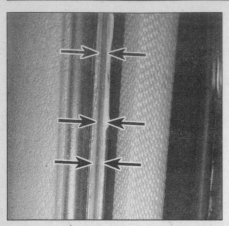

15.2 Apply the grease evenly along the contact surfaces of the full length of the automatic seat belt tracks

15 Automatic seat belt track lubrication (every 6,000 miles or 6 months)

Refer to illustrations 15.2

1   At the specified intervals or whenever the slider makes a chattering noise, lubricate the contact surfaces of the automatic seat belt track with white lithium base grease.

2   With the slider in the parked position, use a small brush to apply the grease sparingly to the entire length of the seat belt track **(see illustration)**.

3   Turn the ignition On and cycle the slider so that it travels the entire length of the track three times by opening and closing the door.

16 Brake check (every 15,000 miles or 12 months)

Warning: *Brake dust produced by lining wear and deposited on brake components may contain asbestos, which is hazardous to your* health. *DO NOT blow it out with compressed air and DO NOT inhale it! DO NOT use gasoline or solvents to remove the dust. Brake system cleaner should be used to flush the dust into a drain pan. After the brake components are wiped clean with a damp rag, dispose of the contaminated rag(s) and solvent in a covered and labeled container. Try to use non-asbestos replacement parts whenever possible.*

Note: *For detailed photographs of the brake system, refer to Chapter 9.*

1   In addition to the specified intervals, the brakes should be inspected every time the wheels are removed or whenever a defect is suspected. Any of the following symptoms could indicate a potential brake system defect: The vehicle pulls to one side when the brake pedal is depressed; the brakes make squealing or dragging noises when applied; brake travel is excessive; the pedal pulsates; brake fluid leaks, usually onto the inside of the tire or wheel.

2   The disc brake pads have built-in wear indicators which should make a high pitched squealing or scraping noise when they are worn to the replacement point. When you hear this noise, replace the pads immediately or expensive damage to the discs can result.

3   Loosen the wheel lug nuts.

4   Raise the vehicle and place it securely on jackstands.

5   Remove the wheels (see *Jacking and towing* at the front of this book, or your owner's manual, if necessary).

Disc brakes

Refer to illustration 16.6

6   There are two pads - an outer and an inner - in each caliper. The pads are visible through small inspection holes in each caliper **(see illustration)**.

7   Check the pad thickness by looking at each end of the caliper and through the inspection hole in the caliper body. If the lining material is less than the thickness listed in this Chapter's Specifications, replace the pads. **Note:** *Keep in mind that the lining material is riveted or bonded to a metal backing plate and the metal portion is not included in this measurement.*

8   If it is difficult to determine the exact thickness of the remaining pad material by the above method, or if you are at all concerned about the condition of the pads, remove the caliper(s), then remove the pads from the calipers for further inspection (refer to Chapter 9).

9   Once the pads are removed from the calipers, clean them with brake cleaner and remeasure them with a small steel pocket ruler or a vernier caliper.

10   Measure the disc thickness with a micrometer to make sure that it still has service life remaining. If any disc is thinner than the specified minimum thickness, replace it (refer to Chapter 9). Even if the disc has service life remaining, check its condition. Look for scoring, gouging and burned spots. If these conditions exist, remove the disc and have it resurfaced (see Chapter 9).

11   Before installing the wheels, check all brake lines and hoses for damage, wear, deformation, cracks, corrosion, leakage, bends and twists, particularly in the vicinity of the rubber hoses at the calipers. Check the clamps for tightness and the connections for leakage. Make sure that all hoses and lines are clear of sharp edges, moving parts and the exhaust system. If any of the above conditions are noted, repair, reroute or replace the lines and/or fittings as necessary (see Chapter 9). Be sure to tighten the lug nuts to the torque listed in this Chapter's Specifications.

Rear drum brakes

Refer to illustration 16.13

12   Refer to Chapter 9 and remove the rear brake drums.

13   Note the thickness of the lining material on the rear brake shoes **(see illustration)** and look for signs of contamination by brake fluid and grease. If the lining material is within 1/16-inch of the recessed rivets or metal

16.6 You will find an inspection hole like this in each caliper - placing a ruler across the hole should enable you to determine the thickness of remaining pad material for both inner and outer pads

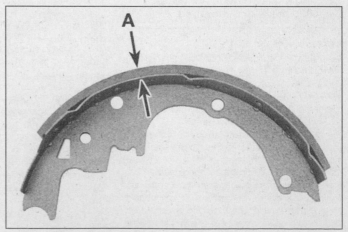

16.13 If the lining is bonded to the brake shoe, measure the lining thickness from the outer surface to the metal shoe, as shown here; if the lining is riveted to the shoe, measure from the lining outer surface to the rivet head

17.1a On single cam models, remove the two air cleaner cover bolts

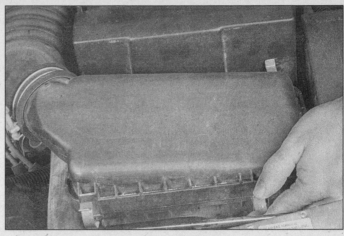

17.1b On all models, release the clips to remove the cover

17.2 Lift the filter out of the housing

shoes, replace the brake shoes with new ones. The shoes should also be replaced if they are cracked, glazed (shiny lining surfaces) or contaminated with brake fluid or grease. See Chapter 9 for the replacement procedure.

14   Check the shoe return and hold-down springs and the adjusting mechanism to make sure they're installed correctly and in good condition. Deteriorated or distorted springs, if not replaced, could allow the linings to drag and wear prematurely.

15   Check the wheel cylinders for leakage by carefully peeling back the rubber boots. If brake fluid is noted behind the boots, the wheel cylinders must be replaced (see Chapter 9).

16   Check the drums for cracks, score marks, deep scratches and hard spots, which will appear as small discolored areas. If imperfections cannot be removed with emery cloth, the drums must be resurfaced by an automotive machine shop (see Chapter 9 for more detailed information).

17   Refer to Chapter 9 and install the brake drums.

18   Install the wheels and snug the wheel lug nuts finger tight.

19   Remove the jackstands and lower the vehicle.

20   Tighten the wheel lug nuts to the torque listed in this Chapter's Specifications.

Brake booster check

21   Sit in the driver's seat and perform the following sequence of tests.

22   With the engine stopped, depress the brake pedal several times- the travel distance should not change.

23   With the brake fully depressed, start the engine - the pedal should move down a little when the engine starts.

24   Depress the brake, stop the engine and hold the pedal in for about 30 seconds - the pedal should neither sink nor rise.

25   Restart the engine, run it for about a minute and turn it off. Then firmly depress the brake several times - the pedal travel should

decrease with each application.

26   If your brakes do not operate as described above when the preceding tests are performed, the brake booster has failed. Refer to Chapter 9 for the removal procedure.

Parking brake

27   Slowly pull up on the parking brake and count the number of clicks you hear until the handle is up as far as it will go. The adjustment is correct if you hear the number of clicks listed in this Chapter's Specifications. If you hear more or fewer clicks, it's time to adjust the parking brake (refer to Chapter 9).

28   An alternative method of checking the parking brake is to park the vehicle on a steep hill with the parking brake set and the transmission in Neutral. If the parking brake cannot prevent the vehicle from rolling, it is in need of adjustment (see Chapter 9).

17   Air filter replacement (every 15,000 miles or 12 months)

Refer to illustrations 17.1a, 17.1b and 17.2

1   The air filter is located inside a housing at the left (drivers) side of the engine compartment. To remove the air filter, remove the

bolts (single cam models) and release the spring clips that keep the two halves of the air cleaner housing together **(see illustrations)**.

2   Lift the cover up and remove the air filter element **(see illustration)**.

3   Inspect the outer surface of the filter element. If it is dirty, replace it. If it is only moderately dusty, it can be reused by blowing it clean from the back to the front surface with compressed air. Because it is a pleated paper type filter, it cannot be washed or oiled. If it cannot be cleaned satisfactorily with compressed air, discard and replace it. **Caution:** *Never drive the vehicle with the air cleaner removed. Excessive engine wear could result and backfiring could even cause a fire under the hood.*

4   Installation is the reverse of removal.

18   Fuel system check (every 15,000 miles or 12 months)

Refer to illustration 18.5

Warning: *Certain precautions should be observed when inspecting or servicing the fuel system components. Work in a well ventilated area and do not allow open flames (cigarettes, appliance pilot lights, etc.) near*

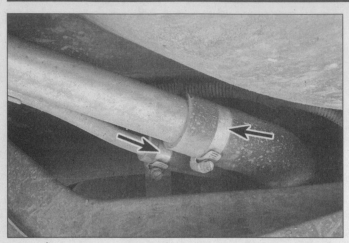

18.5 Make sure the clamps (arrows) are tight and the fuel filler hose is not damaged or split

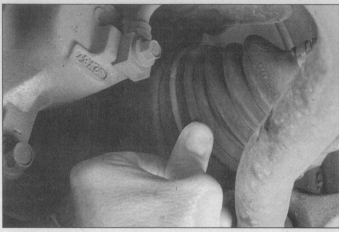

21.2 Flex the driveaxle boots by hand to check for cracks and/or leaking grease

the work area. Mop up spills immediately and do not store fuel soaked rags where they could ignite. It is a good idea to keep a dry chemical (Class B) fire extinguisher near the work area any time the fuel system is being serviced.

1   If you smell gasoline while driving or after the vehicle has been sitting in the sun, inspect the fuel system immediately.

2   Remove the gas filler cap and inspect if for damage and corrosion. The gasket should have an unbroken sealing imprint. If the gasket is damaged or corroded, remove it and install a new one.

3   Inspect the fuel feed and return lines for cracks. Make sure that the threaded flare nut type connectors which secure the metal fuel lines to the fuel injection system and the banjo bolts which secure the banjo fittings to the in-line fuel filter are tight.

4   Since some components of the fuel system - the fuel tank and part of the fuel feed and return lines, for example - are underneath the vehicle, they can be inspected more easily with the vehicle raised on a hoist. If that's not possible, raise the vehicle and support it securely on jackstands.

5   With the vehicle raised and safely supported, inspect the gas tank and filler neck for punctures, cracks and other damage **(see illustration)**. The connection between the filler neck and the tank is particularly critical. Sometimes a rubber filler neck will leak because of loose clamps or deteriorated rubber. These are problems a home mechanic can usually rectify. **Warning:** *Do not, under any circumstances, try to repair a fuel tank (except rubber components). A welding torch or any open flame can easily cause fuel vapors inside the tank to explode.*

6   Carefully check all rubber hoses and metal lines leading away from the fuel tank. Check for loose connections, deteriorated hoses, crimped lines and other damage. Carefully inspect the lines from the tank to the fuel injection system. Repair or replace damaged sections as necessary (see Chapter 4).

19   Manual transaxle lubricant level check (every 15,000 miles or 12 months)

1   A dipstick is used for checking the lubricant level on these models.

2   With the transaxle cold (cool to the touch) and the vehicle parked on a level surface, remove the dipstick from the filler tube located at the left rear (driver's) side of the engine compartment, adjacent to the brake master cylinder.

3   Wipe the dipstick clean, then reinsert and remove it.

4   The level must be even with the Full line on the dipstick.

5   If the level is low, add the specified lubricant through the filler tube, using a funnel.

6   Insert the dipstick into the filler tube and seat it securely.

20   Steering and suspension check (every 15,000 miles or 12 months). **Note:** *For detailed illustrations of the steering and suspension components, refer to Chapter 10.*

With the wheels on the ground

1   With the vehicle stopped and the front wheels pointed straight ahead, rock the steering wheel gently back and forth. If freeplay is excessive, a front wheel bearing, intermediate shaft U-joint, control arm balljoint or tie-rod end is worn or the steering gear is out of adjustment, worn excessively or broken. Refer to Chapter 10 for the appropriate repair procedure.

2   Other symptoms, such as excessive vehicle body movement over rough roads, swaying (leaning) around corners and binding as the steering wheel is turned, may indicate faulty steering and/or suspension components.

3   Check the shock absorbers by pushing down and releasing the vehicle several times at each corner. If the vehicle does not come back to a level position within one or two bounces, the shocks/struts are worn and must be replaced. When bouncing the vehicle up and down, listen for squeaks and noises from the suspension components. Additional information on suspension components can be found in Chapter 10.

Under the vehicle

4   Raise the vehicle with a floor jack and support it securely on jackstands. See *Jacking and towing* at the front of this book for the proper jacking points.

5   Check the tires for irregular wear patterns and proper inflation. See Section 5 in this Chapter for information regarding tire wear and Chapter 10 for the wheel bearing replacement procedures.

6   Inspect the universal joint between the steering shaft and the steering gear housing. Check the steering gear housing for grease leakage or oozing. Make sure that the dust seals and boots are not damaged and that the boot clamps are not loose. Check the steering linkage for looseness or damage. Check the tie-rod ends for excessive play. Look for loose bolts, broken or disconnected parts and deteriorated rubber bushings on all suspension and steering components. While an assistant turns the steering wheel from side to side, check the steering components for free movement, chafing and binding. If the steering components do not seem to be reacting with the movement of the steering wheel, try to determine where the slack is located.

7   Check the balljoint for wear by grasping each front wheel securely and moving it in-and-out to ensure that the balljoint has no play. If any balljoint does have play, replace it. See Chapter 10 for the front balljoint replacement procedure.

8   Inspect the balljoint boots for damage and leaking grease. Replace the balljoints with new ones if they are damaged (see Chapter 10).

21   Driveaxle boot check (every 15,000 miles or 12 months)

Refer to illustration 21.2

1   The driveaxle boots are very important

22.4a Slide the plastic tool up into the hose connection to release the inner clips . . .

22.4b . . . then disconnect the hose from the fuel line

because they prevent dirt, water and foreign material from entering and damaging the constant velocity (CV) joints. Oil and grease can cause the boot material to deteriorate prematurely, so it's a good idea to wash the boots with soap and water.

2   Inspect the boots for tears and cracks as well as loose clamps **(see illustration)**. If there is any evidence of cracks or leaking lubricant, they must be replaced as described in Chapter 8.

22 Fuel filter replacement (every 30,000 miles or 24 months)

Refer to illustrations 22.4a, 22.4b, 22.8 and 22.9

Warning: *Gasoline is extremely flammable, so take extra precautions when you work on any part of the fuel system. Don't smoke or allow open flames or bare light bulbs near the work area, and don't work in a garage where a natural gas-type appliance (such as a water heater or clothes dryer) with a pilot light is present. If you spill any fuel on your skin, rinse it off immediately with soap and water. When*

you perform any kind of work on the fuel system, wear safety glasses and have a Class B type fire extinguisher on hand.

1991 through 1997 vehicles

1   Obtain a replacement filter and hose assembly.

2   Relieve the fuel system pressure (see Chapter 4).

3   In the engine compartment, remove the intake duct from the air cleaner assembly for access to the fuel hose connection.

4   Using the small plastic tool supplied with the replacement filter, disconnect the filter hose fitting at the fuel rail **(see illustrations)**.

6   Raise the vehicle and support it securely on jackstands.

7   Under the vehicle, grasp the fuel line at the filter and turn it a 1/4-turn in each direction to loosen any dirt. Use compressed air or carburetor cleaner to blow or wash any dirt out of the fitting.

8   Squeeze the tabs on the fitting and disconnect the fuel line from the filter, then remove the nuts and detach the filter **(see illustration)**.

9   Push the new disconnect fitting onto the

new filter and lubricate both filter fittings with clean engine oil or spray lubricant **(see illustration)**.

10   Place the new filter in position and install the bracket nuts.

11   Push the quick-connect fitting onto the filter and fuel line, making sure they snap securely into place.

1998 through 1999 vehicles

12   Raise the vehicle and support it securely on jackstands. **Note:** *The fuel filter/fuel pressure regulator assembly is mounted to the bottom of the vehicle near the front left corner of the fuel tank.*

13   Disconnect the negative battery cable. Position it so that it cannot contact the positive cable.

14   Relieve the fuel system pressure (see Chapter 4).

15   Remove the two filter assembly mounting bolts.

16   Carefully open the clip on the fuel tank which retains the fuel lines.

17   Using a special tool such as the one described in Step 4, disconnect the EVAP purge line at the 90-degree quick-connect fitting **(see illustration 22.4a)**.

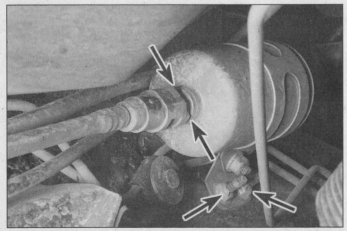

22.8 Press on the two clips (upper arrows) and detach the quick-connect fitting, then remove the nuts (lower arrows) and detach the filter

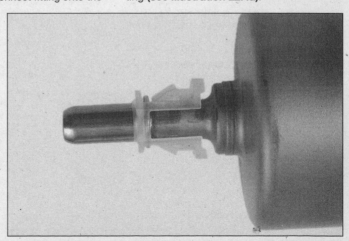

22.9 Install the new quick connect fitting on the filter inlet side

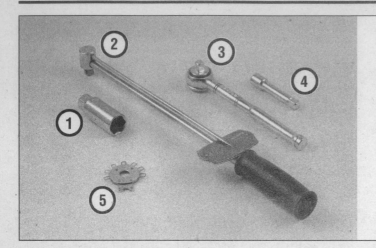

23.1 Tools required for changing spark plugs

1   *Spark plug socket - This will have special padding inside to protect the spark plug's porcelain insulator*
2   *Torque wrench - Although not mandatory, using this tool is the best way to ensure the plugs are tightened properly*
3   *Ratchet - Standard hand tool to fit the spark plug socket*
4   *Extension - Depending on model and accessories, you may need special extensions and universal joints to reach one or more of the plugs*
5   *Spark plug gap gauge - This gauge for checking the gap comes in a variety of styles. Make sure the gap for your engine is included*

23.4a Spark plug manufacturers recommend using a wire-type gauge when checking the gap - if the wire does not slide between the electrodes with a slight drag, adjustment is required

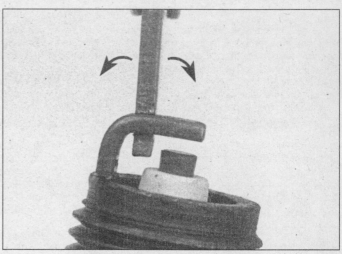

23.4b To change the gap, bend the side electrode only, as indicated by the arrows, and be very careful not to crack or chip the porcelain insulator surrounding the center electrode

18   Disconnect the fuel supply line at the 90-degree quick-connect fitting.
19   Maneuver the filter/regulator and its bracket out from the brake lines.
20   Disconnect the lines from it at the quick-connect fittings.
21   Install a new filter/pressure regulator unit using new fuel line retainers in the female parts of the quick-connect fittings.
22   The remainder of installation is the reverse of removal. Carefully inspect the area for fuel leaks with the engine running before driving the vehicle.

23   Spark plug check and replacement (every 30,000 miles or 24 months)

Refer to illustrations 23.1, 23.4a, 23.4b, 23.6, 23.8, 23.10a and 23.10b
1   Spark plug replacement requires a spark plug socket which fits onto a ratchet wrench. This socket is lined with a rubber grommet to protect the porcelain insulator of the spark plug and to hold the plug while you

insert it into the spark plug hole. You will also need a wire-type feeler gauge to check and adjust the spark plug gap and a torque wrench to tighten the new plugs to the specified torque **(see illustration)**.
2   If you are replacing the plugs, purchase the new plugs, adjust them to the proper gap and then replace each plug one at a time. **Note:** *When buying new spark plugs, it's essential that you obtain the correct plugs for your specific vehicle. This information can be found in the Specifications Section at the beginning of this Chapter, on the Vehicle Emissions Control Information (VECI) label located on the underside of the hood or in the owner's manual. If these sources specify different plugs, purchase the spark plug type specified on the VECI label because that information is provided specifically for your engine.*
3   Inspect each of the new plugs for defects. If there are any signs of cracks in the porcelain insulator of a plug, don't use it.
4   Check the electrode gaps of the new plugs. Check the gap by inserting the wire gauge of the proper thickness between the electrodes at the tip of the plug **(see illustra-**

tion). The gap between the electrodes should be identical to that listed in this Chapter's Specifications or on the VECI label. If the gap is incorrect, use the notched adjuster on the feeler gauge body to bend the curved side electrode slightly **(see illustration)**.
5   If the side electrode is not exactly over the center electrode, use the notched adjuster to align them. **Caution:** *If the gap of a new plug must be adjusted, bend only the base of the ground electrode – do not touch the tip.*

Removal

6   To prevent the possibility of mixing up spark plug wires, work on one spark plug at a time. Remove the wire and boot from one spark plug. Grasp the boot - not the cable - as shown, give it a half twisting motion and pull straight up **(see illustration)**.
7   If compressed air is available, blow any dirt or foreign material away from the spark plug area before proceeding.
8   Remove the spark plug **(see illustration)**.
9   Whether you are replacing the plugs at this time or intend to reuse the old plugs,

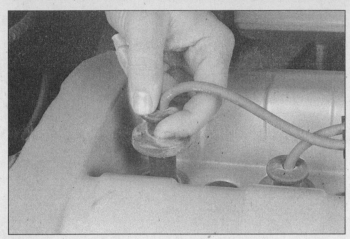

23.6 When removing the spark plug wires, pull only on the boot and twist it back-and-forth

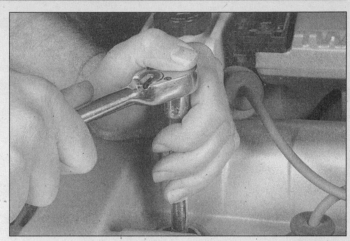

23.8 Use a socket wrench with a long extension to unscrew the spark plug

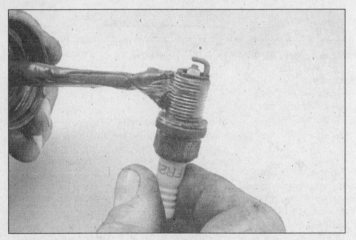

23.10a Apply a coat of anti-seize compound to the spark plug threads

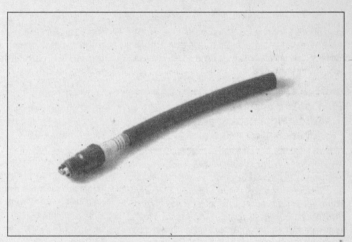

23.10b A length of 3/8-inch ID rubber hose will save time and prevent damaged threads when installing the spark plugs

compare each old spark plug with those shown in color on the inside rear cover of this manual to determine the overall running condition of the engine.

Installation

10   Prior to installation, apply a coat of anti-seize compound to the spark plug threads **(see illustration)**. It's often difficult to insert spark plugs into their holes without cross-threading them. To avoid this possibility, fit a short piece of 3/8-inch ID rubber hose over the end of the spark plug **(see illustration)**. The flexible hose acts as a universal joint to help align the plug with the plug hole. Should the plug begin to cross-thread, the hose will slip on the spark plug, preventing thread damage. Tighten the plug to the torque listed in this Chapter's Specifications.

11   Attach the plug wire to the new spark plug, again using a twisting motion on the boot until it is firmly seated on the end of the spark plug.

12   Follow the above procedure for the remaining spark plugs, replacing them one at a time to prevent mixing up the spark plug wires.

24   Spark plug and coil wire check and replacement (every 30,000 miles or 24 months)

1   The spark plug wires should be checked whenever new spark plugs are installed.

2   Begin this procedure by making a visual check of the spark plug wires while the engine is running. In a darkened area (make sure there is adequate ventilation) start the engine and observe each plug wire. Be careful not to come into contact with any moving engine parts. If there is a break in the wire, you will see arcing or a small spark at the damaged area. If arcing is noticed, make a note to obtain new wires.

3   The spark plug wires should be inspected one at a time to prevent mixing up the order, which is essential for proper engine operation. Each original plug wire should be numbered to help identify its location. If the number is illegible, a piece of tape can be marked with the correct number and wrapped around the plug wire.

4   Disconnect the plug wire from the spark plug. A removal tool can be used for this pur-

pose or you can grasp the rubber boot, twist the boot half a turn and pull the boot free. Do not pull on the wire itself.

5   Check inside the boot for corrosion, which will look like a white crusty powder.

6   Push the wire and boot back onto the end of the spark plug. It should fit tightly onto the end of the plug. If it doesn't, remove the wire and use pliers to carefully crimp the metal connector inside the wire boot until the fit is snug.

7   Using a clean rag, wipe the entire length of the wire to remove built-up dirt and grease. Once the wire is clean, check for burns, cracks and other damage. Do not bend the wire sharply, because the conductor might break.

8   Disconnect the wire from the coil pack. Again, pull only on the rubber boot. Check for corrosion and a tight fit. Reinstall the wire on the coil.

9   Inspect the remaining spark plug wires, making sure that each one is securely fastened at the coil and spark plug when the check is complete.

10   If new spark plug wires are required, purchase a set for your specific engine

25.4 Use pliers to remove the radiator drain plug - be careful, the coolant will gush out of the plug opening

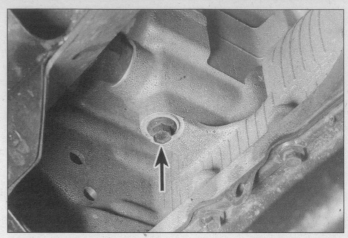

25.5 Remove the engine block drain plug (arrow)

model. Pre-cut wire sets with the boots already installed are available. Remove and replace the wires one at a time to avoid mix-ups in the firing order.

25   Cooling system servicing (draining, flushing and refilling) (every 30,000 miles or 24 months)

Warning: *Do not allow engine coolant (antifreeze) to come in contact with your skin or painted surfaces of the vehicle. Rinse off spills immediately with plenty of water. Antifreeze is highly toxic if ingested. Never leave antifreeze laying around in an open container or in puddles on the floor; children and pets are attracted by it's sweet smell and may drink it. Check with local authorities about disposing of used antifreeze. Many communities have collection centers which will see that antifreeze is disposed of safely.*

1   Periodically, the cooling system should be drained, flushed and refilled to replenish the antifreeze mixture and prevent formation of rust and corrosion, which can impair the performance of the cooling system and cause engine damage. When the cooling system is serviced, all hoses and the radiator cap should be checked and replaced if necessary.

Draining

Refer to illustrations 25.4 and 25.5
2   Apply the parking brake and block the wheels. If the vehicle has just been driven, wait several hours to allow the engine to cool down before beginning this procedure.
3   Once the engine is completely cool, remove the coolant reservoir cap.
4   Move a large container under the radiator drain to catch the coolant. Unscrew the drain fitting (a pair of pliers may be required to turn it) **(see illustration)**. Be careful, as the coolant will flow with considerable force out of the drain plug opening.
5   After the coolant stops flowing out of

the radiator, move the container under the engine block drain plug **(see illustration)**. Loosen the plug and allow the coolant in the block to drain.
6   While the coolant is draining, check the condition of the radiator hoses, heater hoses and clamps (refer to Section 13 if necessary).
7   Replace any damaged clamps or hoses (see Chapter 3).

Flushing

8   Once the system is completely drained, flush the radiator with fresh water from a garden hose until water runs clear at the drain. The flushing action of the water will remove sediments from the radiator but will not remove rust and scale from the engine and cooling tube surfaces.
9   These deposits can be removed by the chemical action of a cleaner. Follow the procedure outlined in the manufacturer's instructions. If the radiator is severely corroded, damaged or leaking, it should be removed (see Chapter 3) and taken to a radiator repair shop.
10   Remove the overflow hose from the coolant recovery reservoir. Drain the reservoir and flush it with clean water, then reconnect the hose.

Refilling

11   Install and tighten the radiator drain plug. Install and tighten the engine block drain plug.
12   Place the heater temperature control in the maximum heat position.
13   Slowly add new coolant (a 50/50 mixture of water and phosphate-free antifreeze) to the radiator until it's full. Add coolant to the reservoir up to the lower mark.
14   Run the engine in a well-ventilated area for two or three minutes.
15   Turn the engine off and add more coolant mixture to bring the level back up to the Full Cold line on the reservoir.
16   Start the engine, allow it to reach normal operating temperature and check for leaks.

26   Exhaust system check (every 30,000 miles or 24 months)

Refer to illustration 26.4
1   With the engine cold (at least three hours after the vehicle has been driven), check the complete exhaust system from its starting point at the engine to the end of the tailpipe. This should be done on a hoist where unrestricted access is available.
2   Check the pipes and connections for evidence of leaks, severe corrosion or damage. Make sure that all brackets and hangers are in good condition and tight.
3   At the same time, inspect the underside of the body for holes, corrosion, open seams, etc. which may allow exhaust gases to enter the passenger compartment. Seal all body openings with silicone or body putty.
4   Rattles and other noises can often be traced to the exhaust system, especially the mounts and hangers **(see illustration)**. Try to move the pipes, muffler and catalytic converter. If the components can come in contact with the body or suspension parts,

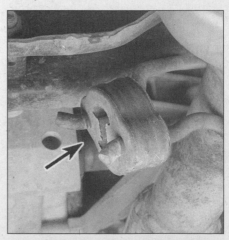

26.4 Check the exhaust system rubber hangers (arrow) for damage

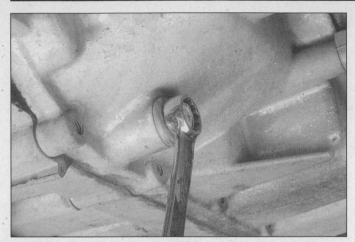

27.7 Remove the transaxle drain plug with a box-end wrench

27.11 Since the transaxle fluid pressure filter (accessible from below) is on very tight, you'll need a special wrench for removal - DO NOT use the wrench to tighten the new filter

secure the exhaust system with new mounts.

5   Check the running condition of the engine by inspecting inside the end of the tailpipe. The exhaust deposits here are an indication of engine state-of-tune. If the pipe is black and sooty or coated with white deposits, the engine is in need of a tune-up, including a thorough fuel system inspection.

27   Automatic transaxle fluid and filter change (every 30,000 miles or 24 months)

Refer to illustrations 27.7 and 27.11
Note: *The automatic transaxle fluid pressure filter looks just like the engine oil filter. Be sure you are removing the proper filter when changing the oil or transaxle fluid as they are not interchangeable and installing the wrong one will result in damage to the engine and/or transaxle.*

1   At the specified time intervals, the automatic transaxle fluid should be drained and replaced.

2   Before beginning work, purchase the specified transmission fluid (see *Recommended fluids and lubricants* at the front of this Chapter).

3   Other tools necessary for this job include jackstands to support the vehicle in a raised position, a wrench, filter wrench, a drain pan capable of holding at least eight pints, newspapers and clean rags.

4   The fluid should be drained immediately after the vehicle has been driven. Hot fluid is more effective than cold fluid at removing built up sediment. **Warning**: *Fluid temperature can exceed 350-degrees F in a hot transaxle. Wear protective gloves!*

5   After the vehicle has been driven to warm up the fluid, raise it and place it on jackstands for access to the transaxle drain plug and fluid pressure filter.

6   Move the necessary equipment under the vehicle, being careful not to touch any of the hot exhaust components.

7   With the drain pan in place, remove the drain plug and allow the fluid to drain **(see illustration)**.

8   After all the fluid has drained, wipe off the drain plug with a clean rag. Even minute metal particles clinging to the plug would immediately contaminate the new fluid.

9   Clean the area around the drain plug opening, reinstall the plug and tighten it securely, but do not strip the threads.

10   Move the drain pan into position under the transaxle pressure filter, which looks just like the engine oil filter.

11   Loosen the transaxle pressure filter **(see illustration)** by turning it counterclockwise with the filter wrench. Any standard filter wrench will work. Sometimes the filter is screwed on so tightly that it cannot be loosened. If this situation occurs, punch a metal bar or long screwdriver directly through the side of the canister and use it as a T-bar to turn the filter. Be prepared for oil to spurt out of the canister as it is punctured. Once the filter is loose, use your hands to unscrew it from the transaxle housing. Just as the filter is detached from the block, immediately tilt the open end up to prevent the fluid inside the filter from spilling out. Move the drain pan into position under the oil filter.

12   With a clean rag, wipe off the mounting surface on the transaxle. If a residue of old fluid is allowed to remain, it will prevent the new filter from seating properly. Also make sure that the none of the old gasket remains

stuck to the mounting surface. It can be removed with a scraper if necessary.

13   Compare the old filter with the new one to make sure they are the same type. It will probably be labeled *Transmission*. Smear some transmission fluid on the rubber gasket of the new filter and screw it into place. Because overtightening the filter will damage the gasket, do not use a filter wrench to tighten the filter. Tighten it by hand until the gasket contacts the seating surface. Then seal the filter by giving it an additional 3/4-turn.

14   Lower the vehicle.

15   With the engine off, add new fluid to the transaxle through the dipstick tube (see *Recommended fluids and lubricants* for the recommended fluid type and capacity). Use a funnel to prevent spills. It is best to add a little fluid at a time, continually checking the level with the dipstick (see Section 7). Allow the fluid time to drain into the pan.

16   Start the engine and shift the selector into all positions from P through L, then shift into P and apply the parking brake.

17   With the engine idling, check the fluid level. Add fluid up to the hatched area on the dipstick (see Section 7).

28   Manual transaxle lubricant change (every 30,000 miles or 24 months)

1   Remove the drain plug and drain the fluid.

2   Reinstall the drain plug securely.

3   Use a funnel to add the specified amount of new lubricant through the dipstick opening and check the fluid level (Section 19). See *Recommended lubricants and fluids* for the specified lubricant amount and type.

Notes

Chapter 2 Part A Engines

Contents

Specifications

General

Firing order	1-3-4-2
Cylinder numbers (timing chain end-to-transaxle end)	1-2-3-4

Cylinder locations and coil-pack terminal diagram

Camshaft(s)

SOHC engine

Journal diameter	
Standard	1.7480 to 1.7490 inches
Service limit	1.7470 inches minimum
Lobe lift (intake and exhaust)	0.2531 to 0.2556 inch
Runout (all journals)	
Standard	0.0020 inch
Service limit	0.0028 inch maximum
Journal oil clearance	
Standard	0.0020 to 0.0040 inch
Service limit	0.0054 inch maximum
Endplay (thrust)	
Standard	0.0028 to 0.0079 inch
Service limit	0.0098 inch maximum

Camshaft(s) (continued)

DOHC engine (both camshafts)

   Journal diameter

      1991 through 1998

         Standard .... 1.1398 to 1.1406 inches

         Service limit .... 1.139 inches minimum

      1999

         Standard .... 1.7480 to 1.1.7490 inches

         Service limit .... 1.7470 inches minimum

   Lobe lift

      1991 through 1998

         Intake .... 0.3528 to 0.3559 inch

         Exhaust .... 0.3409 to 0.3441 inch

      1999 .... 0.2531 to 0.2556 inch

   Runout

      1991 through 1998

         Standard .... 0.0020 inch

         Service limit .... 0.004 inch maximum

      1999

         Standard .... 0.0020 inch

         Service limit .... 0.0028 inch maximum

   Journal oil clearance

      1991 through 1998

         Standard .... 0.0021 to 0.0030 inch

         Service limit .... 0.005 inch maximum

      1999

         Standard .... 0.0020 to 0.0040 inch

         Service limit .... 0.0054 inch maximum

   Endplay (thrust)

      1991 through 1998

         Standard .... 0.0020 to 0.0080 inch

         Service limit .... 0.010 inch maximum

      1999

         Standard .... 0.0028 to 0.0079 inch

         Service limit .... 0.0098 inch maximum

Warpage limits (all engines)

Cylinder head

   Length .... 0.0028 inch

   Width .... 0.0012 inch

   Intake flange .... 0.004 inch maximum

   Exhaust flange .... 0.004 inch

Intake manifold .... 0.008 inch maximum

Exhaust manifold .... 0.014 inch maximum

Oil pump clearances (all engines)

Gear tip clearance .... 0.006 inch maximum

End-to end clearance .... 0.005 inch maximum

Driven gear-to-housing clearance .... 0.011 inch maximum

Rocker arm shaft (SOHC engine)

Outside diameter

   Standard .... 0.624 to 0.625 inch

   Service limit .... 0.620 inch minimum

Timing chain stretch

1991 through 1998

   SOHC

      Standard .... 16.50 to 16.61 inches

      Service limit .... 16.73 inches maximum

   DOHC

      Standard .... 22.87 to 22.99 inches

      Service limit .... 23.15 inches maximum

Timing chain tensioner

Plunger extension

   1991 through 1998

      SOHC (standard range - depressed to full extension) .... 0.059 to 0.413 inch

      DOHC (standard range - depressed to full extension) .... 0.059 to 0.374 inch

      Service limit (all) .... 0.8626 inch maximum

Plunger extension (continued)
  1999 (standard range - depressed to full extension)
    Standard .... 0.045 to 0.388 inch
    Service limit... 0.7092 inch maximum

Timing chain guide groove wear
1991 through 1998
  Standard.. 0.0 inch
  Service limit .... 0.0984 inch maximum
1999
  Standard.. 0.0 inch
  Service limit .... 0.0394 inch maximum

Valve lifter/tappet
SOHC engine
  Lifter body diameter
    Standard .... 0.8420 to 0.8427 inch
    Service limit... 0.8417 minimum
  Bore diameter
    Standard .... 0.8434 to 0.8444
    Service limit... 0.8445 inch
  Lifter body oil clearance
    Standard .... 0.0007 to 0.0024 inch
    Service limit... 0.0028 maximum
  Lifter body guide flat
    Standard .... 0.7551 to 0.7598 inch
    Service limit... 0.7539 inch
  Lifter guide plate width
    Standard .... 0.7634 to 0.7685 inch
    Service limit... 0.7632 inch
  Lifter/roller total length
    Standard .... 2.0563 to 2.0697 inches
    Service limit... 2.0531 inches minimum
1991 through 1998 DOHC engine
  Lifter body diameter
    Standard .... 1.2976 to 1.2982 inches
    Service limit... 1.30 inches
  Lifter bore diameter
    Standard .... 1.2992 to 1.3002 inches
    Service limit... 1.3004 inches
  Lifter body oil clearance
    Standard .... 0.0010 to 0.0026 inch
    Service limit... 0.003
1999 DOHC
  Lash adjuster body diameter
    Standard .... 0.4722 to 0.4728 inch
    Service limit... 0.4672 inch minimum
  Lash adjuster body clearance
    Standard .... 0.0003 to 0.0017 inch
    Service limit... 0.0024 inch maximum

Torque specifications
                                                                              Ft-lbs (unless otherwise indicated)
Camshaft bearing cap bolts (DOHC engine only)................................... 124 in-lbs
Camshaft sprocket(s)-to-camshaft(s) bolt 75
Camshaft thrust plate bolts-to-head (SOHC engine) 19
Crankshaft pulley/vibration damper bolt (all).. 159
Cylinder head bolts (in sequence **see illustration 12.24a or 12.24b**)
  Initial installation (clean and dry) 48
  Step 2.. Remove the bolts, apply engine oil to the threads, then reinstall them
  Step 3.. 22
  Step 4
    SOHC engine 33
    DOHC engine 37
  Step 5.. Turn an additional 1/4-turn (90-degrees)
Engine mounts
  Right (front) mount-to-block bolts(1991).............................. 52
  Right (front) mount-to-block bolts (1992 and later).............. 37
  Rear mount-to-block (both sides) 35
  SOHC and all TBI models 16

Torque specifications (continued)   Ft-lbs (unless otherwise indicated)

Engine mounts
  DOHC
    1995 23
    1996 through 1997 19
    1998 through 1999 .... 150 in-lbs
Flywheel-to-crankshaft 59
Driveplate-to-crankshaft 44
Idler pulley bolts to front cover 33
Intake manifold-to-cylinder head bolts
  TBI models .... 188 in-lbs
  MPFI models 22
Oil baffle plate-to-block (DOHC engine) 41
Oil pan bolts .... 80 in-lbs
Oil pick-up tube-to-block .... 133 in-lbs
Oil pipe bracket-to-baffle plate (DOHC engine) ... 133 in-lbs
Oil pipe bracket-to-block (SOHC engine) 41
Oil pump cover screws .... 97 in-lbs
Rear main oil seal retainer bolts .... 97 in-lbs
Rocker arm shaft-to-head 19
Serpentine belt tensioner pulley 22
Serpentine belt tensioner-to-block 22
Timing chain cover lower center bolt .... 89 in-lbs
Timing chain cover upper bolts 22
Timing chain guide bolts 19
Timing chain tensioner (all engines) .... 168 in-lbs
Valve cover bolts
  SOHC engine 22
  DOHC engine .... 89 in-lbs

1 General information

This Part of Chapter 2 is devoted to in-vehicle repair procedures for all engines. All information concerning engine removal and installation and engine block and cylinder head overhaul can be found in Part B of this Chapter.

The following repair procedures are based on the assumption that the engine is installed in the vehicle. If the engine has been removed from the vehicle and mounted on a stand, many of the steps outlined in this Part of Chapter 2 will not apply.

The Specifications included in this Part of Chapter 2 apply only to the procedures contained in this Part. Part B of Chapter 2 contains the Specifications necessary for cylinder head and engine block rebuilding.

2 Repair operations possible with the engine in the vehicle

Many major repair operations can be accomplished without removing the engine from the vehicle.

Clean the engine compartment and the exterior of the engine with some type of degreaser before any work is done. It will make the job easier and help keep dirt out of the internal areas of the engine.

Depending on the components involved, it may be helpful to remove the hood to improve access to the engine as repairs are performed (refer to Chapter 11 if necessary). Cover the fenders to prevent damage to the paint. Special pads are available, but an old bedspread or blanket will also work.

If vacuum, exhaust, oil or coolant leaks develop, indicating a need for gasket or seal replacement, the repairs can generally be made with the engine in the vehicle. The intake and exhaust manifold gaskets, oil pan gasket, crankshaft oil seals and cylinder head gasket are all accessible with the engine in place.

Exterior engine components, such as the intake and exhaust manifolds, the oil pan, the oil pump, the water pump, the starter motor, the alternator and the fuel system components can be removed for repair with the engine in place.

Since the cylinder head can be removed without pulling the engine, camshaft and valve component servicing can also be accomplished with the engine in the vehicle. Replacement of the timing chain and sprockets is also possible with the engine in the vehicle.

In extreme cases caused by a lack of necessary equipment, repair or replacement of piston rings, pistons, connecting rods and rod bearings is possible with the engine in the vehicle. However, this practice is not recommended because of the cleaning and prepa-ration work that must be done to the components involved.

3 Top Dead Center (TDC) for number one piston - locating

1   Top Dead Center (TDC) is the highest point in the cylinder that each piston reaches as it travels up-and-down when the crankshaft turns. Each piston reaches TDC on the compression stroke and again on the exhaust stroke, but TDC generally refers to piston position on the compression stroke.
2   Positioning the piston(s) at TDC is an essential part of many procedures such as timing chain/sprocket removal and camshaft removal.
3   Before beginning this procedure, be sure to place the transmission in Neutral and apply the parking brake or block the rear wheels. Also, disable the ignition system by detaching the primary (low voltage) electrical connectors at the ignition coil pack/module (see Chapter 5). Remove the spark plugs (see Chapter 1).
4   When looking at the drivebelt end of the engine, normal crankshaft rotation is clockwise. In order to bring any piston to TDC, the crankshaft must be turned with a socket and ratchet attached to the bolt threaded into the center of the vibration damper on the crankshaft.

4.3 Remove the two valve cover bolts (SOHC engine)

4.5 On DOHC models remove the EGR valve solenoid from the valve cover (arrow)

5   Have an assistant turn the crankshaft with a socket and ratchet as described in the previous Step while you hold a finger over the number one spark plug hole (the one closest to the timing chain end of the engine). **Note:** *Since the spark plugs on DOHC models are deeply recessed, thread an adapter hose from a compression tester into the spark plug hole and feel for compression at the end of the hose.*

6   When the piston approaches TDC, air pressure will be felt at the spark plug hole. At this time instruct your assistant to stop turning the crankshaft.

7   Insert a drinking straw into the spark plug hole and have your assistant turn the crankshaft slowly. As the piston rises the straw will be pushed out. Note the point where the straw stops moving out - this is TDC.

8   After the number one piston has been positioned at TDC on the compression stroke, TDC for any of the remaining pistons

can be located by turning the crankshaft clockwise, 180-degrees at a time, following the firing order. Or, you can simply repeat Steps 5 through 7 for any particular cylinder.

4   Valve cover - removal and installation

Refer to illustrations 4.3, 4.5, 4.6, 4.7a, 4.7b, 4.9, 4.14a and 4.14b
Caution: *On models equipped with a theft-deterrent radio, make sure you have the correct activation code, or the theft deterrent system is turned off, before disconnecting the battery.*
Removal
1   Disconnect the negative battery cable. Position it so that it cannot contact the positive cable.
2   Disconnect the PCV hose and the breather hose from the valve cover. Remove the accelerator cable bracket and any other

components which interfere with valve cover removal.

SOHC engine
3   Remove the two valve cover hold-down bolts. Inspect the bolt insulators for damage and replace them if necessary.
4   Remove the valve cover. If it sticks, use a soft-face hammer or a hammer and a block of wood to break the bond between the gasket and the cover.

DOHC engines
5   Unbolt and remove the EGR valve solenoid from the valve cover **(see illustration)**.
6   Disconnect the spark plug wires from the clips and remove the wires from the spark plugs **(see illustration)**.
7   Using a no. 40 Torx driver, remove the bolts holding the valve cover in place **(see illustration)**. **Note:** *Inspect the insulators for cracks* **(see illustration)**. *Replace the insula-*

4.6 Disconnect the spark plug wires from their clips, remove them from the spark plugs and set them away from the valve cover

4.7a On DOHC engines remove the Torx bolts that secure the valve cover to the cylinder head

2A

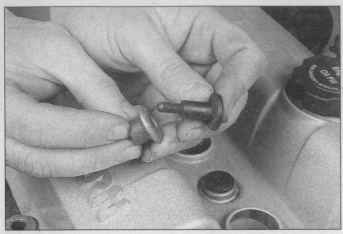

4.7b Inspect the insulators on each bolt and replace them if deterioration is found

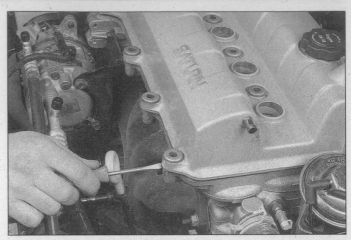

4.9 If the valve cover is stuck carefully pry the valve cover up to break the gasket seal, but pry no more than necessary to avoid gasket surface damage

5.2 Working from the ends, loosen the rocker arm shaft bolts a little at a time

5.7 Snap one end of each lifter guide plate retaining spring (arrows) onto the rocker arm shaft between the number one and two cylinder rocker arms and also between number three and four cylinder rocker arms

tors if deterioration exists.

8   Remove the EGR valve (see Chapter 6).
9   Remove the cover. If it sticks, knock it loose with a rubber mallet or a hammer and a block of wood. If you must pry the cover loose be certain to pry at a point near the corner and only enough to break the gasket seal (see illustration). Caution: *Be extremely careful not to damage the gasket surfaces of either the cover or the cylinder head - both are aluminum and can be easily cracked or gouged.*

Installation

10   The mating surfaces of the cylinder head and valve cover must be clean when the cover is installed. Use a gasket scraper to remove all traces of sealant, then clean the mating surfaces with carburetor spray cleaner or brake system cleaner. Don't use petroleum-based solvents. If there's residue or oil on the mating surfaces when the cover is installed, oil leaks may develop.
11   The original gasket may be reused if it is not damaged. Make sure that it is clean and

that it is installed in its groove properly.
12   Apply a small amount of RTV sealant to the line where the front cover contacts the cylinder head. Install the valve cover.
13   On SOHC engines, tighten the two bolts progressively and evenly to the torque listed in the Specifications in this Chapter.
14   On DOHC engines, tighten the bolts in several increments following the correct sequence (see illustrations). Refer to the Specifications in this Chapter for the proper final torque.

5   Rocker arm assembly (SOHC models only) - removal, inspection and installation

Removal

1   Remove the valve cover (see Section 4).
2   Loosen the rocker arm assembly bolts a little at a time, working from the ends of the shaft towards the center. Remove the bolts and lift the shaft(s) off the cylinder head.

Inspection

3   Clean the rocker shafts and rocker arms with solvent and dry them thoroughly. Keep the rocker arms in order so they can be returned to their original positions. Do the components from one shaft at a time to avoid mixing up the parts.

Rocker arms

4   Inspect all the rocker arms for excessive wear on the tips that contact the valve stem and camshaft.
5   Slide each rocker arm, one at a time, onto the shaft in its proper position. Wiggle the arm on the shaft and note what the freeplay feels like. Now slide the rocker arm onto an unworn portion of the rocker arm shaft and check the freeplay again. If there is a noticeable difference between these two checks, either the rocker arm or the shaft is worn excessively. Proceed to the next Step and check the rocker arm shaft. If it isn't worn excessively, replace the rocker arm(s).

Rocker arm shafts

6   Check the shafts for scoring, excessive

5.8 Install the rocker arm and shaft assembly on the cylinder head. Make sure the rocker arm slots (notches) are squarely seated on the hydraulic lifter plunger and that the retaining springs are properly located in the lifter guide plate

6.3 Raise the engine and support it in this position by placing a wood block between the frame and the mount bracket (1992 and later models)

2A

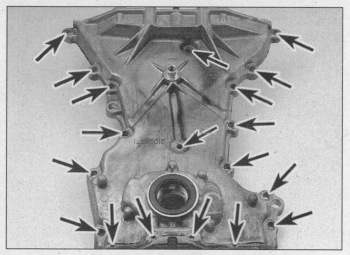

6.5 Locations of the timing chain cover bolts (arrows) (DOHC models)

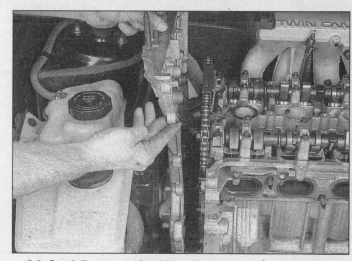

6.6 Carefully remove the timing chain cover from the engine

wear and other damage. The areas where the rocker arms contact the shafts should be smooth and dull (not shiny). If there is a visible ridge at the edge of where the rocker arm rides, the shaft is probably worn excessively. Measure the diameter of the shaft, in the areas where the rocker arms ride, with a micrometer. If the diameter of the shaft is less than specified, replace the shaft.

Installation

Refer to illustrations 5.7 and 5.8

7   Lubricate the rocker arm shaft with clean engine oil or engine assembly lube and install the rocker arms on both shafts **(see illustration)**. Snap one end of each lifter guide plate retaining spring onto the rocker arm shaft between number one and two cylinder rocker arms and also between number three and four cylinder rocker arms.

8   Install the rocker arm and shaft assembly on the cylinder head. Make sure the rocker arm slots (notches) are squarely seated on the hydraulic lifter plungers **(see illustration)**. Also be certain that the retaining

spring is positioned in the guide plate slot.

9   Tighten the shaft bolts, a little at a time, to the torque listed in this Chapter's Specifications.

10   The remainder of installation is the reverse procedure of removal.

---

6   Timing chain cover, chain and sprockets - removal, inspection and installation

Removal

All engines

Refer to illustrations 6.3, 6.5, 6.6, 6.8 and 6.9

1   Position the engine at TDC compression for the number one cylinder (see Section 3).

2   Remove the crankshaft damper/pulley (see Section 16) and remove the accessory drivebelt tensioner. Remove the right wheel and the inner splash shield if required for access.

3   Raise the engine, placing a jack and

block of wood under the oil pan to gain clearance to remove the timing chain cover. Place a block of wood between the cradle and the mount bracket to support it in the raised position **(see illustration)**.

4   Remove the water pump pulley (see Chapter 3). **Note:** *It may not be necessary on all models to remove this pulley, but it makes access to the bolts easier. Remove the serpentine belt idler pulley and the tensioner.*

5   Remove the four oil pan-to-timing chain cover bolts and the remaining timing chain cover bolts **(see illustration)**. **Note:** *On 1992 and later engines, there is a front timing cover attachment bolt located under the engine mount flange and above the accessory drivebelt tensioner. It may be necessary to cut the oil pan RTV sealant using a sharp razor blade.*

6   Remove the timing chain cover **(see illustration)**. If it sticks, knock it loose with a soft-faced hammer or a hammer and a block of wood.

7   Reinstall the vibration damper bolt in the front of the crankshaft. Rotate the crankshaft

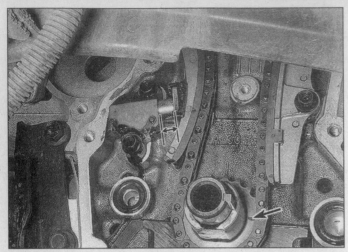

6.8 Measure the extension of the tensioner plunger (also note the position of the mark on the crankshaft sprocket)

6.9 Place a wrench on the flats of the camshaft to prevent it from turning when loosening the sprocket bolt(s)

6.11 Release the plunger lock (arrow) and make sure the piston moves in-and-out freely

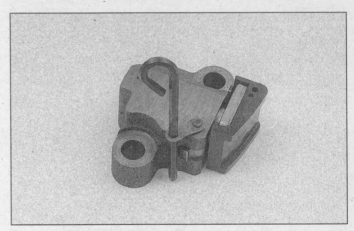

6.24 Before reinstalling the tensioner, compress the plunger and lock it in this position by putting an Allen wrench or drill bit in the locking arm

90-degrees (1/4-turn) clockwise, lining up the mark on the crankshaft sprocket with the main bearing cap split at the block. This will ensure the valves won't hit the pistons if the camshaft(s) turn slightly when the camshaft sprocket bolts are removed and installed.

8   Before removal of the chain, tensioner and guides, measure the distance of tensioner plunger extension **(see illustration)**. Compare your measurement with the value listed in this Chapter's Specifications. If the plunger extension exceeds the specified limit, the timing chain is probably stretched past its wear limit (see Step 12).

9   Remove the camshaft sprocket(s) **(see illustration)**, timing chain guides, tensioner, and crankshaft sprockets from the front of the engine block.

Inspection

Refer to illustration 6.11

10   Inspect the individual sprocket teeth and keyway for wear and damage. Check the chain for cracked plates and pitted or worn rollers. Check the chain tensioner contact shoe for wear. Check the chain guides for

wear and damage. On SOHC models, check the camshaft trust plate and sprocket thrust surface for wear. Replace any excessively worn or defective parts with new ones.

11   Check the tensioner for proper operation:

a) *Release the plunger lock* **(see illustration)** *and make sure the piston moves freely.*

b) *Submerge the tensioner in a can of oil or solvent, remove and depress the plunger and make sure the oil feed hole is not plugged and will release the oil.* **Note:** *Also inspect the oil feed hole in the block to be certain it's not plugged.*

12   Lay the timing chain on a flat surface, pull it taut and measure its inside diameter. Compare your measurement to the Specifications at the beginning of this Chapter. Replace the chain if necessary.

Installation

Refer to illustrations 6.24 and 6.26

13   If the camshaft(s) have been removed, reinstall them at this time (see Section 7).

SOHC engines

14   Reinstall the camshaft sprocket temporarily (don't install the bolt). Place the camshaft sprocket in the TDC position. The dimple in the sprocket should be in the 12-o'clock position.

15   Install an alignment pin through the camshaft sprocket and into the hole in the cylinder head to ensure the camshaft is in the proper position. You can use a 3/16-inch drill bit for an alignment pin. The marks are at the 12-o'clock position.

16   Reinstall the crankshaft sprocket, if removed. Turn the crankshaft counterclockwise 90-degrees (1/4-turn) and align the mark on the sprocket with the mark on the front of the engine.

17   Remove the camshaft sprocket and install the timing chain on the sprocket, aligning the bright link with the dimple in the 12-o'clock position on the cam sprocket. Loop the timing chain under the crankshaft sprocket and install the cam sprocket on the camshaft. Another bright link, this one at the crankshaft sprocket, should be in alignment with the tooth on the sprocket that is in the 6-

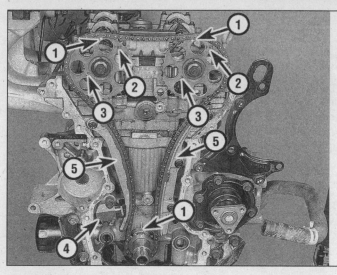

6.26 Assembled view of the timing chain, gears, tensioner and guides (DOHC engine)

1 TDC alignment marks
2 Camshaft sprocket assembly reference
3 Alignment pin holes
4 Timing chain tensioner
5 Timing chain guides

o'clock position.

18 Install the camshaft sprocket bolt and, holding the camshaft stationary with a wrench (positioned on the flats of the camshaft), tighten the bolt to the torque listed in this Chapter's Specifications. Verify that all marks are in alignment. Proceed to Step 23.

DOHC engines

19 Reinstall the camshaft sprockets and tighten the bolts to the torque listed in this Chapter's Specifications. Be sure to prevent the camshaft from turning by placing a wrench on the flats of the camshaft.

20 Place the camshafts in the TDC position. The dimple in the sprockets should be in the 12-o'clock position. Install alignment pins through the camshaft sprockets and into the holes in the cylinder head to ensure the sprockets are in proper alignment. You can use 3/16-inch drill bits for alignment pins.

21 Reinstall the crankshaft sprocket, if removed. Turn the crankshaft counterclockwise 90-degrees (1/4-turn) and align the mark on the sprocket with the mark on the front of the engine. The marks are at the 12-o'clock position.

22 Install the timing chain. There are two bright links in the chain, paired together - the tooth on the crankshaft sprocket that points in the 6-o'clock position should mesh between these links. The other two bright links should be in alignment with the dimples in the camshaft sprockets.

All engines

23 Lubricate the timing chain guides with moly-based grease. Install the chain guides, tightening the bolts to the torque values listed in this Chapter's Specifications.

24 Before reinstalling the tensioner, compress the tensioner and lock it in this position by putting an Allen wrench or drill bit in the locking arm **(see illustration)**. Bolt the tensioner to the block in this retracted position.

25 Remove the Allen wrench or drill bit so the tensioner fully extends against the chain guide.

26 Recheck all the timing marks to make sure they haven't moved **(see illustration)**.

27 Slowly rotate the crankshaft two complete revolutions in the normal direction of rotation (clockwise) and again bring the engine to TDC. If you feel any resistance, stop and find out why. Check all sprocket and block alignment marks to verify that everything is properly assembled.

28 Carefully clean the cover with acetone or lacquer thinner. Caution: It is very important that all traces of old RTV sealant are removed from the block and the timing chain cover prior to applying a fresh 2 mm bead of RTV sealant. Use special care to ensure that no sealant remains in the bolt holes or the oil drain-back hole. Apply a 4 mm bead of RTV sealer to the front of the oil pan. Apply extra sealer to all joints and gaps. Also apply sealer around the center top bolt hole.

29 Install the cover and gradually tighten all bolts starting from the middle and working towards each end. The center bolt(s) should be tightened last in each sequence. Note: The lower center bolt requires much less torque than the others, refer to the torques listed in the Specifications in this Chapter. Install the oil pan bolts.

30 Make certain that the oil drain hole in the bottom of the front seal bore is clear by pumping oil through it.

31 The remainder of installation is the reverse of removal. **Note:** It is recommended that the timing chain cover oil seal be replaced before re-installation of the cover (see Section 16).

7 Camshaft(s) - removal, inspection and installation

SOHC engine

Removal

Note 1: The camshaft can be removed out either end of the cylinder head. If the cylinder head has been removed, it can be removed from the front by removing the thrust plate. With the cylinder head in place, the camshaft

must be removed out the rear. The following procedure describes removal of the camshaft with the cylinder head installed.

Note 2: If the cylinder head has been removed, it's advisable to remove the camshaft out the front end of the head. This is accomplished by removing the camshaft thrust plate. Always use new Torx head screws when installing the camshaft thrust plate.

1 Remove the valve cover (see Section 4). Remove the rocker arm shaft assemblies (see Section 5).

2 Remove the hydraulic valve lifters and guide plates (see Section 8).

3 Before removing the camshaft, measure the camshaft endplay by mounting a dial indicator to the front end of the cylinder head, with the probe resting on the camshaft. Pry the camshaft back-and-forth in the cylinder head. The reading is the camshaft endplay. Compare the reading to this Chapter's Specifications. If the endplay is greater than specified, the camshaft thrust plate is probably worn.

4 Remove the timing chain and sprockets (see Section 6).

5 Remove the battery (see Chapter 5).

6 Using a hammer and a punch, knock the camshaft plug inward, then retrieve it with a magnet or pliers. Discard the plug - a new one must be used upon installation.

7 Carefully slide the camshaft out the rear of the cylinder head.

Inspection

Refer to illustrations 7.9a, 7.9b and 7.9c

8 Visually examine the cam lobes and bearing journals for score marks, pitting, galling and evidence of overheating (blue, discolored areas). Look for flaking away of the hardened surface of each lobe.

9 Using a micrometer, measure the diameter of each camshaft journal and the lobe lift of each camshaft lobe **(see illustrations)**. If the journal diameter or lobe lift of any one of these is less than specified, replace the camshaft.

7.9a Using a micrometer, measure the camshaft journals and compare the measurements to Specifications

2A

7.9b Measure the camshaft lobes at their greatest dimension, write down the measurements . . .

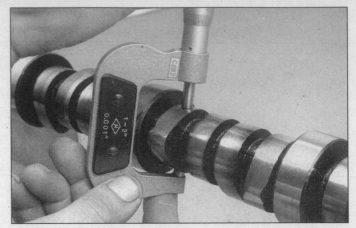

7.9c . . . and subtract the measurement of the lobe diameters at their smallest dimension to obtain the lobe lift specification

Installation

Refer to illustration 7.10

10   Apply moly-base grease or camshaft installation lube to the cam lobes and journals **(see illustration)**.

11   Install the camshaft in the cylinder head, making sure that the dowel pin is in the 12-

7.10 Coat the lobes and journals with cam lube

o'clock position.

12   Apply a thread locking compound (Saturn part no. 21485277, Loctite 242 or equivalent) to the outer sealing surface of the new camshaft plug. Drive the plug into its bore using a bushing driver or a large socket.

13   Install the valve lifters and guide plates (see Section 8). Install the rocker arm assemblies (see Section 5).

14   Install the timing chain, sprockets and timing chain cover (see Section 6).

15   The remainder of installation is the reverse of removal.

DOHC engines

Removal

Refer to illustrations 7.20 and 7.21

16   Remove the valve cover (see Section 4).

17   Measure the thrust clearance (endplay) of the camshaft(s) with a dial indicator **(see illustration 7.3)**. If the endplay is greater than the value listed in this Chapter's Specifications, replace the camshaft and/or the cylinder head.

18   Remove the timing chain and camshaft sprockets (see Section 6).

19   The camshaft bearing caps are num-

bered from front to rear, to ensure they are reinstalled in the same locations when reassembled.

20   Working on one camshaft at a time, loosen the bearing cap bolts **(see illustration)** 1/4-turn at a time.

21   Remove the caps and lift the camshaft off the cylinder head **(see illustration)**. Refer to Section 8 and remove the lifters (this will prevent the lifters from becoming filled with oil - if they're not removed, you'll have to let the engine sit for a few hours after the camshafts have been installed to let the lifters bleed down). **Caution:** *Keep all bearing caps in the correct order of removal. They MUST be reinstalled in the same locations. Also, don't mix up the camshafts.*

22   Repeat this procedure for removal of the other camshaft.

Inspection

Refer to illustrations 7.24a and 7.24b

23   Perform the checks described in Steps 8 and 9.

24   Check the oil clearance for each camshaft journal as follows:

   a)   Clean the bearing caps and the cam-

7.20 Loosen the camshaft bearing cap bolts, 1/4-turn at a time (DOHC engine)

7.21 Remove the caps and lift the camshaft off the cylinder head

7.24a Lay a strip of Plastigage on each camshaft journal

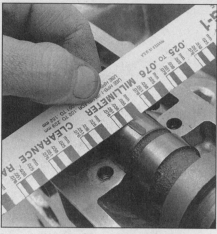

7.24b Compare the width of the crushed Plastigage to the scale on the envelope to determine the oil clearance

shaft journals with lacquer thinner or acetone.

b) Carefully lay the camshaft(s) in place in the head. Don't install the lifters and don't use any lubrication.

c) Lay a strip of Plastigage on each journal **(see illustration)**.

d) Install the bearing caps with the arrows pointing toward the front (timing chain end) of the engine.

e) Tighten the bearing cap bolts, a little at a time, to the torque listed in this Chapter's Specifications. **Note:** Don't turn the camshaft while the Plastigage is in place.

f) Remove the bolts and detach the caps.

g) Compare the width of the crushed Plastigage (at its widest point) to the scale on the Plastigage envelope **(see illustration)**.

h) If the clearance is greater than specified, and the diameter of any journal is less than specified, replace the camshaft. If the journal diameters are within specifications but the oil clearance is too great, the cylinder head is worn and must be replaced.

i) Scrape off the Plastigage with your fingernail or the edge of a credit card - don't scratch or nick the journals or bearing caps.

Installation

25 If the lifters have been removed, install them (see Section 8).

26 Apply moly-base grease or camshaft installation lube to the camshaft lobes and bearing journals **(see illustration 7.10)**. Install each camshaft with the dowel pin in the 12-o'clock position. Make sure the intake and exhaust camshafts are installed in their correct side of the cylinder head. **Caution:** Each camshaft is identified with either a IN (intake) or EX (exhaust); this mark is made between the first and second journal (counting from the timing chain end of the engine). The camshaft(s) can be broken if they are installed incorrectly.

27 Install the camshaft bearing caps in numerical order with the arrows pointing toward the timing chain end of the engine, then tighten the bolts, a little at a time, to the torque listed in this Chapter's Specifications.

Start with the center bolts and work towards the outer bolts.

28 Install the camshaft sprockets, timing chain and timing chain cover (see Section 6).

29 The remainder of installation is the reverse of the removal procedure.

8 Hydraulic valve lifters - removal, inspection and installation

Removal

1 Remove the valve cover (see Section 4).

SOHC engines

Refer to illustration 8.4

2 Remove the rocker arm shaft assembly (see Section 5).

3 Remove the two retaining clips that hold the roller lifter guide plates in place.

4 Lift off the guide plates and remove the lifters **(see illustration)**. Using a magnet if necessary, lift out each valve lifter and set them in numbered boxes, plastic sandwich bags or other containers so they can be reinstalled in the same position during reassembly.

DOHC engines

Refer to illustration 8.6

5 Remove the camshafts (see Section 7).

6 Remove the lifters **(see illustration)**. **Caution:** Using a magnet if necessary, lift out the valve lifters and set them in numbered boxes, plastic sandwich bags or other containers so they can be reinstalled in the same position during reassembly.

Inspection

7 Inspect each lifter for score marks, pitting, galling and evidence of overheating (blue, discolored areas). Look for flaking away of the hardened surface of each lifter.

8 Measure each lifter for wear and compare your measurements with the values

2A

8.4 Lift off the guide plates and remove the lifters from the cylinder head (SOHC engine)

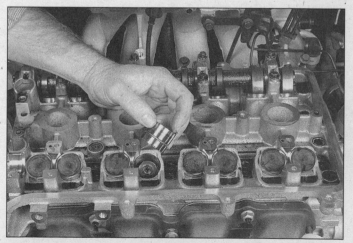

8.6 Remove the lifters (use a magnet if necessary) and set them in numbered boxes or other type of container to keep them organized for reassembly in the exact same location

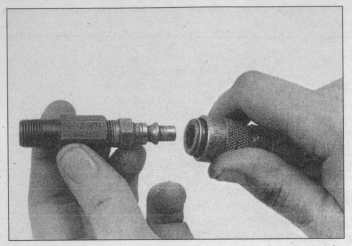

9.5 This is what the air hose adapter that threads into the spark plug hole looks like - they're commonly available from auto parts stores

9.10 Use a valve spring compressor to compress the springs, then remove the keepers from the valve stem with a magnet or small needle-nose pliers

listed in this Chapter's Specifications. Replace any lifter that is worn excessively.

Installation

9   Lubricate the lifters with clean engine oil or engine assembly lube and install them in their original positions in the cylinder head.

10   On SOHC engines, rotate the lifters until the flat sides are parallel to the intake and exhaust manifold flanges and install the lifter guide plates. **Caution:** *Make sure the guide plates are properly seated (with the retaining spring slot up) on the cylinder head and all lifters fit squarely in the guide plate.*

11   The remainder of installation is the reverse of removal.

9   Valve springs, retainers and seals - replacement

Caution: *On models equipped with a theft-deterrent radio, make sure you have the correct activation code, or the theft deterrent system is turned off, before disconnecting the battery.*

Note: *Broken valve springs and defective valve stem seals can be replaced without removing the cylinder head. Two special tools and a compressed air source are normally required to perform this operation, so read through this Section carefully and rent or buy the tools before beginning the job. If compressed air isn't available, a length of nylon rope can be used to keep the valves from falling into the cylinder during this procedure.*

Disassembly

Refer to illustrations 9.5, 9.10 and 9.11

1   Remove the spark plugs.

2   Disconnect the cable from the negative terminal of the battery.

3   On DOHC engines remove the camshaft(s) (see Section 7). On SOHC engines remove the rocker arm and shaft assembly (see Section 5).

4   Turn the crankshaft until the piston in the affected cylinder is at Top Dead Center (TDC) on the compression stroke (refer to Section 3 for instructions). If you're replacing all of the valve stem seals, begin with cylinder number one and work on the valves for one cylinder at a time.

5   Thread an adapter into the spark plug hole **(see illustration)** and connect an air hose from a compressed air source to it. Most auto parts stores can supply the air hose adapter. **Note:** *Many cylinder compression gauges utilize a screw-in fitting that may work with your air hose quick-disconnect fitting.*

6   Apply compressed air to the cylinder. **Warning:** *The piston may be forced down by compressed air, causing the crankshaft to turn suddenly. If the wrench used when positioning the number one piston at TDC is still attached to the bolt in the crankshaft nose, it could cause damage or injury when the crankshaft moves.*

7   The valves should be held in place by the air pressure. If the valve faces or seats are in poor condition, leaks may prevent air pressure from retaining the valves - refer to the alternative procedure below.

8   If you don't have access to compressed air, an alternative method can be used. Position the piston at a point approximately 45-degrees before TDC on the compression stroke, then feed a long piece of nylon rope through the spark plug hole until it fills the combustion chamber. Be sure to leave the end of the rope hanging out of the engine so it can be removed easily.

9   Use a large ratchet and socket to rotate the crankshaft in the normal direction of rotation (clockwise, viewed from the front) until slight resistance is felt.

10   Stuff shop rags into the cylinder head holes around the valves to prevent parts and tools from falling into the engine, then use a valve spring compressor to compress the spring. Remove the keepers with small needle-nose pliers or a magnet **(see illustration)**.

9.11 Remove the valve guide seal with a pair of pliers

Note: *A couple of different types of tools are available for compressing the valve springs with the head in place. One type grips the lower spring coils and presses on the retainer as the knob is turned, while the other type utilizes a bolt or stud and nut for leverage. Both types work very well, although the lever type is usually less expensive.*

11   Remove the spring retainer and valve spring, then remove the stem oil seal **(see illustration)**.

12   Wrap a rubber band or tape around the top of the valve stem so the valve won't fall into the combustion chamber, then release the air pressure. **Note:** *If a rope was used instead of air pressure, turn the crankshaft slightly in the direction opposite normal rotation.*

13   Inspect the valve stem for damage. Rotate the valve in the guide and check the end for eccentric movement, which would indicate that the valve is bent.

14   Move the valve up-and-down in the guide and make sure it doesn't bind. If the valve stem binds, either the valve is bent or the guide is damaged. In either case, the head will have to be removed for repair.

9.16 Gently tap the seal into place with a hammer and a deep socket

9.18 Apply a small dab of grease to each keeper as shown here before installation - it'll hold them in place on the valve stem as the spring is released

Reassembly

Refer to illustrations 9.16 and 9.18

15   Reapply air pressure to the cylinder to retain the valve in the closed position, then remove the tape or rubber band from the valve stem. If a rope was used instead of air pressure, rotate the crankshaft in the normal direction of rotation until slight resistance is felt.
16   Lubricate the valve stem with engine oil and install a new oil seal **(see illustration)**.
17   Install the spring in position over the valve.
18   Install the valve spring retainer. Compress the valve spring and carefully position the keepers in the groove. Apply a small dab of grease to the inside of each keeper to hold it in place if necessary **(see illustration)**.
19   Remove the pressure from the spring tool and make sure the keepers are seated.
20   Disconnect the air hose and remove the adapter from the spark plug hole. If a rope was used in place of air pressure, pull it out of the cylinder.
21   Refer to Section 7 and install the camshaft(s).
22   Install the rest of the parts in the reverse

order of the removal procedure.
23   Start and run the engine, then check for oil leaks and unusual sounds coming from the valve cover area.

10   Intake manifold - removal and installation

Refer to illustrations 10.6 and 10.10
Caution: *On models equipped with a theft-deterrent radio, make sure you have the correct activation code, or the theft deterrent system is turned off, before disconnecting the battery.*

Removal

1   Disconnect the negative cable from the battery.
2   Drain the cooling system (see Chapter 1).
3   Disconnect the accelerator cable from the throttle body or TBI unit. Detach the cable housing from its bracket (see Chapter 4).
4   Disconnect the electrical connectors for the injectors, idle air control valve, throttle position switch and MAP sensor.

5   Disconnect the vacuum hose at the canister purge valve, throttle body connector, fuel regulator, EGR valve solenoid and PCV hose at the valve cover.
6   Disconnect the heater hose at the intake manifold **(see illustration)**.
7   Unbolt the power steering pump (see Chapter 10) and set it aside without disconnecting any of the hoses.
8   Remove the intake manifold nuts.
9   Raise the vehicle and support it securely on jackstands and remove the right side tire and the splash shield.
10   From under the vehicle, remove the intake manifold support brace attached to the intake next to the alternator **(see illustration)**.
11   Label and detach all wire harnesses, control cables and hoses still connected to the intake manifold.
12   Remove the intake manifold.

Installation

13   Use a scraper to remove all traces of old gasket material and sealant from the manifold and cylinder head, then clean the mating surfaces with lacquer thinner or acetone. **Caution:** *Be careful not to gouge the soft aluminum.* If the gasket was leaking, check the manifold for warpage using a straightedge and feeler gauge. Compare your measurement with this Chapter's Specifications. If it's warped beyond the maximum allowable limit, have it resurfaced at an automotive machine shop.
14   Install a new gasket, then position the manifold on the head and install the nuts.
15   Tighten the nuts in three or four equal steps to the torque listed in this Chapter's Specifications. Work from the center out towards the ends to avoid warping the manifold.
16   Install the remaining parts in the reverse order of removal.
17   Refill the cooling system with the proper type and mixture of antifreeze (see Chapter 1).
18   Before starting the engine, check the throttle linkage for smooth operation.

2A

10.6 Disconnect the heater hose at the left (driver's) side of the intake manifold

10.10 Remove the support brace, which can be reached from under the vehicle and is attached to the intake next to the alternator

11.3 Unplug the electrical connector and, if you're installing a new manifold, unscrew the oxygen sensor (arrow)

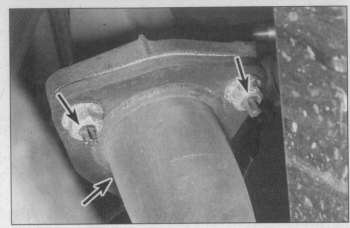

11.5 Remove the three nuts (arrows) and disconnect the exhaust pipe from the exhaust manifold

19 Run the engine and check for coolant and vacuum leaks.

20 Road test the vehicle and check for proper operation of all accessories, including the cruise control system (if equipped).

11 Exhaust manifold - removal and installation

Refer to illustrations 11.3 and 11.5

Warning: *The engine must be completely cool before beginning this procedure.*

Caution: *On models equipped with a theft-deterrent radio, make sure you have the correct activation code, or the theft deterrent system is turned off, before disconnecting the battery.*

Removal

1 Disconnect the negative cable from the battery.

2 Remove the heat shield from the manifold, if equipped.

3 Unplug the oxygen sensor electrical connector **(see illustration)**. If you're installing a new manifold, remove the sensor (see Chapter 6).

4 Apply penetrating oil to the exhaust manifold mounting nuts and the exhaust pipe flange nuts.

5 Disconnect the exhaust pipe from the exhaust manifold **(see illustration)**.

6 Remove the nuts and detach the manifold from the cylinder head.

Installation

7 Use a scraper to remove all traces of old gasket material and carbon deposits from the manifold and cylinder head mating surfaces. **Caution:** *When scraping, be very careful not to gouge or scratch the delicate aluminum cylinder head.* If the gasket was leaking, check the manifold for warpage using a straightedge and feeler gauge. Compare your measurement with this Chapter's Specifications. If it's warped beyond the maximum allowable limit, have it resurfaced at an automotive machine shop.

8 Position a new exhaust manifold gasket over the studs on the cylinder head.

9 Install the manifold and thread the mounting nuts into place.

10 Working from the center out, tighten the nuts/bolts to the torque listed in this Chapter's Specifications in several equal steps.

11 Reinstall the remaining parts in the reverse order of removal.

12 Run the engine and check for exhaust leaks.

12 Cylinder head - removal and installation

Caution 1: *The engine must be completely cool before beginning this procedure.*

Caution 2: *On models equipped with a theft-deterrent radio, make sure you have the correct activation code, or the theft deterrent system is turned off, before disconnecting the battery.*

Removal

1 Disconnect the negative cable from the battery.

2 Drain the coolant from the engine block and radiator (see Chapter 1).

3 Drain the engine oil and remove the oil filter (see Chapter 1).

4 Remove the intake manifold (see Section 10).

5 Remove the exhaust manifold (see Section 11).

6 Remove the valve cover (see Section 4).

7 Remove the timing chain (see Section 6).

8 Remove the alternator (see Chapter 5).

9 Unbolt the power steering pump and set it aside without disconnecting the hoses (see Chapter 10).

10 Label and detach any remaining wiring or hoses that would interfere with head removal.

11 On 1992 and later models, support the engine with a jack and block of wood positioned under the oil pan. Remove the right engine mount (see Section 17).

12 Using a breaker bar and a 6-point socket, loosen the cylinder head bolts in 1/4-turn increments until they can be removed by hand. Work in a sequence opposite that of the tightening sequence to avoid warping or cracking the head **(see illustrations 12.25a or 12.25b)**.

13 Lift the cylinder head off the engine block. If it's stuck, very carefully pry up at the transaxle end, beyond the gasket surface. Set the head on two wood blocks to avoid damaging the valves.

14 Remove all external components from the head to allow for thorough cleaning and inspection. See Chapter 2, Part B, for cylinder head servicing procedures.

Installation

Refer to illustrations 12.19, 12.24a and 12.24b

15 The mating surfaces of the cylinder head and block must be perfectly clean when the head is installed. **Note:** *Replacement 1991 and 1992 cylinder head gaskets have a 0.157 inch (4 mm) orifice and the cylinder block orifice plug is no longer required. The oil galley plug in the block with the orifice can be removed with a slide hammer and a screw puller - this should be removed whenever a cylinder head gasket is replaced.*

16 Use a gasket scraper to remove all traces of carbon and old gasket material, then clean the mating surfaces with lacquer thinner or acetone. If there's oil on the mating surfaces when the head is installed, the gasket may not seal correctly and leaks could develop. When working on the block, stuff the cylinders with clean shop rags to keep out debris. Use a vacuum cleaner to remove material that falls into the cylinders. **Caution:** *Be careful not to gouge the soft aluminum of the cylinder head.*

17 Check the block and head mating surfaces for nicks, deep scratches and other damage. If damage is slight, it can be removed with a file; if it's excessive, machining may be the only alternative. Also check the cylinder head for warpage (see Chapter 2, Part B).

18 Use a tap of the correct size to chase (clean up) the threads in the head bolt holes,

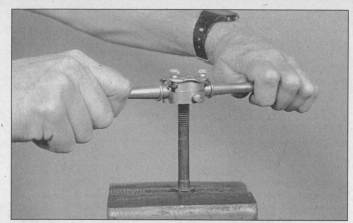

12.19 A die should be used to remove corrosion from the head bolt threads prior to installation

12.24a Cylinder head bolt TIGHTENING sequence (SOHC engine)

12.24b Cylinder head bolt TIGHTENING sequence (DOHC engine)

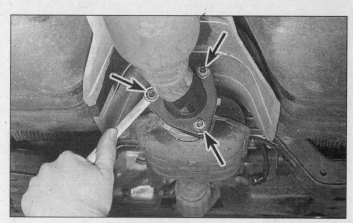

13.5 Disconnect the exhaust pipe from the catalytic converter (arrows)

then clean the holes with compressed air - make sure that nothing remains in the holes. **Warning:** *Wear eye protection when using compressed air!*

19   Mount each bolt in a vise and run a die down the threads to remove corrosion and restore the threads **(see illustration).** Dirt, corrosion, sealant and damaged threads will affect torque readings.

20   Install the components that were removed from the head.

21   Position the new gasket over the dowel pins in the block.

22   Carefully set the head on the block without disturbing the gasket.

23   Before installing the head bolts, clean the threads of all corrosion.

24   Install the bolts (clean and dry) and tighten them, in sequence, to 48 ft-lbs **(see illustrations).**

25   Remove the bolts (using the opposite of the tightening sequence), coat the threads with engine oil and reinstall the bolts, tightening them following the procedure below:

   a) *In sequence, tighten the bolts to 22 ft-lbs.*
   b) *In sequence, tighten the bolts to 37 ft-lbs on DOHC models and 33 ft-lbs on SOHC models.*
   c) *Finally, in sequence, tighten all bolts an additional 90-degrees (1/4-turn).* **Note:** *Gauges that connect to your torque*

wrench and measure angles are available at most auto parts stores. Another method is to make a small paint mark on each head bolt and, after the beginning position of the bolt is noted, turn each an additional 1/4-turn (90-degrees).

26   The remaining installation steps are the reverse of removal.

27   Refill the cooling system, install a new oil filter and add oil to the engine (see Chapter 1).

28   Run the engine and check for leaks. Set the ignition timing (see Chapter 5) and road test the vehicle.

13   Oil pan - removal and installation

Refer to illustrations 13.5, 13.6, 13.8, 13.9, 13.13 and 13.16
Caution: *On models equipped with a theft-deterrent radio, make sure you have the correct activation code, or the theft deterrent system is turned off, before disconnecting the battery.*

1   Disconnect the negative cable from the battery.

2   Set the parking brake and block the rear wheels. Raise the front of the vehicle and support it securely on jackstands.

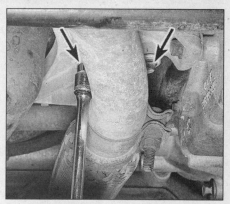

13.6 Remove the two bracket bolts (arrows) that hold the exhaust pipe up to the bottom of the engine

3   Remove the splash shields under the engine.

4   Drain the engine oil and remove the oil filter (see Chapter 1).

5   Disconnect the exhaust pipe from the catalytic converter **(see illustration).**

6   Remove the two bracket bolts **(see illustration)** that hold the exhaust pipe up to the bottom of the engine.

7   Unbolt the exhaust pipe from the exhaust manifold (see Section 11) and remove the exhaust pipe assembly.

13.8   Remove the six bolts (arrows) and the engine-to-transaxle reinforcement plate

13.9   Remove the three bolts (arrows) and the flywheel access cover

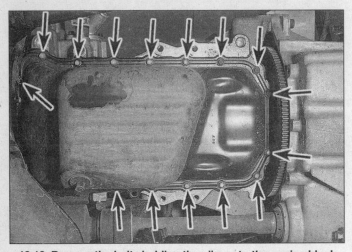

13.13   Remove the bolts holding the oil pan to the engine block (some of the bolts aren't visible in this photo)

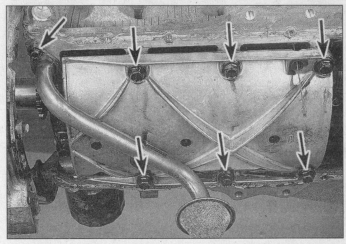

13.16   If necessary, remove the crankcase oil baffle (DOHC models only) and oil pickup tube from the engine block (arrows)

8   Remove the engine-to transaxle reinforcement plate **(see illustration)**.

9   Remove the flywheel/driveplate access cover **(see illustration)**.

10   Remove the right-side wheel and splash shield.

11   Remove crankshaft damper/pulley (see Section 16).

12   Loosen and back out the four front engine mount bolts approximately 13 mm (1/2-inch).

13   Remove all the oil pan bolts **(see illustration)**.

14   Using a suitable tool pry the front engine mount away from the cylinder block to allow for oil pan removal.

15   Detach the oil pan. If it's stuck, pry it loose carefully with a screwdriver or putty knife. Don't damage the mating surfaces of the pan and block or leaks could develope.

16   If necessary, remove the crankcase oil baffle (windage tray) and oil pickup tube **(see illustration)**.

17   Use a scraper to remove all traces of old gasket material and sealant from the block and oil pan. Clean the mating surfaces with lacquer thinner or acetone.

18   Make sure the threaded bolt holes in the block are clean.

19   Check the oil pan flange for distortion, particularly around the bolt holes. If necessary, place the pan on a block of wood and use a hammer to flatten and restore the gasket surface.

20   Apply a 4 mm (5/32-inch) bead of sealant to the oil pan flange. **Note:** *The oil pan must be installed within 3 minutes once the sealant has been applied.*

21   Carefully position the oil pan on the engine block and install the bolts. Working from the center out, tighten them to the torque listed in this Chapter's Specifications in three or four steps.

22   The remainder of installation is the reverse of removal. Be sure to add oil and install a new oil filter.

23   Run the engine and check for oil leaks.

---

14   Oil pump - removal, inspection and installation

Caution: *On models equipped with a theft-deterrent radio, make sure you have the cor-*

rect activation code, or the theft deterrent system is turned off, before disconnecting the battery.

Removal

Refer to illustration 14.4

1   Disconnect the negative cable from the battery.

2   Set the parking brake and block the rear wheels. Raise the front of the vehicle and support it securely on jackstands.

3   Remove the timing chain cover (which is also the housing for the oil pump assembly) (see Section 6).

4   Using a number 30 Torx driver, remove the oil pump housing cover bolts and cover **(see illustration)**.

5   If necessary, take the timing chain cover to a dealer service department, automotive machine shop or other repair facility to have the oil pressure relief valve removed (special tools are required to do this, but it is recommended that it not be removed unless there is some damage indicated, such as a low oil pressure problem). **Caution:** *Whenever the pressure relief valve is removed from the pump it must be replaced with a new one*

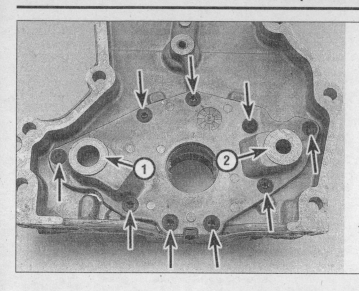

14.4 Remove the Torx screws that attach the pump cover to the timing chain cover

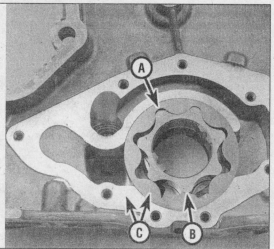

14.9 Once the cover is removed inspect all the individual parts as well as the housing for galling, pitting, score marks or indications overheating (blue, discolored areas) and perform the following checks:

a) *Measure the clearance between the oil pump driven gear and the pump housing*

b) *Measure the tip clearance between the drive and driven gear*

c) *Place a straightedge across the pump housing and measure the clearance between the gearset and the pump cover*

because removal causes damage to the valve sealing seat.

6   Use a felt pen to mark the gears so that they can be installed in the same positions. **Note:** *The chamfers on the gears face towards the front cover.* Remove the oil pump gears from the housing and set them aside.

7   Now would be a good time to replace the seal in the timing chain cover (see Section 16).

Inspection and reassembly

Refer to illustration 14.9

8   Inspect the gears and the pump housing for signs of overheating (a blue color), nicks, scoring or other signs of wear. Replace any damaged parts.

9   Reinstall the gears into the oil pump housing **(see illustration)** and, using feeler gauges, measure the clearances of:

   The driven gear-to-pump housing clearance
   The drive gear-to-driven gear tip clearance
   The gearset-to-oil pump housing clearance (or endplay)

10   Compare your measurements to the values listed in this Chapter's Specifications.

11   Be sure the surfaces of the pump hous-

ing are clean and dry before reassembly.

12   If necessary, have a new oil pressure relief valve installed by a dealer service department, automotive machine shop or other repair facility.

13   Lubricate the gear set with clean engine oil. Reinstall the gears.

14   Pack the pump cavities with petroleum jelly (this will prime the pump and ensure good suction when the engine is started).

15   Install the cover and tighten the screws to the torque listed in this Chapter's Specifications. **Note:** *The factory recommends that the cover screws be replaced whenever they are removed because they are coated with a sealant to prevent oil leakage.*

16   It's a good idea to remove the oil pan and inspect the screen at the end of the oil pick-up tube for any debris that might be plugging it. Either clean the tube and screen completely or replace it with a new one at this time.

Installation

17   Install new oil pressure and suction seals into the engine block **(see illustration 14.4).**

18   Install the front cover (see Section 6).

19   Reinstall the remaining parts in the

reverse order of removal.

20   Install a new oil filter, add oil (see Chapter 1), start the engine and check for oil pressure and leaks.

15   Flywheel/driveplate - removal, inspection and installation

Refer to illustration 15.3

Removal

1   Remove the transaxle (see Chapter 7).

2   If you're working on a model with a manual transaxle, remove the pressure plate and clutch disc (see Chapter 8). Now is a good time to check/replace the clutch components.

3   Use a center-punch or paint to make alignment marks on the flywheel/driveplate and crankshaft to ensure correct alignment during installation **(see illustration).**

4   Remove the bolts that secure the flywheel/driveplate to the crankshaft. If the crankshaft turns, wedge a screwdriver in the ring gear teeth to jam the flywheel.

5   Remove the flywheel/driveplate from the crankshaft. Since the flywheel is fairly heavy, be sure to support it while removing the last bolt.

Inspection

6   Clean the flywheel to remove grease and oil. Inspect the surface for cracks, rivet grooves, burned areas and score marks. Light scoring can be removed with emery cloth. Check for cracked and broken ring gear teeth. Lay the flywheel on a flat surface and use a straightedge to check for warpage. If necessary, take the flywheel to an automotive machine shop to have it resurfaced.

7   Clean and inspect the mating surfaces of the flywheel/driveplate and the crankshaft. If the crankshaft rear seal is leaking, replace it before reinstalling the flywheel/driveplate.

Installation

8   Position the flywheel/driveplate against the crankshaft. Be sure to align the marks

15.3 Mark the flywheel/driveplate and the crankshaft so they can be reassembled in the same relative position

2A

16.5 Remove the bolt from the crankshaft pulley using a large socket and ratchet or breaker bar to break the bolt loose after the pulley has been held in place with a prybar or large screwdriver

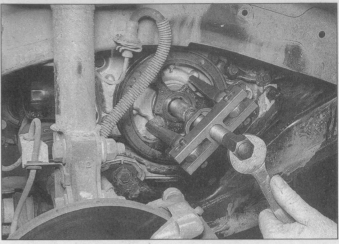

16.6 Once the bolt has been removed, remove the damper/pulley by using a two or three jaw puller

16.7 Carefully pry the crankshaft seal out of the bore (arrow) - DO NOT nick or scratch the camshaft or seal bore

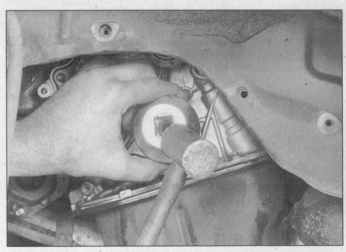

16.9 Gently tap the new seal into place with the spring side toward the engine

made during removal. Note that some engines have an alignment dowel or staggered bolt holes to ensure correct installation. Before installing the bolts, apply thread locking compound to the threads.

9   Wedge a screwdriver in the ring gear teeth to keep the flywheel/driveplate from turning as you tighten the bolts to the torque listed in this Chapter's Specifications. Follow a criss-cross pattern and work up to the final torque in three or four steps.

10   The remainder of installation is the reverse of the removal procedure.

---

16   Crankshaft oil seals - replacement

Caution: *On models equipped with a theft-deterrent radio, make sure you have the correct activation code, or the theft deterrent system is turned off, before disconnecting the battery.*

Front oil seal

Refer to illustration 16.5, 16.6, 16.7 and 16.9

1   Disconnect the negative cable from the battery.

2   Set the parking brake and block the rear wheels. Raise the front of the vehicle and support it securely on jackstands.

3   Drain the engine oil and remove the oil filter (see Chapter 1).

4   Remove the right-side wheel and splash shield.

5   Remove the vibration damper retaining bolt by placing a prybar or large screwdriver through the damper and wedging it under the block **(see illustration)**. This will hold the damper in place so the center bolt can be loosened and removed.

6   Use two or three jaw puller **(see illustration)** to remove the damper.

7   Carefully pry the seal out of the timing chain cover **(see illustration)** with a screwdriver or seal removal tool. Don't scratch the bore or damage the crankshaft in the process (if the crankshaft is damaged, the new seal

will end up leaking).

8   Clean the bore in the housing and coat the outer edge of the new seal with multi-purpose grease. Apply multi-purpose grease to the seal lip.

9   Using a seal driver or a socket with an outside diameter slightly smaller than the outside diameter of the seal, carefully drive the new seal into place with a hammer **(see illustration)**. Make sure it's installed squarely and driven in to the same depth as the original. Check the seal after installation to make sure the garter spring didn't pop out of place.

10   The rest of the installation is the reverse of the removal procedure. Apply a small amount of RTV sealer between the crankshaft pulley and the large pulley washer to prevent oil leaks at this point.

11   Run the engine and check for oil leaks at the front seal.

Rear main oil seal

Refer to illustrations 16.13, 16.16 and 16.17

12   The transaxle must be removed from the

16.13 Lubricate the crankshaft journal and the lip of the new seal with engine oil and tap the new seal into place - the seal lip is stiff and can be easily damaged during installation if you're not careful

16.16 After removing the retainer from the engine, support it on wood blocks and drive out the old seal with a punch and hammer

vehicle for this procedure (see Chapter 7).

13   The seal can be replaced without dropping the oil pan or removing the seal retainer. However, this method can be more difficult because the lip of the seal is quite stiff and it's possible to cock the seal in the retainer bore or damage it during installation. If you want to take the chance, pry out the old seal with a screwdriver or seal removal tool. **Note:** *Slots are provided in the seal retainer for prying - don't pry between the seal and the crankshaft.* Apply engine oil to the crankshaft seal journal and the lip of the new seal and carefully push the new seal into place. The lip is stiff so carefully work it onto the seal journal of the crankshaft with a smooth object like the end of a socket extension as you tap the seal into place **(see illustration)**.

14   The following method is recommended but requires removal of the oil pan (see Section 13) and the seal retainer.

15   After the oil pan has been removed, remove the bolts, detach the seal retainer and peel off all the old gasket material.

16   Position the seal and retainer assembly on a couple of wood blocks and drive the old seal out from the back side with a punch **(see illustration)**.

17   Drive the new seal into the retainer with a block of wood **(see illustration)** or a section of pipe slightly smaller in diameter than the outside diameter of the seal.

18   Lubricate the crankshaft seal journal and the lip of the new seal with engine oil. Position a new gasket on the engine block.

19   Slowly and carefully push the seal onto the crankshaft. The seal lip is stiff, so work it onto the crankshaft with a smooth object such as the end of a socket extension as you push the retainer against the block.

20   Install and tighten the retainer bolts to the torque listed in this Chapter's Specifications. The bottom sealing flange of the retainer must not extend below the bottom sealing flange (oil pan rail) of the block.

21   The remaining steps are the reverse of removal.

16.17 Drive the new seal into the retainer with a wood block or a section of pipe, if you have one large enough - make sure you don't cock the seal in the retainer bore

17   Engine mounts - check and replacement

Caution: *On models equipped with a theft-deterrent radio, make sure you have the correct activation code, or the theft deterrent system is turned off, before disconnecting the battery.*

1   Engine mounts seldom require attention, but broken or deteriorated mounts should be replaced immediately or the added strain placed on the driveline components may cause damage or wear.

Check

2   During the check, the engine must be raised slightly to remove the weight from the mounts.

3   Raise the vehicle and support it securely on jackstands, then position a jack under the engine oil pan. Place a block of wood between the jack head and the oil pan, then carefully raise the engine just enough to take the weight off the mounts. **Warning:** *DO NOT*

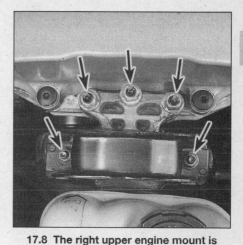

17.8 The right upper engine mount is taken off by removing the five nuts and lifting it straight up

place any part of your body under the engine when it's supported only by a jack!

4   Check the mounts to see if the rubber is cracked, hardened or separated from the metal portion. Sometimes the rubber will split right down the center.

5   Check for relative movement between the mounts and the engine or frame (use a large screwdriver or prybar to attempt to move the mounts). If movement is noted, lower the engine and tighten the mount fasteners.

6   Rubber preservative should be applied to the mounts to slow deterioration.

Replacement

Refer to illustrations 17.8, 17.10a and 17.10b

7   Disconnect the negative battery cable from the battery, then raise the vehicle and support it securely on jackstands (if not already done). Support the engine as described in Step 3.

8   The upper engine mount is taken off by removing the five nuts and lifting it straight up **(see illustration)**.

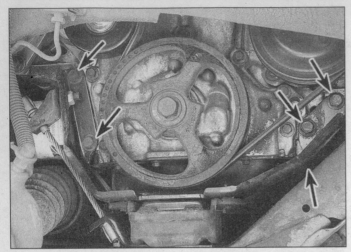

17.10a Details of the right-side lower engine mount (1991 models)

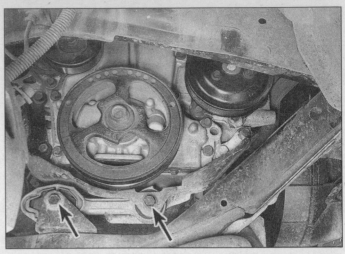

17.10b Details of the right-side lower engine mount (1992 and later models)

9   To remove the right lower engine mount raise the vehicle and support it securely on jackstands, then position a jack under the engine oil pan. Place a block of wood between the jack head and the oil pan, then carefully raise the engine just enough to take the weight off the mounts. **Warning:** *DO NOT place any part of your body under the engine when it's supported only by a jack!*

10   Remove the bolts that attach the mount to the engine block and the nut that hold the mount to the frame cradle **(see illustrations).**

11   Installation is the reverse of removal. Use thread locking compound on the mount bolts/nuts and be sure to tighten them securely.

12   See Chapter 7A for the torque rod replacement procedure.

Chapter 2 Part B
General engine overhaul procedures

Contents

Specifications

General

Cylinder compression pressure (all engines)
Standard	190 psi minimum
Service Limit	180 psi minimum

Oil pressure
Idle (normal operating temperature)	13.0 psi minimum
2000 RPM (hot)	36.0 psi minimum
2000 RPM (cold)	80.0 psi maximum

Engine block

Deck surface warpage
Standard	0.0009 inch
Service limit	0.0020 inch

Bore diameter
Cylinder Nos. 1, 2 and 3	3.2280 to 3.2287 inches
Service limit	3.2297 inches maximum
Cylinder No. 4	3.2283 to 3.2291 inches
Service limit	3.2301 inches maximum

Out-of-round
Standard	0.0004 inch
Service limit	0.0020 inch maximum

Engine block (continued)
Taper
    Standard.. 0.0004 inch
    Service limit.. 0.0020 inch maximum

Crankshaft and connecting rods
Crankshaft runout.. 0.0020 inch maximum
Endplay
    Standard.. 0.0020 to 0.0079 inch
    Maximum... 0.0098 inch maximum
Main bearing journal
    Diameter
        Standard .... 2.2438 to 2.2444 inches
        Service limit .... 2.2437 inches maximum
    Taper .... 0.0004 inch maximum
    Out-of-round .... 0.0004 inch maximum
Main bearing oil clearance
    Standard.. 0.0002 to 0.002 inch
    Maximum... 0.0024 inch maximum

Connecting rod journal
Diameter
    Standard.. 1.8500 to 1.8508 inches
    Minimum.. 1.8500 inches minimum
Taper .... 0.0004 inch maximum
Out-of-round... 0.0004 inch maximum
Connecting rod oil clearance
    1991 through 1995
        Standard .... 0.00039 to 0.0025 inch
        Service limit .... 0.0029 inch maximum
    1996 through 1999
        Standard .... 0.0001 to 0.0021 inch
        Service limit .... 0.0025 inch maximum
Connecting rod side clearance (endplay)
    1991 through 1994 .... 0.0126 inch maximum
    1995 through 1999
        Standard .... 0.0065 to 0.01713 inch
        Service limit .... 0.01850 inch maximum

Piston and rings
Piston diameter
    Standard
        1991 through 1998.. 3.2270 to 3.2277 inches
        1999 .... 3.2270 to 3.2280 inches
    Service limit .... 3.2266 minimum
Piston-to-bore clearance
    1991 through 1998
        Cylinders No. 1, 2 and 3
            Standard... 0.0002 to 0.0017 inch
            Service limit .... 0.0028 inch maximum
        Cylinder 4
            Standard... 0.0006 to 0.0021 inch
            Service limit .... 0.0028 inch maximum
    1999
        Cylinders No. 2 and 3
            Standard... 0.0000 to 0.0017 inch
            Service limit .... 0.0028 inch maximum
        Cylinders No. 1 and 4
            Standard... 0.0003 to 0.0021 inch
            Service limit .... 0.0028 inch maximum
Piston rings
    End gap
        1991 through 1998
            No. 1 (top ring)
                Standard .... 0.0098 to 0.0157 inch
                Service limit... 0.0197 inch maximum
            No. 2 (middle ring)
                 Standard .... 0.0098 to 0.0197 inch
                Service limit... 0.0236 inch maximum

Oil ring
    Standard .... 0.0098 to 0.0492 inch
    Service limit .... 0.0551 inch maximum
1999
  No. 1 (top ring)
    Standard .... 0.0071 to 0.0130 inch
    Service limit... 0.0177 inch maximum
  No. 2 (middle ring)
    Standard .... 0.0138 to 0.0216 inch
    Service limit .... 0.0256 inch maximum
  Oil ring
    Standard .... 0.0039 to 0.0197 inch
    Service limit .... 0.0256 inch maximum
Piston ring side clearance
  1991 through 1998
    No. 1 (top ring)
      Standard .... 0.0016 to 0.0032 inch
      Service limit .... 0.0035 inch maximum
    No. 2 (middle ring)
      Standard .... 0.0012 to 0.0031 inch
      Service limit .... 0.0031 inch maximum
  1999
    No. 1 (top ring)
      Standard .... 0.0016 to 0.0028 inch
      Service limit .... 0.0035 inch maximum
    No. 2 (middle ring)
      Standard .... 0.0012 to 0.0031 inch
      Service limit .... 0.0031 inch maximum

Valves and related components

Valve lash .... Not adjustable
Valve spring free length
  SOHC engine
    Intake and exhaust.. 1.8898 to 1.9134 inches
  DOHC
    Intake and exhaust
      1991 through 1997 .... 1.61 inches
      1998... 1.65 inches
      1999... 1.56 inches
Valve spring squareness
  SOHC engine
    Standard .... 0.080 inch
    Service limit... 0.100 inch maximum
  DOHC engine
    Standard .... 0.080 inch
    Service limit... 0.100 inch maximum
Valve margin width
  Intake
    Standard .... 0.0378 to 0.0496 inch
    Service limit... 0.035 inch minimum
  Exhaust
    Standard .... 0.0441 to 0.0559 inch
    Service limit .... 0.039 inch minimum
Valve seat width
  SOHC engine
    Intake
      Standard... 0.0394 to 0.0512 inch
      Service limit .... 0.063 inch maximum
    Exhaust
      Standard... 0.0512 to 0.0630 inch
      Service limit .... 0.075 inch maximum
  DOHC engine
    Intake
      Standard... 0.0394 to 0.0512 inch
      Service limit .... 0.0653 inch maximum
    Exhaust
      Standard... 0.0512 to 0.0630 inch
      Service limit .... 0.0756 inch maximum
Valve face angle (intake and exhaust) .... 44.5 to 45.5 degrees

Valves and related components (continued)

Valve stem diameter
  1991 through 1998
    Intake
      Standard... 0.2736 to 0.274 inch
      Service limit.. 0.273 inch minimum
    Exhaust
      Standard... 0.2726 to0.2736 inch
      Service limit.. 0.272 inch minimum
  1999
    Intake
      Standard... 0.2347 to 0.2363 inch
      Service limit.. 0.2342 inch minimum
    Exhaust
      Standard... 0.2342 to 0.2349 inch
      Service limit.. 0.2336 inch minimum
  Exhaust
    Standard... 0.2736 to 0.2740 inch
    Service limit.. 0.272 inch minimum
Valve stem-to-guide clearance
  Intake
    Standard... 0.001 to 0.0025 inch
    Service limit.. 0.004 inch maximum
  Exhaust
    Standard... 0.0015 to 0.0032 inch
    Service limit.. 0.005 inch maximum
Valve spring pressure
  SOHC engine
    Spring height
      1.28 inches
        Standard... 202 to 211 lbs
        Service limit.. 196 lbs minimum
      1.61 inches
        Standard... 76 to 87 lbs
        Service limit.. 75 lbs minimum
  1991 through 1997
    Spring height
      0.984 inch
        Standard... 163 to 180 lbs
        Service limit.. 157 lbs minimum
      1.34 inches
        Standard... 66 to 77 lbs
        Service limit.. 63 lbs minimum
  1998
    Spring height
      0.984 inch
        Standard... 163 to 177 lbs
        Service limit.. 155 lbs minimum
      1.388 inches
        Standard... 66 to 74 lbs
        Service limit.. 63 lbs minimum
  1999
    Spring height
      0.847 inch
        Standard... 99 to 109 lbs
        Service limit.. 94 lbs minimum
      1.221 inches
        Standard... 42 to 48 lbs
        Service limit.. 40 lbs minimum

Torque specifications*   **Ft-lbs** (unless otherwise indicated)

Main bearing cap bolts.. 37
Connecting rod cap nuts
  1991 through 1998.. 33
  1999
    Step 1... 19
    Step 2... Tighten an additional 75-degrees

* **Note:** *Refer to Part A for additional torque specifications*

1 General information

Included in this portion of Chapter 2 are the general overhaul procedures for the cylinder head and internal engine components.

The information ranges from advice concerning preparation for an overhaul and the purchase of replacement parts to detailed, step-by-step procedures covering removal and installation of internal engine components and the inspection of parts.

The following Sections have been written based on the assumption that the engine has been removed from the vehicle. For information concerning in-vehicle engine repair, as well as removal and installation of the external components necessary for the overhaul, see Part A of this Chapter and Section 7 of this Part.

The Specifications included in this Part are only those necessary for the inspection and overhaul procedures which follow. Refer to Part A for additional Specifications.

2 Engine overhaul - general information

It's not always easy to determine when, or if, an engine should be completely overhauled, as a number of factors must be considered.

High mileage is not necessarily an indication that an overhaul is needed, while low mileage doesn't preclude the need for an overhaul. Frequency of servicing is probably the most important consideration. An engine that's had regular and frequent oil and filter changes, as well as other required maintenance, will most likely give many thousands of miles of reliable service. Conversely, a neglected engine may require an overhaul very early in its life.

Excessive oil consumption is an indication that piston rings, valve seals and/or valve guides are in need of attention. Make sure that oil leaks aren't responsible before deciding that the rings and/or guides are bad. Perform a cylinder compression check to determine the extent of the work required (see Section 3).

Check the oil pressure with a gauge installed in place of the oil pressure sending unit and compare it to the Specifications. If it's extremely low, the bearings and/or oil pump are probably worn out.

Loss of power, rough running, knocking or metallic engine noises, excessive valve train noise and high fuel consumption rates may also point to the need for an overhaul, especially if they're all present at the same time. If a complete tune-up doesn't remedy the situation, major mechanical work is the only solution.

An engine overhaul involves restoring the internal parts to the specifications of a new engine. During an overhaul, the piston rings are replaced and the cylinder walls are reconditioned (rebored and/or honed). If a rebore is done by an automotive machine shop, new oversize pistons will also be installed. The main bearings, connecting rod bearings and camshaft bearings are generally replaced with new ones and, if necessary, the crankshaft may be reground to restore the journals. Generally, the valves are serviced as well, since they're usually in less-than-perfect condition at this point. The end result should be a like-new engine that will give many trouble free miles. **Note:** *Critical cooling system components such as the hoses, drivebelts, thermostat and water pump MUST be replaced with new parts when an engine is overhauled. The radiator should be checked carefully to ensure that it isn't clogged or leaking* (see Chapter 3).

Before beginning the engine overhaul, read through the entire procedure to familiarize yourself with the scope and requirements of the job. Overhauling an engine isn't difficult, but it is time consuming. Plan on the vehicle being tied up for a minimum of two weeks, especially if parts must be taken to an automotive machine shop for repair or reconditioning. Check on availability of parts and make sure that any necessary special tools and equipment are obtained in advance. Most work can be done with typical hand tools, although a number of precision measuring tools are required for inspecting parts to determine if they must be replaced. Often an automotive machine shop will handle the inspection of parts and offer advice concerning reconditioning and replacement. **Note:** *Always wait until the engine has been completely disassembled and all components, especially the engine block, have been inspected before deciding what service and repair operations must be performed by an automotive machine shop. Since the block's condition will be the major factor to consider when determining whether to overhaul the original engine or buy a rebuilt one, never purchase parts or have machine work done on other components until the block has been thoroughly inspected. As a general rule, time is the primary cost of an overhaul, so it doesn't pay to install worn or substandard parts.*

As a final note, to ensure maximum life and minimum trouble from a rebuilt engine, everything must be assembled with care in a spotlessly clean environment.

3 Compression check

Refer to illustration 3.6

1 A compression check will tell you what mechanical condition the upper end (pistons, rings, valves, head gasket[s]) of your engine is in. Specifically, it can tell you if the compression is down due to leakage caused by worn piston rings, defective valves and seats or a blown head gasket. **Note:** *The engine must be at normal operating temperature and the battery must be fully charged for this check.*

3.6 A compression gauge with a threaded fitting for the spark plug hole is preferred over the type that requires hand pressure to maintain the seal - be sure to open the throttle valve as far as possible during the compression check

2 Begin by cleaning the area around the spark plugs before you remove them (compressed air should be used, if available, otherwise a small brush or even a bicycle tire pump will work). The idea is to prevent dirt from getting into the cylinders as the compression check is being done.

3 Remove all of the spark plugs from the engine (see Chapter 1).

4 Block the throttle wide open.

5 Detach the primary (low voltage) wiring from the coil pack (see Chapter 5). The fuel pump circuit should also be disabled (see Chapter 4).

6 Install the compression gauge in the spark plug hole **(see illustration)**.

7 Crank the engine over at least seven compression strokes and watch the gauge. The compression should build up quickly in a healthy engine. Low compression on the first stroke, followed by gradually increasing pressure on successive strokes, indicates worn piston rings. A low compression reading on the first stroke, which doesn't build up during successive strokes, indicates leaking valves or a blown head gasket (a cracked head could also be the cause). Deposits on the undersides of the valve heads can also cause low compression. Record the highest gauge reading obtained.

8 Repeat the procedure for the remaining cylinders and compare the results to the Specifications.

9 Add some engine oil (about three squirts from a plunger-type oil can) to each cylinder, through the spark plug holes, and repeat the test.

10 If the compression increases after the oil is added, the piston rings are definitely worn. If the compression doesn't increase significantly, the leakage is occurring at the valves or head gasket. Leakage past the valves may be caused by burned valve seats and/or faces or warped, cracked or bent valves.

11 If two adjacent cylinders have equally low compression, there's a strong possibility

2B

that the head gasket between them is blown. The appearance of coolant in the combustion chambers or the crankcase would verify this condition.

12   If one cylinder is 20-percent lower than the others, and the engine has a slightly rough idle, a worn exhaust lobe on the camshaft could be the cause.

13   If the compression is unusually high, the combustion chambers are probably coated with carbon deposits. If that's the case, the cylinder head(s) should be removed and decarbonized.

14   If compression is way down or varies greatly between cylinders, it would be a good idea to have a leak-down test performed by an automotive repair shop. This test will pinpoint exactly where the leakage is occurring and how severe it is.

4   Engine removal - methods and precautions

If you've decided that an engine must be removed for overhaul or major repair work, several preliminary steps should be taken.

Locating a suitable place to work is extremely important. Adequate work space, along with storage space for the vehicle, will be needed. If a shop or garage isn't available, at the very least a flat, level, clean work surface made of concrete or asphalt is required.

Cleaning the engine compartment and engine before beginning the removal procedure will help keep tools clean and organized.

An engine hoist or A-frame will also be necessary. Make sure the equipment is rated in excess of the combined weight of the engine and transaxle. Safety is of primary importance, considering the potential hazards involved in lifting the engine out of the vehicle.

If the engine is being removed by a novice, a helper should be available. Advice and aid from someone more experienced would also be helpful. There are many instances when one person cannot simultaneously perform all of the operations required when lifting the engine out of the vehicle.

Plan the operation ahead of time. Arrange for or obtain all of the tools and equipment you'll need prior to beginning the job. Some of the equipment necessary to perform engine removal and installation safely and with relative ease are (in addition to an engine hoist) a heavy duty floor jack, complete sets of wrenches and sockets as described in the front of this manual, wooden blocks and plenty of rags and cleaning solvent for mopping up spilled oil, coolant and gasoline. If the hoist must be rented, make sure that you arrange for it in advance and perform all of the operations possible without it beforehand. This will save you money and time.

Plan for the vehicle to be out of use for quite a while. A machine shop will be required to perform some of the work which the do-it-yourselfer can't accomplish without

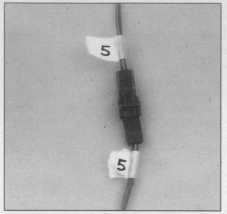

5.7   Label each wire before unplugging the connector

special equipment. These shops often have a busy schedule, so it would be a good idea to consult them before removing the engine in order to accurately estimate the amount of time required to rebuild or repair components that may need work.

Always be extremely careful when removing and installing the engine. Serious injury can result from careless actions. Plan ahead, take your time and a job of this nature, although major, can be accomplished successfully.

5   Engine - removal and installation

Refer to illustrations 5.7, 5.14, 5.15, 5.16a, 5.16b, 5.17, 5.20a, 5.20b, 5.21a and 5.21b
Warning: *Gasoline is extremely flammable, so take extra precautions when disconnecting any part of the fuel system. Don't smoke or allow open flames or bare light bulbs in or near the work area and don't work in a garage where a natural gas appliance (such as a clothes dryer or water heater) is installed. If you spill gasoline on your skin, rinse it off immediately. Have a fire extinguisher rated for gasoline fires handy and know how to use it.*
Caution: *On models equipped with a theft-deterrent radio, make sure you have the correct activation code, or the theft deterrent system is turned off, before disconnecting the battery.*
Note: *Read through the entire Section before beginning this procedure. On automatic transaxle models, the engine may be removed separately, or the engine and transaxle may be removed as a unit. On manual transaxle models, the engine and transaxle must be removed together as a unit. Refer to Chapter 7A for additional information regarding removal of the engine and manual transaxle as a unit.*

Removal

1   Relieve the fuel system pressure (see Chapter 4).

2   Place protective covers on the fenders and cowl and remove the hood (see Chapter 11).

5.14   Remove the nuts (arrows) and detach the exhaust pipe from the manifold (the third nut is not visible in this photo)

3   Remove the battery and battery tray (see Chapter 5).

4   Remove the air cleaner assembly (see Chapter 4).

5   Remove the cruise control actuator and bracket, if equipped.

6   Raise the vehicle and support it securely on jackstands. Drain the cooling system and engine oil and remove the drivebelts (see Chapter 1). **Note:** *Don't raise the vehicle any higher than necessary.*

7   Clearly label and disconnect all vacuum lines, coolant and emissions hoses, electrical connectors, ground straps and fuel lines (for fuel line removal see Chapter 4). Masking tape and/or a touch up paint applicator work well for marking items **(see illustration)**. Take instant photos or sketch the locations of components and brackets.

8   Remove the cooling fan, shroud and radiator (see Chapter 3).

9   Release the residual fuel pressure in the tank by removing the gas cap, then disconnect the fuel lines between the engine and chassis (see Chapter 4). Plug or cap all open fittings.

10   Disconnect the throttle linkage (and transaxle shift control cable, if equipped) from the throttle body (see Chapter 4).

11   Remove the alternator and starter (see Chapter 5).

12   On power steering equipped vehicles, unbolt the power steering pump (see Chapter 10). Tie the pump aside without disconnecting the hoses.

13   On air conditioned models, unbolt the compressor and set it aside (see Chapter 3). **Warning:** *Do not disconnect the refrigerant hoses.*

14   Unbolt the exhaust pipe from the exhaust manifold **(see illustration)**.

15   Unbolt and remove the engine-to-transmission reinforcement plate **(see illustration)**.

16   On automatic transaxle equipped models, detach the torque converter cover from the lower bellhousing **(see illustration)**. Remove the torque converter-to-driveplate fasteners **(see illustration)** and push the

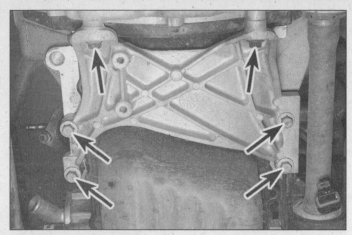

5.15 Unbolt and remove the engine-to-transaxle reinforcement plate

5.16a Remove the bolts (arrows) and take off the torque converter cover

5.16b Remove the bolts that attach the torque converter to the driveplate

5.17 Attach a chain or an engine removal sling (available at the rental shop you got the hoist from) and take up the slack in the chain to support the weight of the engine

2B

converter back slightly into the bellhousing.
17   Attach a lifting sling or chain to the brackets on the engine **(see illustration)**. Position a hoist and connect the sling or chain to it. Take up the slack until there is slight tension on the hoist. **Warning:** *Do not place any part of your body under the*

engine/transaxle when it's supported only by a hoist or other lifting device.
18   Recheck to be sure nothing except the mounts are still connecting the engine to the vehicle. Disconnect anything still remaining.
19   On automatic transaxle models, place a floor jack under the transmission, under the

bellhousing area, to support the transaxle when the engine is unbolted and removed. Be sure to place a block of wood on the jack head to serve as a cushion.
20   Remove the remaining engine mounts **(see illustrations)**.
21   Slowly lift the engine or engine/transaxle

5.20a With the weight of the engine support by the hoist, disconnect the lower front mount (arrow) . . .

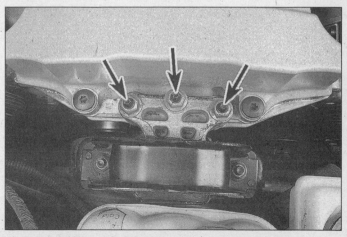

5.20b . . . and the front upper mount (arrows)

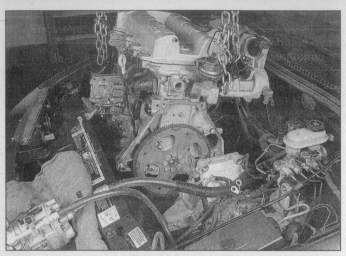

5.21a Slowly raise the engine from the vehicle - keep checking for anything still attached to the engine assembly

5.21b Guide the engine/transaxle assembly as it comes up and out to clear any components in the engine compartment

assembly out of the vehicle (see illustrations). It may be necessary to pry the mounts away from the frame brackets.

Installation

22   Check the engine/transaxle mounts. If they're worn or damaged, replace them.
23   On manual transaxle equipped models, inspect the clutch components (see Chapter 8) and, on automatic transaxle models, inspect the converter seal and bushing.
24   On automatic transaxle equipped models, apply a dab of grease to the nose of the torque converter and to the seal lips.
25   Carefully guide the transaxle into place, following the procedure outlined in Chapter 7A or 7B. **Caution:** *Do not use the bolts to force the engine and transaxle into alignment. It may crack or damage major components.* Install the engine-to-transaxle bolts and tighten them securely.
26   If you're working on a model equipped with an automatic transaxle, install the torque converter-to-driveplate fasteners and tighten them to the torque listed in the Chapter 7B Specifications.
27   Attach the hoist to the engine and carefully lower the engine assembly into the engine compartment.
28   Install the mount bolts and tighten them securely.
29   Reinstall the remaining components and fasteners in the reverse order of removal.
30   Add coolant, oil, power steering and transmission fluids as needed (see Chapter 1).
31   Run the engine and check for proper operation and leaks. Shut off the engine and recheck the fluid levels.

6   Engine rebuilding alternatives

The do-it-yourselfer is faced with a number of options when performing an engine overhaul. The decision to replace the engine block, piston/connecting rod assemblies and

crankshaft depends on a number of factors, with the number one consideration being the condition of the block. Other considerations are cost, access to machine shop facilities, parts availability, time required to complete the project and the extent of prior mechanical experience on the part of the do-it-yourselfer.

Some of the rebuilding alternatives include:

Individual parts - If the inspection procedures reveal that the engine block and most engine components are in reusable condition, purchasing individual parts may be the most economical alternative. The block, crankshaft and piston/connecting rod assemblies should all be inspected carefully. Even if the block shows little wear, the cylinder bores should be surface honed.

Short block - A short block consists of an engine block with a crankshaft and piston/connecting rod assemblies already installed. All new bearings are incorporated and all clearances will be correct. The existing camshaft, valve train components, cylinder head(s) and external parts can be bolted to the short block with little or no machine shop work necessary.

Long block - A long block consists of a short block plus an oil pump, oil pan, cylinder head, camshaft and valve train components, timing pulleys and belt. All components are installed with new bearings, seals and gaskets incorporated throughout. The installation of manifolds and external parts is all that's necessary.

Give careful thought to which alternative is best for you and discuss the situation with local automotive machine shops, auto parts dealers and experienced rebuilders before ordering or purchasing replacement parts.

7   Engine overhaul - disassembly sequence

1   It's much easier to disassemble and work on the engine if it's mounted on a

portable engine stand. A stand can often be rented quite cheaply from an equipment rental yard. Before the engine is mounted on a stand, the flywheel/driveplate and rear oil seal retainer should be removed from the engine.
2   If a stand isn't available, it's possible to disassemble the engine with it blocked up on the floor. Be extra careful not to tip or drop the engine when working without a stand.
3   If you're going to obtain a rebuilt engine, all external components must come off first, to be transferred to the replacement engine, just as they will if you're doing a complete engine overhaul yourself. These include:

> Emissions control components
> Coil pack, spark plug wires and spark plugs
> Thermostat and housing
> Water pump
> EFI components
> Intake/exhaust manifolds
> Oil filter (always use a new filter)
> Engine mounts
> Clutch and flywheel/driveplate
> Engine rear plate

Note: *When removing the external components from the engine, pay close attention to details that may be helpful or important during installation. Note the installed position of gaskets, seals, spacers, pins, brackets, washers, bolts and other small items.*
4   If you're obtaining a short block, which consists of the engine block, crankshaft, pistons and connecting rods all assembled, then the cylinder head, oil pan, front timing cover and oil pump will have to be removed as well. See *Engine rebuilding alternatives* for additional information regarding the different possibilities to be considered.
5   If you're planning a complete overhaul, the engine must be disassembled and the internal components removed in the following order.

> Valve cover
> Oil pan and pick-up tube

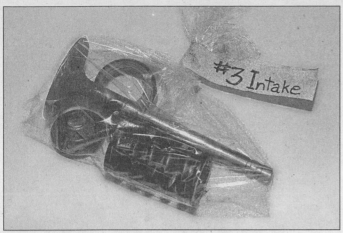

8.2 A small plastic bag, with an appropriate label, can be used to store the valve train components so they can be kept together and reinstalled in the original position

8.3 Use a valve spring compressor to compress the spring, then remove the keepers from the valve stem

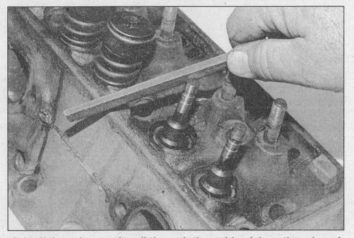

8.4a If the valve won't pull through the guide, deburr the edge of the stem end and the area around the top of the keeper groove with a file or whetstone

8.4b Remove the valve guide seal with a pair of pliers

2B

Timing chain cover
Oil pump
Timing chain and sprockets
Rocker arm and shaft assemblies
Camshaft(s) and lifters
Cylinder head
Piston/connecting rod assemblies
Crankshaft rear oil seal retainer
Crankshaft and main bearings

6　Before beginning the disassembly and overhaul procedures, make sure the following items are available. Also, refer to *Engine overhaul - reassembly sequence* for a list of tools and materials needed for engine reassembly.

Common hand tools
Small cardboard boxes or plastic bags
　for storing parts
Gasket scraper
Ridge reamer
Vibration damper puller
Micrometers
Telescoping gauges
Dial indicator set
Valve spring compressor
Cylinder surfacing hone

Piston ring groove cleaning tool
Electric drill motor
Tap and die set
Wire brushes
Oil gallery brushes
Cleaning solvent

8　Cylinder head - disassembly

Refer to illustrations 8.2, 8.3, 8.4a and 8.4b
Note: *New and rebuilt cylinder heads are commonly available for most engines at dealerships and auto parts stores. Due to the fact that some specialized tools are necessary for the disassembly and inspection procedures, and replacement parts may not be readily available, it may be more practical and economical for the home mechanic to purchase a replacement head rather than taking the time to disassemble, inspect and recondition the original.*

1　Cylinder head disassembly involves removal of the intake and exhaust valves and related components. It's assumed that the lifters or rocker arms and camshaft(s) have already been removed (see Part A as needed).

2　Before the valves are removed, arrange to label and store them, along with their related components, so they can be kept separate and reinstalled in the same valve guides they are removed from **(see illustration)**.

3　Compress the springs on the first valve with a spring compressor and remove the keepers **(see illustration)**. Carefully release the valve spring compressor and remove the retainer, the spring and the spring seat (if used).

4　Pull the valve out of the head and remove the oil seal from the guide (in either order). If the valve binds in the guide (won't pull through), push it back into the head and deburr the area around the keeper groove with a fine file or whetstone **(see illustrations)**.

5　Repeat the procedure for the remaining valves. Remember to keep all the parts for each valve together so they can be reinstalled in the same locations.

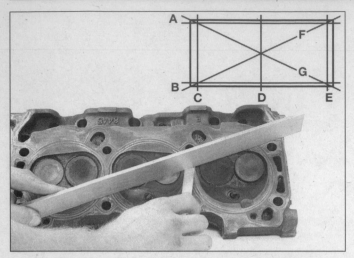

9.12 Check the cylinder head gasket surface for warpage by trying to slip a feeler gauge under the straightedge (see this Chapter's Specifications for the maximum warpage allowed and use a feeler gauge of that thickness)

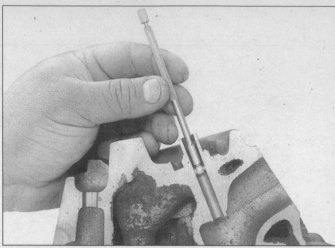

9.14 Use a small hole gauge to determine the inside diameter of the valve guides (the gauge is then measured with a micrometer)

6   Once the valves and related components have been removed and stored in an organized manner, the head should be thoroughly cleaned and inspected. If a complete engine overhaul is being done, finish the engine disassembly procedures before beginning the cylinder head cleaning and inspection process.

9   Cylinder head - cleaning and inspection

1   Thorough cleaning of the cylinder head and related valve train components, followed by a detailed inspection, will enable you to decide how much valve service work must be done during the engine overhaul. **Note:** *If the engine was severely overheated, the cylinder head is probably warped* (see Step 12).

Cleaning

2   Scrape all traces of old gasket material and sealing compound off the head gasket, intake manifold and exhaust manifold sealing surfaces. Be very careful not to gouge the cylinder head. Special gasket removal solvents that soften gaskets and make removal much easier are available at auto parts stores.
3   Remove all built-up scale from the coolant passages.
4   Run a stiff wire brush through the various holes to remove deposits that may have formed in them.
5   Run an appropriate size tap into each of the threaded holes to remove corrosion and thread sealant that may be present. If compressed air is available, use it to clear the holes of debris produced by this operation. **Warning:** *Wear eye protection when using compressed air!*
6   Clean the exhaust and intake manifold stud threads with a wire brush.

7   Clean the cylinder head with solvent and dry it thoroughly. Compressed air will speed the drying process and ensure that all holes and recessed areas are clean. **Note:** *Decarbonizing chemicals are available and may prove very useful when cleaning cylinder heads and valve train components. They are very caustic and should be used with caution. Be sure to follow the instructions on the container.*
8   Clean the lifters and rocker arms (if used) with solvent and dry them thoroughly (don't mix them up during the cleaning process). Compressed air will speed the drying process and can be used to clean out the oil passages.
9   Clean all the valve springs, spring seats, keepers and retainers with solvent and dry them thoroughly. Do the components from one valve at a time to avoid mixing up the parts.
10   Scrape off any heavy deposits that may have formed on the valves, then use a motorized wire brush to remove deposits from the valve heads and stems. Again, make sure the valves don't get mixed up.

Inspection

Note: *Be sure to perform all of the following inspection procedures before concluding that machine shop work is required. Make a list of the items that need attention. The inspection procedures for the lifters and rocker arms, as well as the camshaft(s), can be found in Part A.*

Cylinder head
Refer to illustrations 9.12, 9.14, 9.15 and 9.16
11   Inspect the head very carefully for cracks, evidence of coolant leakage and other damage. If cracks are found, check with an automotive machine shop concerning repair. If repair isn't possible, a new cylinder head should be obtained.

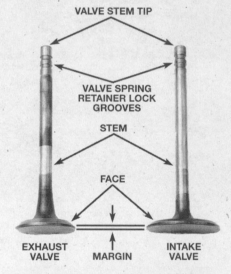

9.15 Check for valve wear at the points shown here

12   Using a straightedge and feeler gauge, check the head gasket mating surface for warpage **(see illustration)**. If the warpage exceeds the specified limit, it can be resurfaced at an automotive machine shop. Excessive warping will require that the head be straightened prior to machining to ensure that the camshaft bores remain straight. **Note:** *Before attempting to recondition a warped head, have a machine shop inspect it for cracks.*
13   Examine the valve seats in each of the combustion chambers. If they're pitted, cracked or burned, the head will require valve service that's beyond the scope of the home mechanic.
14   Measure the valve guide inside diameter with a small hole gauge and micrometer **(see illustration)**, then measure the valve stem diameter and subtract it from the valve guide diameter to obtain the stem-to-guide clear-

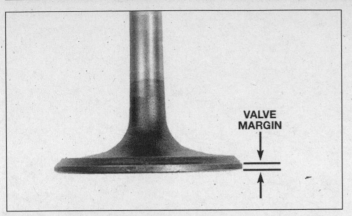

9.16 The margin width on each valve must be as specified (if no margin exists, the valve cannot be reused)

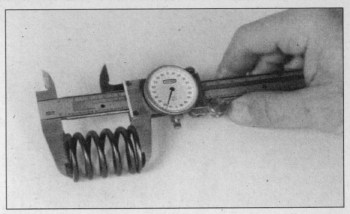

9.17 Measure the free length of each valve spring with a dial or vernier caliper

ance. Also, check the valve stem deflection crosswise (parallel to the rocker arm) with a dial indicator attached securely to the head. The valve must be in the guide and approximately 1/16-inch off the seat. The total valve stem movement indicated by the gauge needle must be noted. If it exceeds the specified stem-to-guide clearance limit, the valve guides should be replaced. After this is done, if there's still some doubt regarding the condition of the valve guides they should be checked by an automotive machine shop (the cost should be minimal).

Valves

15   Carefully inspect each valve face for uneven wear, deformation, cracks, pits and burned areas. Check the valve stem for scuffing and galling and the neck for cracks. Rotate the valve and check for any obvious indication that it's bent. Look for pits and excessive wear on the end of the stem **(see illustration)**. The presence of any of these conditions indicates the need for valve service by an automotive machine shop.
16   Measure the margin width on each valve **(see illustration)**. Any valve with a margin narrower than specified will have to be replaced with a new one.

Valve components

Refer to illustrations 9.17 and 9.18
17   Check each valve spring for wear (on the ends) and pits. Measure the free height and compare it to the Specifications **(see illustration)**. Any springs that are shorter than specified have sagged and should not be reused. The tension of all springs should be checked with a special fixture before deciding that they're suitable for use in a rebuilt engine (take the springs to an automotive machine shop for this check).
18   Stand each spring on a flat surface and check it for squareness **(see illustration)**. If any of the springs are distorted or sagged, replace all of them with new parts.
19   Check the spring retainers and keepers for obvious wear and cracks. Any questionable parts should be replaced with new ones, as extensive damage will occur if they fail

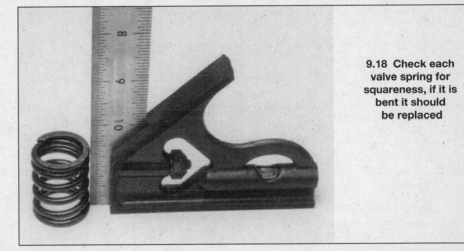

9.18 Check each valve spring for squareness, if it is bent it should be replaced

during engine operation.
20   Any damaged or excessively worn parts must be replaced with new ones.
21   If the inspection process indicates that the valve components are in generally poor condition and worn beyond the limits specified, which is usually the case in an engine that's being overhauled, reassemble the valves in the cylinder head and refer to valve servicing recommendations (see Section 10).

10   Valves - servicing

1   Because of the complex nature of the job and the special tools and equipment needed, servicing of the valves, the valve seats and the valve guides, commonly known as a valve job, should be done by a professional.
2   The home mechanic can remove and disassemble the head, do the initial cleaning and inspection, then reassemble and deliver it to a dealer service department or an automotive machine shop for the actual service work. Doing the inspection will enable you to see what condition the head and valvetrain components are in and will ensure that you know what work and new parts are required when dealing with an automotive machine shop.

3   The dealer service department, or automotive machine shop, will remove the valves and springs, recondition or replace the valves and valve seats, check and replace the valve guides (as necessary), check and replace the valve springs, spring retainers and keepers (as necessary), replace the valve seals with new ones, reassemble the valve components and make sure the installed spring height is correct. The cylinder head gasket surface will also be resurfaced (within specification limits) if it's warped.
4   After the valve job has been performed by a professional, the head will be in like-new condition. When the head is returned, be sure to clean it again before installation on the engine to remove any metal particles and abrasive grit that may still be present from the valve service or head resurfacing operations. Use compressed air, if available, to blow out all the oil holes and passages.

11   Cylinder head - reassembly

Refer to illustration 11.3
1   Regardless of whether or not the head was sent to an automotive repair shop for valve servicing, make sure it's clean before beginning reassembly.

11.3 Using either the special tool made for installation of valve seals or a small deep socket, lightly tap the seal into place until it's seated

12.1 A ridge reamer is required to remove the ridge from the top of each cylinder - do this before removing the pistons!

12.3 Remove the oil baffle bolts (arrows) and baffle plate that covers the crankshaft

12.4 Check the connecting rod side clearance (endplay) with a feeler gauge as shown here

2   If the head was sent out for valve servicing, the valves and related components will already be in place.

3   Install new seals on each of the valve guides. Gently tap each intake valve seal into place until it's seated on the guide **(see illustration)**. **Caution:** *Don't hammer on the valve seals once they're seated or you may damage them. Don't twist or cock the seals during installation or they won't seat properly on the valve stems.*

4   Beginning at one end of the head, lubricate and install the first valve. Apply moly-base grease or clean engine oil to the valve stem.

5   Drop the spring seat or shim(s) (if equipped) over the valve guide and set the valve spring and retainer in place.

6   Compress the springs with a valve spring compressor and carefully install the keepers in the upper groove, then slowly release the compressor and make sure the keepers seat properly. Apply a small dab of grease to each keeper to hold it in place if necessary.

7   Repeat the procedure for the remaining valves. Be sure to return the components to their original locations - don't mix them up!

12 Pistons/connecting rods - removal

Refer to illustrations 12.1, 12.3, 12.4, 12.5 and 12.7

Note: *Prior to removing the piston/connecting rod assemblies, remove the cylinder head, the oil pan and the oil pump pick-up tube by referring to the appropriate Sections in Part A.*

1   Use your fingernail to feel if a ridge has formed at the upper limit of ring travel (about 1/4-inch down from the top of each cylinder). If carbon deposits or cylinder wear have produced ridges, they must be completely removed with a special tool **(see illustration)**. Follow the manufacturer's instructions provided with the tool. Failure to remove the ridges before attempting to remove the piston/connecting rod assemblies may result in piston breakage. **Caution:** *When removing a ridge in the cylinder bore, DO NOT damage the top of the cylinder bore or the block deck*

surface with the ridge removal tool.

2   After the cylinder ridges have been removed, turn the engine upside-down so the crankshaft is facing up.

3   On DOHC models only, remove the six bolts and the oil baffle plate **(see illustration)**

4   Before the connecting rods are removed, check the endplay (side clearance) with feeler gauges. Slide them between the first connecting rod and the crankshaft throw until the play is removed **(see illustration)**. The endplay is equal to the thickness of the feeler gauge(s). If the endplay exceeds the service limit, new connecting rods will be required. If new rods (or a new crankshaft) are installed, the endplay may fall under the specified minimum (if it does, the rods will have to be machined to restore it - consult an automotive machine shop for advice if necessary). Repeat the procedure for the remaining connecting rods.

5   Check the connecting rods and caps for identification marks. **Note:** *Saturn has color coded the rods and caps at the time of assembly to prevent the interchanging of similar parts. If they aren't plainly marked, use a*

12.5 Mark the rod bearing caps in order from the front of the engine to the rear (one mark for the front cap, two for the second one and so on)

12.7 To prevent damage to the crankshaft journals and cylinder walls, slip sections of hose over the rod bolts before removing the piston/rod assemblies

13.1 Checking crankshaft endplay with a dial indicator

13.3 Checking crankshaft endplay with a feeler gauge

2B

small center-punch to make the appropriate number of indentations on each rod and cap (1, 2, 3, etc., depending on the engine type and cylinder they're associated with) **(see illustration)**.

6　Loosen each of the connecting rod cap nuts 1/2-turn at a time until they can be removed by hand. Remove the number one connecting rod cap and bearing insert. Don't drop the bearing insert out of the cap.

7　Slip a short length of plastic or rubber hose over each connecting rod cap bolt to protect the crankshaft journal and cylinder wall as the piston is removed **(see illustration)**.

8　Remove the bearing insert and push the connecting rod/piston assembly out through the top of the engine. Use a wooden hammer handle to push on the upper bearing surface in the connecting rod. If resistance is felt, double-check to make sure that all of the ridge was removed from the cylinder. **Caution:** *DO NOT pull the piston assembly down through the cylinder bore beyond its normal travel. The oil control rings can lodge between the bottom of the cylinder liner and the machined step in the block, and act as a snap-ring, holding the piston firmly in place.*

Cylinder liner removal will be required to free the piston.

9　Repeat the procedure for the remaining cylinders.

10　After removal, reassemble the connecting rod caps and bearing inserts in their respective connecting rods and install the cap nuts finger tight. Leaving the old bearing inserts in place until reassembly will help prevent the connecting rod bearing surfaces from being accidentally nicked or gouged.

11　Don't separate the pistons from the connecting rods (see Section 17).

13　Crankshaft - removal

Refer to illustrations 13.1, 13.3 and 13.5
Note: *The crankshaft can be removed only after the engine has been removed from the vehicle. It's assumed that the flywheel or driveplate, vibration damper, timing chain, oil pan, oil pick-up tube, oil pump and piston/connecting rod assemblies have already been removed. The rear main oil seal retainer must be unbolted and separated*

from the block before proceeding with crankshaft removal.

1　Before the crankshaft is removed, check the endplay. Mount a dial indicator with the stem in line with the crankshaft and touching the crankshaft end **(see illustration)**.

2　Pry the crankshaft all the way to the rear and zero the dial indicator. Next, pry the crankshaft to the front as far as possible and check the reading on the dial indicator. The distance that it moves is the endplay. If it's greater than specified, check the crankshaft thrust surfaces for wear. If no wear is evident, new thrust washers (on early engines) or bearings (on later engines) should correct the endplay.

3　If a dial indicator isn't available, feeler gauges can be used. Gently pry or push the crankshaft all the way to the front of the engine. Slip feeler gauges between the crankshaft and the front face of the thrust main bearing to determine the clearance **(see illustration)**. The thrust bearing is number three (center).

4　Check the main bearing caps to see if they're marked to indicate their locations. They should be numbered consecutively

13.5 If the main bearing caps are difficult to remove, work them forward and back using the main bearing cap bolts as levers

14.1a A hammer and a large punch can be used to knock the core plugs sideways in their bores

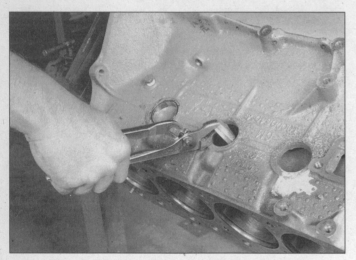

14.1b Pull the core plugs from the block with pliers

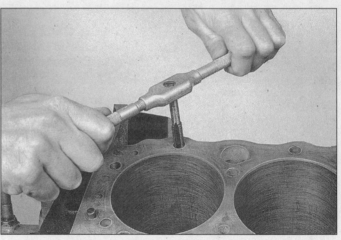

14.8 All bolt holes in the block - particularly the main bearing cap and head bolt holes - should be cleaned and restored with a tap (be sure to remove debris from the holes after this is done)

from the front of the engine to the rear. If they aren't, mark them with number stamping dies or a center-punch. Main bearing caps generally have a cast-in arrow, which points to the front of the engine. Loosen the main bearing cap bolts 1/4-turn at a time each, starting with the front and rear caps and working toward the center, until they can be removed by hand. Note if any stud bolts are used and make sure they're returned to their original locations when the crankshaft is reinstalled.

5   Gently tap the caps with a soft-face hammer, then separate them from the engine block. If necessary, use the bolts as levers to remove the caps **(see illustration)**. Try not to drop the bearing inserts if they come out with the caps.

6   Carefully lift the crankshaft out of the engine. It may be a good idea to have an assistant available, since the crankshaft is quite heavy. With the bearing inserts in place in the engine block and main bearing caps or cap assembly, return the caps to their respective locations on the engine block and tighten the bolts finger tight.

14   Engine block - cleaning

Refer to illustrations 14.1a, 14.1b, 14.8 and 14.10

Caution: *The core plugs (also known as freeze or soft plugs) may be difficult or impossible to retrieve if they're driven into the block coolant passages.*

1   Using the wide end of a punch **(see illustration)** tap in on the outer edge of the core plug to turn the plug sideways in the bore. Then, using a pair of pliers, pull the core plug from the engine block **(see illustration)**. Don't worry about the condition of the old core plugs as they are being removed because they will be replaced on reassembly with new plugs.

2   Using a gasket scraper, remove all traces of gasket material from the engine block. **Caution:** *The engine block is made from aluminum, as is the cylinder head - be EXTREMELY CAREFUL not to nick or gouge the gasket sealing surfaces.*

3   Remove the main bearing caps or cap

assembly and separate the bearing inserts from the caps and the engine block. Tag the bearings, indicating which cylinder they were removed from and whether they were in the cap or the block, then set them aside.

4   Remove all of the threaded oil gallery plugs from the block. The plugs are usually very tight - they may have to be drilled out and the holes retapped. Use new plugs when the engine is reassembled.

5   If the engine is extremely dirty it should be taken to an automotive machine shop for cleaning.

6   After the block is returned, clean all oil holes and oil galleries one more time. Brushes specifically designed for this purpose are available at most auto parts stores. Flush the passages with warm water until the water runs clear, dry the block thoroughly and wipe all machined surfaces with a light, rust preventive oil. If you have access to compressed air, use it to speed the drying process and to blow out all the oil holes and galleries. **Warning:** *Wear eye protection when using compressed air!*

14.10 A large socket on an extension can be used to drive the new core plugs into the bores

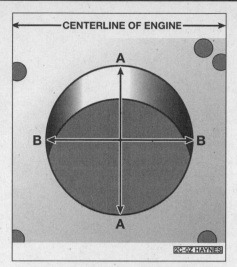

15.4a Measure the diameter of each cylinder at a right angle to the engine centerline (A), and parallel to the engine centerline (B) - out-of-round is the difference between A and B; taper is the difference between A and B at the top of the cylinder and A and B at the bottom of the cylinder

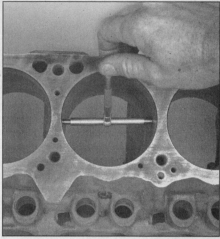

15.4b The ability to feel when the telescoping gauge is at the correct point will be developed over time, so work slowly and repeat the check until you're satisfied the bore measurement is accurate

7   If the block isn't extremely dirty or sludged up, you can do an adequate cleaning job with hot soapy water and a stiff brush. Take plenty of time and do a thorough job. Regardless of the cleaning method used, be sure to clean all oil holes and galleries very thoroughly, dry the block completely and coat all machined surfaces with light oil.

8   The threaded holes in the block must be clean to ensure accurate torque readings during reassembly. Run the proper size tap into each of the holes to remove rust, corrosion, thread sealant or sludge and restore damaged threads **(see illustration)**. If possible, use compressed air to clear the holes of debris produced by this operation. Now is a good time to clean the threads on the head bolts and the main bearing cap bolts as well.

9   Reinstall the main bearing caps and tighten the bolts finger tight.

10   After coating the sealing surfaces of the new core plugs with Permatex no. 2 sealant, install them in the engine block **(see illustration)**. Make sure they're driven in straight or leakage could result. Special tools are available for this purpose, but a large socket, with an outside diameter that will just slip into the core plug, a 1/2-inch drive extension and a hammer will work just as well.

11   Apply non-hardening sealant (such as Permatex no. 2 or Teflon pipe sealant) to the new oil gallery plugs and thread them into the holes in the block. Make sure they're tightened securely.

12   If the engine isn't going to be reassembled right away, cover it with a large plastic trash bag to keep it clean.

15   Engine block - inspection

Refer to illustrations 15.4a, 15.4b, 15.4c, 15.12a and 15.12b

1   Before the block is inspected, it should be cleaned (see Section 14).

2   Visually check the block for cracks, rust and corrosion. Look for stripped threads in the threaded holes. It's also a good idea to have the block checked for hidden cracks by an automotive machine shop that has the special equipment to do this type of work. If defects are found, have the block repaired, if possible, or replaced.

3   Check the cylinder bores for scuffing and scoring.

4   Measure the diameter of each cylinder at the top (just under the ridge area), center and bottom of the cylinder bore, parallel to the crankshaft axis **(see illustrations)**.

5   Next, measure each cylinder's diameter at the same three locations across the crankshaft axis. Compare the results to the Specifications.

6   If the required precision measuring tools aren't available, the piston-to-cylinder clearances can be obtained, though not quite as accurately, using feeler gauge stock. Feeler gauge stock comes in 12-inch lengths and various thicknesses and is generally available at auto parts stores.

7   To check the clearance, select a feeler gauge and slip it into the cylinder along with the matching piston. The piston must be positioned exactly as it normally would be. The feeler gauge must be between the piston and cylinder on one of the thrust faces (90-degrees to the piston pin bore).

8   The piston should slip through the cylinder (with the feeler gauge in place) with moderate pressure.

9   If it falls through or slides through easily, the clearance is excessive and a new piston will be required. If the piston binds at the lower end of the cylinder and is loose toward the top, the cylinder is tapered. If tight spots are encountered as the piston/feeler gauge is rotated in the cylinder, the cylinder is out-of-round.

10   Repeat the procedure for the remaining pistons and cylinders.

11   If the cylinder walls are badly scuffed or scored, or if they're out-of-round or tapered beyond the limits given in the Specifications, have the engine block rebored and honed at an automotive machine shop. If a rebore is

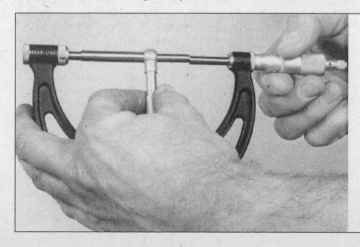

15.4c The gauge is then measured with a micrometer to determine the bore size

2B

15.12a Check the block deck for distortion with a precision straightedge and a feeler gauge

15.12b Lay the straightedge across the block, diagonally and from end-to-end when making the check

16.3a A "bottle brush" hone will produce better results if you've never honed cylinders before

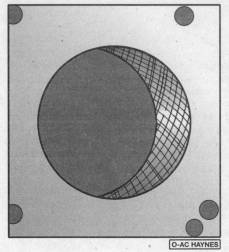

16.3b The cylinder hone should leave a smooth, crosshatch pattern with the lines intersecting at approximately a 60-degree angle

O-AC HAYNES

done, oversize pistons and rings will be required.

12   Using a precision straightedge and a feeler gauge, check the block deck (the surface that mates with the cylinder head) for distortion **(see illustrations)**. If it's distorted beyond the specified limit, it can be resurfaced by an automotive machine shop.

13   If the cylinders are in reasonably good condition and not worn to the outside of the limits, and if the piston-to-cylinder clearances can be maintained properly, then they don't have to be rebored. Honing is all that's necessary (see Section 16).

16 Cylinder honing

Refer to illustrations 16.3a and 16.3b

1   Prior to engine reassembly, the cylinder bores must be honed so the new piston rings will seat correctly and provide the best possible combustion chamber seal. **Note:** *If you don't have the tools or don't want to tackle the honing operation, most automotive machine shops will do it for a reasonable fee.*

2   Before honing the cylinders, install the main bearing caps (without bearing inserts) and tighten the bolts to the torque listed in this Chapter's Specifications.

3   Two types of cylinder hones are commonly available - the flex hone or "bottle brush" type and the more traditional surfacing hone with spring-loaded stones. Both will do the job, but for the less experienced mechanic the "bottle brush" hone will probably be easier to use. You'll also need some kerosene or honing oil, rags and an electric drill motor. Proceed as follows:

a) *Mount the hone in the drill motor, compress the stones and slip it into the first cylinder* **(see illustration)**. *Be sure to wear safety goggles or a face shield!*

b) *Lubricate the cylinder with plenty of honing oil, turn on the drill and move the hone up-and-down in the cylinder at a pace that will produce a fine crosshatch pattern on the cylinder walls. Ideally, the crosshatch lines should intersect at approximately a 60-degrees angle* **(see illustration)**. *Be sure to use plenty of lubricant and don't take off any more*

material than is absolutely necessary to produce the desired finish. **Note:** *Piston ring manufacturers may specify a smaller crosshatch angle than the traditional 60-degrees - read and follow any instructions included with the new rings.*

c) *Don't withdraw the hone from the cylinder while it's running. Instead, shut off the drill and continue moving the hone up-and-down in the cylinder until it comes to a complete stop, then compress the stones and withdraw the hone. If you're using a "bottle brush" type hone, stop the drill motor, then turn the chuck in the normal direction of rotation while withdrawing the hone from the cylinder.*

d) *Wipe the oil out of the cylinder and repeat the procedure for the remaining cylinders.*

4   After the honing job is complete, chamfer the top edges of the cylinder bores with a small file so the rings won't catch when the pistons are installed. Be very careful not to nick the cylinder walls with the end of the file.

5   The entire engine block must be washed

17.4a The piston ring grooves can be cleaned with a special tool, as shown here . . .

17.4b . . . or a section of a broken ring

again very thoroughly with warm, soapy water to remove all traces of the abrasive grit produced during the honing operation. **Note:** *The bores can be considered clean when a lint-free white cloth - dampened with clean engine oil - used to wipe them out doesn't pick up any more honing residue, which will show up as gray areas on the cloth. Be sure to run a brush through all oil holes and galleries and flush them with running water.*

6   After rinsing, dry the block and apply a coat of light rust preventive oil to all machined surfaces. Wrap the block in a plastic trash bag to keep it clean and set it aside until reassembly.

17 Pistons/connecting rods - inspection

Refer to illustrations 17.4a, 17.4b, 17.10 and 17.11

1   Before the inspection process can be carried out, the piston/connecting rod assemblies must be cleaned and the original piston rings removed from the pistons. **Note:** *Always use new piston rings when the engine is reassembled.*

2   Using a piston ring installation tool, carefully remove the rings from the pistons. Be careful not to nick or gouge the pistons in the process.

3   Scrape all traces of carbon from the top of the piston. A hand-held wire brush or a piece of fine emery cloth can be used once the majority of the deposits have been scraped away. Do not, under any circumstances, use a wire brush mounted in a drill motor to remove deposits from the pistons. The piston material is soft and may be eroded away by the wire brush.

4   Use a piston ring groove cleaning tool to remove carbon deposits from the ring grooves. If a tool isn't available, a piece broken off the old ring will do the job. Be very careful to remove only the carbon deposits - don't remove any metal and do not nick or scratch the sides of the ring grooves **(see illustrations)**.

5   Once the deposits have been removed, clean the piston/rod assemblies with solvent and dry them with compressed air (if available). **Warning:** *Wear eye protection when using compressed air!* Make sure the oil return holes in the back sides of the ring grooves and the oil hole in the lower end of

each rod are clear.

6   If the pistons and cylinder walls aren't damaged or worn excessively, and if the engine block is not rebored, new pistons won't be necessary. Normal piston wear appears as even vertical wear on the piston thrust surfaces and slight looseness of the top ring in its groove. New piston rings, however, should always be used when an engine is rebuilt.

7   Carefully inspect each piston for cracks around the skirt, at the pin bosses and at the ring lands.

8   Look for scoring and scuffing on the thrust faces of the skirt, holes in the piston crown and burned areas at the edge of the crown. If the skirt is scored or scuffed, the engine may have been suffering from overheating and/or abnormal combustion, which caused excessively high operating temperatures. The cooling and lubrication systems should be checked thoroughly. A hole in the piston crown is an indication that abnormal combustion (preignition) was occurring. Burned areas at the edge of the piston crown are usually evidence of spark knock (detonation). If any of the above problems exist, the causes must be corrected or the damage will occur again. The causes may include intake air leaks, incorrect fuel/air mixture, incorrect ignition timing and EGR system malfunctions.

9   Corrosion of the piston, in the form of small pits, indicates that coolant is leaking into the combustion chamber and/or the crankcase. Again, the cause must be corrected or the problem may persist in the rebuilt engine.

10   Measure the piston ring side clearance by laying a new piston ring in each ring groove and slipping a feeler gauge in beside it . Check the clearance at three or four locations around each groove. Be sure to use the correct ring for each groove - they are different. If the side clearance is greater than specified, new pistons will have to be used.

11   Check the piston-to-bore clearance by measuring the bore (see Section 15) and the piston diameter. Make sure the pistons and bores are correctly matched. Measure the

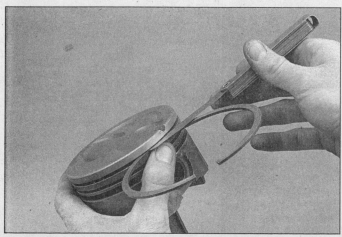

17.10 Check the ring side clearance with a feeler gauge at several points around the groove

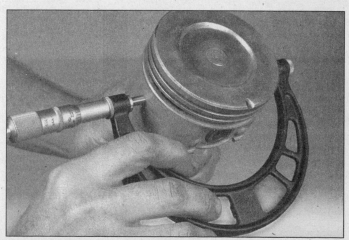

17.11 Measure the piston diameter at a 90-degree angle to the piston pin and in line with it

2B

18.1 The oil holes should be chamfered so sharp edges don't gouge or scratch the new bearings

18.2 Use a wire or stiff plastic bristle brush to clean the oil passages in the crankshaft

piston across the skirt, at a 90-degree angle to the piston pin, the specified distance down from the top of the piston or the lower edge of the oil ring groove **(see illustration)**. Subtract the piston diameter from the bore diameter to obtain the clearance. If it's greater than specified, the block will have to be rebored and new pistons and rings installed.

12   Check the piston-to-rod clearance by twisting the piston and rod in opposite directions. Any noticeable play indicates excessive wear, which must be corrected. The piston/connecting rod assemblies should be taken to an automotive machine shop to have the pistons and rods re-sized and new pins installed.

13   If the pistons must be removed from the connecting rods for any reason, they should be taken to an automotive machine shop. While they are there have the connecting rods checked for bend and twist, since automotive machine shops have special equipment for this purpose. **Note:** *Unless new pistons and/or connecting rods must be installed, do not disassemble the pistons and connecting rods.*

14   Check the connecting rods for cracks and other damage. Temporarily remove the rod caps, lift out the old bearing inserts, wipe the rod and cap bearing surfaces clean and inspect them for nicks, gouges and scratches. After checking the rods, replace the old bearings, slip the caps into place and tighten the nuts finger tight. **Note:** *If the engine is being rebuilt because of a connecting rod knock, be sure to install new rods.*

18   Crankshaft - inspection

Refer to illustrations 18.1, 18.2, 18.4, 18.6 and 18.8

1   Remove all burrs from the crankshaft oil holes with a stone, file or scraper **(see illustration)**.

2   Clean the crankshaft with solvent and dry it with compressed air (if available). Be sure to clean the oil holes with a stiff brush and flush them with solvent **(see illustration)**.

3   Check the main and connecting rod bearing journals for uneven wear, scoring, pits and cracks.

4   Rub a penny across each journal several times **(see illustration)**. If a journal picks up copper from the penny, it's too rough and must be reground.

5   Check the rest of the crankshaft for cracks and other damage. It should be magnafluxed to reveal hidden cracks - an automotive machine shop will handle the procedure.

6   Using a micrometer, measure the diameter of the main and connecting rod journals and compare the results to the Specifications **(see illustration)**. By measuring the diameter at a number of points around each journal's circumference, you'll be able to determine whether or not the journal is out-of-round. Take the measurement at each end of the journal, near the crank throws, to determine if the journal is tapered. Crankshaft runout

18.4 Rubbing a penny lengthwise on each journal will reveal its condition - if copper rubs off and is embedded in the crankshaft, the journals should be reground

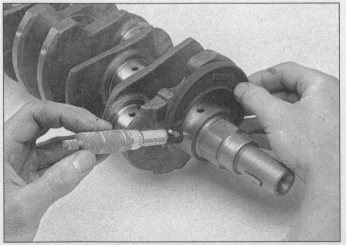

18.6 Measure the diameter of each crankshaft journal at several points to detect taper and out-of-round conditions

18.8 If the seals have worn grooves in the crankshaft journals, or if the seal contact surfaces are nicked or scratched, the new seals will leak

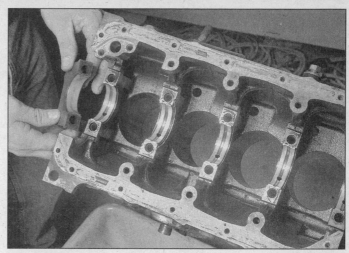

19.1a Before discarding the used bearings examine them for indications of any possible problems with the crankshaft, noting the bearing location it came from

should be checked also, but large V-blocks and a dial indicator are needed to do it correctly. If you don't have the equipment, have a machine shop check the runout.

7 If the crankshaft journals are damaged, tapered, out-of-round or worn beyond the limits given in the Specifications, have the crankshaft reground by an automotive machine shop. Be sure to use the correct size bearing inserts if the crankshaft is reconditioned.

8 Check the oil seal journals at each end of the crankshaft for wear and damage (see illustration). If the seal has worn a groove in the journal, or if it's nicked or scratched, the new seal may leak when the engine is reassembled. In some cases, an automotive machine shop may be able to repair the journal by pressing on a thin sleeve. If repair isn't feasible, a new or different crankshaft should be installed.

9 Examine the main and rod bearing

inserts (see Section 19).

10 If the crankshaft requires replacement make certain that you examine the ignition (DIS system) reluctor ring, this is how you can be certain that the crankshaft you get will be a proper match for your model year. The 1991 and 1992 crankshafts can be identified from 1993 and later crankshafts by measuring the distance between the reluctor slot and the casting parting line. 1993 and later crankshafts have ten degrees advanced reluctor ring slots and CANNOT be used with 1991 and 1992 engines. The number three crankshaft rod journal must be at TDC (top dead center) and the view shown is looking from the timing chain end of the engine towards the transaxle end of the engine. Be sure to match the original crankshaft with any replacement purchased.

19 Main and connecting rod bearings - inspection and selection

Inspection

Refer to illustrations 19.1a and 19.1b

1 Even though the main and connecting rod bearings should be replaced with new ones during the engine overhaul, the old bearings should be retained for close examination, as they may reveal valuable information about the condition of the engine (see illustrations).

2 Bearing failure occurs because of lack of lubrication, the presence of dirt or other foreign particles, overloading the engine and corrosion. Regardless of the cause of bearing failure, it must be corrected before the engine is reassembled to prevent it from happening again.

3 When examining the bearings, remove them from the engine block, the main bearing caps, the connecting rods and the rod caps

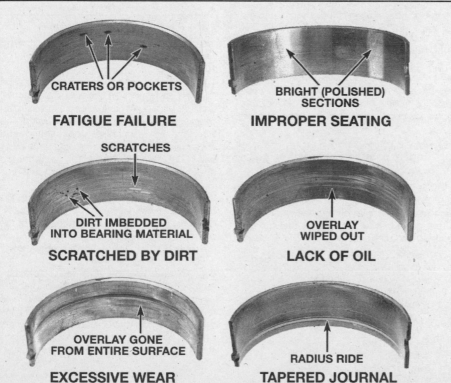

CRATERS OR POCKETS
FATIGUE FAILURE

BRIGHT (POLISHED) SECTIONS
IMPROPER SEATING

SCRATCHES
DIRT IMBEDDED INTO BEARING MATERIAL
SCRATCHED BY DIRT

OVERLAY WIPED OUT
LACK OF OIL

OVERLAY GONE FROM ENTIRE SURFACE
EXCESSIVE WEAR

RADIUS RIDE
TAPERED JOURNAL

19.1b Typical bearing failures

and lay them out on a clean surface in the same general position as their location in the engine. This will enable you to match any bearing problems with the corresponding crankshaft journal.

4   Dirt and other foreign particles get into the engine in a variety of ways. It may be left in the engine during assembly, or it may pass through filters or the PCV system. It may get into the oil, and from there into the bearings. Metal chips from machining operations and normal engine wear are often present. Abrasives are sometimes left in engine components after reconditioning, especially when parts are not thoroughly cleaned using the proper cleaning methods. Whatever the source, these foreign objects often end up embedded in the soft bearing material and are easily recognized. Large particles will not embed in the bearing and will score or gouge the bearing and journal. The best prevention for this cause of bearing failure is to clean all parts thoroughly and keep everything spotlessly clean during engine assembly. Frequent and regular engine oil and filter changes are also recommended.

5   Lack of lubrication (or lubrication breakdown) has a number of interrelated causes. Excessive heat (which thins the oil), overloading (which squeezes the oil from the bearing face) and oil leakage or throw off (from excessive bearing clearances, worn oil pump or high engine speeds) all contribute to lubrication breakdown. Blocked oil passages, which usually are the result of misaligned oil holes in a bearing shell, will also oil starve a bearing and destroy it. When lack of lubrication is the cause of bearing failure, the bearing material is wiped or extruded from the steel backing of the bearing. Temperatures may increase to the point where the steel backing turns blue from overheating.

6   Driving habits can have a definite effect on bearing life. Full throttle, low speed operation (lugging the engine) puts very high loads on bearings, which tends to squeeze out the oil film. These loads cause the bearings to flex, which produces fine cracks in the bearing face (fatigue failure). Eventually the bearing material will loosen in pieces and tear away from the steel backing. Short trip driving leads to corrosion of bearings because insufficient engine heat is produced to drive off the condensed water and corrosive gases. These products collect in the engine oil, forming acid and sludge. As the oil is carried to the engine bearings, the acid attacks and corrodes the bearing material.

7   Incorrect bearing installation during engine assembly will lead to bearing failure as well. Tight fitting bearings leave insufficient bearing oil clearance and will result in oil starvation. Dirt or foreign particles trapped behind a bearing insert result in high spots on the bearing which lead to failure.

Selection

8   If the original bearings are worn or damaged, or if the oil clearances are incorrect

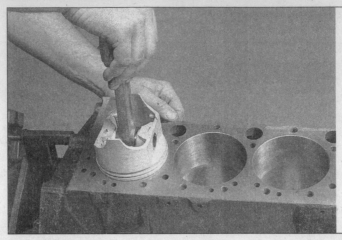

21.3 When checking piston ring end gap, the ring must be square in the cylinder bore (this is done by pushing the ring down with the top of a piston as shown)

(see Section 22 or 24) and the crankshaft is undamaged or worn excessively, new bearings should be purchased for engine reassembly. However, if the crankshaft has been reground, new undersize bearings must be installed. The automotive machine shop that reconditions the crankshaft will provide or help you select the correct size bearings. Regardless of how the bearing sizes are determined, measure the oil clearance with Plastigage to ensure the bearings are the right size.

Main bearings

9   If you need to use a STANDARD size main bearing, install one that has the same number as the original bearing.

10   If you need to use a STANDARD size rod bearing, install one that has the same number as the number stamped into the connecting rod cap.

All bearings

11   Remember, the oil clearance is the final judge when selecting new bearing sizes. If you have any questions or are unsure which bearings to use, get help from a dealer parts or service department or other parts supplier.

20   Engine overhaul - reassembly sequence

1   Before beginning engine reassembly, make sure you have all the necessary new parts, gaskets and seals as well as the following items on hand:

Common hand tools
A torque wrench
Piston ring installation tool
Piston ring compressor
Short lengths of rubber or plastic hose to fit over connecting rod bolts
Plastigage
Feeler gauges
A fine-tooth file
New engine oil
Engine assembly lube or moly-base grease
Gasket sealant
Thread locking compound

2   In order to save time and avoid problems, engine reassembly must be done in the following general order:

Piston rings
Crankshaft and main bearings
Piston/connecting rod assemblies
Rear main oil seal
Cylinder head and rocker arms or lifters (Part A)
Camshaft(s) (Part A)
Oil pump (Part A)
Timing chain, sprockets and cover (Part A)
Oil pick-up tube
Oil pan (Part A)
Valve cover (Part A)
Intake and exhaust manifolds (Part A)
Flywheel/driveplate (Part A)

21   Piston rings - installation

Refer to illustrations 21.3, 21.4, 21.7a, 21.7b and 21.10

1   Before installing the new piston rings, the ring end gaps must be checked. It's assumed that the piston ring side clearance has been checked and verified correct (see Section 17).

2   Lay out the piston/connecting rod assemblies and the new ring sets so the ring sets will be matched with the same piston and cylinder during the end gap measurement and engine assembly.

3   Insert the top (number one) ring into the first cylinder and square it up with the cylinder walls by pushing it in with the top of the piston **(see illustration)**. The ring should be near the bottom of the cylinder, at the lower limit of ring travel.

4   To measure the end gap, slip feeler gauges between the ends of the ring until a gauge equal to the gap width is found **(see illustration)**. The feeler gauge should slide between the ring ends with a slight amount of drag. Compare the measurement to the Specifications. If the gap is larger or smaller than specified, double-check to make sure you have the correct rings before proceeding.

5   Repeat the procedure for each ring that will be installed in the first cylinder and for each ring in the remaining cylinders. Remem-

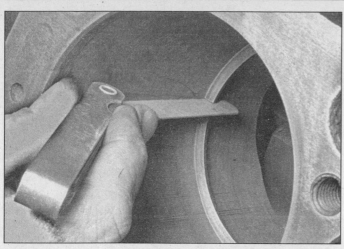

21.4 With the ring square in the cylinder, measure the end gap with a feeler gauge

21.7a Installing the spacer/expander in the oil control ring groove

21.7b DO NOT use a piston ring installation tool when installing the oil ring side rails

21.10 Installing the compression rings with a ring expander - the mark (arrow) must face up

2B

ber to keep rings, pistons and cylinders matched up.

6 Once the ring end gaps have been checked/corrected, the rings can be installed on the pistons.

7 The oil control ring (lowest one on the piston) is usually installed first. It's composed of three separate components. Slip the spacer/expander into the groove **(see illustration)**. If an anti-rotation tang is used, make sure it's inserted into the drilled hole in the ring groove. Next, install the lower side rail. Don't use a piston ring installation tool on the oil ring side rails, as they may be damaged. Instead, place one end of the side rail into the groove between the spacer/expander and the ring land, hold it firmly in place and slide a finger around the piston while pushing the rail into the groove **(see illustration)**. Next, install the upper side rail in the same manner.

8 After the three oil ring components have been installed, check to make sure that both the upper and lower side rails can be turned smoothly in the ring groove.

9 The number two (middle) ring is installed next. It's usually stamped with a mark which must face up, toward the top of the piston.

Note: *Always follow the instructions printed on the ring package or box - different manufacturers may require different approaches. Do not mix up the top and middle rings, as they have different cross sections.*

10 Use a piston ring installation tool and make sure the identification mark is facing the top of the piston, then slip the ring into the middle groove on the piston. Don't expand the ring any more than necessary to slide it over the piston **(see illustration)**.

11 Install the number one (top) ring in the same manner. Make sure the mark is facing up. Be careful not to confuse the number one and number two rings.

12 Repeat the procedure for the remaining pistons and rings.

22 Crankshaft - installation and main bearing oil clearance check

Refer to illustrations 22.10, 22.12 and 22.14

1 Crankshaft installation is the first major step in engine reassembly. It's assumed at this point that the engine block and

crankshaft have been cleaned, inspected and repaired or reconditioned.

2 Position the engine with the bottom facing up.

3 Remove the main bearing cap bolts and lift out the caps. Lay the caps out in the proper order to ensure correct installation.

4 If they're still in place, remove the old bearing inserts from the block and the main bearing caps. Wipe the main bearing surfaces of the block and caps with a clean, lint-free cloth. They must be kept spotlessly clean!

Main bearing oil clearance check

5 Clean the back sides of the new main bearing inserts and lay the bearing half with the oil groove and hole in each main bearing saddle in the block (on all engines covered by this manual, the bearings with oil holes go in the block and those without oil holes go in the caps). Lay the other bearing half from each bearing set in the corresponding main bearing cap. Make sure the tab on each bearing insert fits into the recess in the block or

22.10 Lay the Plastigage strips (arrow) on the main bearing journals, parallel to the crankshaft centerline

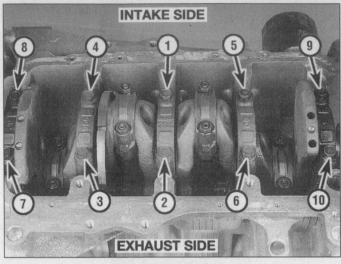

22.12 Main bearing cap tightening sequence

22.14 Compare the width of the crushed Plastigage to the scale on the envelope to determine the main bearing oil clearance (always take the measurement at the widest point of the Plastigage); be sure to use the correct scale - standard and metric ones are included

cap. Also, the oil holes in the block must line up with the oil holes in the bearing insert. **Caution:** *Do not hammer the bearings into place and don't nick or gouge the bearing faces. No lubrication should be used at this time.*

6   The thrust bearings (washers) must be installed in the number three main bearing saddle (on the cylinder block side). Be sure to install them with the oil grooves facing out (away from the main bearing saddle). Apply a thin film of moly-based grease or engine assembly lube to the back sides of the thrust bearings to hold them in place. **Note:** *The thrust washers are used only on early engines. Later engines use conventional thrust bearings in the number three bearing location.*

7   Clean the faces of the bearings in the block and the crankshaft main bearing journals with a clean, lint-free cloth. Check or clean the oil holes in the crankshaft, as any dirt here can go only one way - straight through the new bearings.

8   Once you're certain the crankshaft is clean, carefully lay it in position in the main bearings. No lubricant should be used at this time.

9   Before the crankshaft can be permanently installed, the main bearing oil clearance must be checked.

10   Trim several pieces of the appropriate size Plastigage (they must be slightly shorter than the width of the main bearings) and place one piece on each crankshaft main bearing journal, parallel with the journal axis **(see illustration)**.

11   Clean the faces of the bearings in the caps and install the caps in their respective positions (don't mix them up) with the arrows pointing toward the front of the engine. Don't disturb the Plastigage. Apply a light coat of oil to the bolt threads and the under sides of the bolt heads, then install them.

12   Following the recommended sequence **(see illustration)**, tighten the main bearing cap bolts, in three steps, to the torque listed in this Chapter's Specifications. Don't rotate

the crankshaft at any time during this operation!

13   Remove the bolts and carefully lift off the main bearing caps. Keep them in order. Don't disturb the Plastigage or rotate the crankshaft. If any of the main bearing caps are difficult to remove, tap them gently from side-to-side with a soft-face hammer to loosen them.

14   Compare the width of the crushed Plastigage on each journal to the scale printed on the Plastigage envelope to obtain the main bearing oil clearance **(see illustration)**. Check the Specifications to make sure it's correct.

15   If the clearance is not as specified, the bearing inserts may be the wrong size which means different ones will be required (see Section 19). Before deciding that different inserts are needed, make sure that no dirt or oil was between the bearing inserts and the caps or block when the clearance was measured. If the Plastigage is noticeably wider at one end than the other, the journal may be tapered (see Section 18).

16   Carefully scrape all traces of the Plastigage material off the main bearing journals and/or the bearing faces. Don't nick or scratch the bearing faces.

Final crankshaft installation

17   Carefully lift the crankshaft out of the engine. Clean the bearing faces in the block, then apply a thin, uniform layer of clean moly-base grease or engine assembly lube to each of the bearing surfaces. Coat the thrust washers as well.

18   Lubricate the crankshaft surfaces that contact the oil seals with moly-base grease, engine assembly lube or clean engine oil.

19   Make sure the crankshaft journals are clean, then lay the crankshaft back in place in the block. Clean the faces of the bearings in the caps or cap assembly, then apply the same lubricant to them. Install the caps in their respective positions with the arrows pointing toward the front of the engine. **Note:** *If you haven't already done so, be sure to install the thrust washers in the number three main bearing saddle.*

20   Apply a light coat of oil to the bolt threads and the under sides of the bolt heads, then install them. Tighten all except the number three cap bolts (the one with the thrust washers on the block side) to the torque listed in this Chapter's Specifications (work from the center out and approach the final torque in three steps). Tighten the num-

23.3 After removing the retainer from the block, support it on two wood blocks and drive out the old seal with a punch and hammer

23.5 Drive the new seal into the retainer with a wood block or a section of pipe, if you have one large enough - make sure you don't cock the seal in the retainer bore

ber three cap bolts to 10-to-12 ft-lbs. Tap the ends of the crankshaft forward and backward with a lead or brass hammer to line up the thrust washer and crankshaft surfaces. Re-tighten all main bearing cap bolts to the specified torque, following the recommended sequence.

21 Rotate the crankshaft a number of times by hand to check for any obvious binding.

22 Check the crankshaft endplay with a feeler gauge or a dial indicator (see Section 13). The endplay should be correct if the crankshaft thrust faces aren't worn or damaged and new thrust washers have been installed.

23 Install a new rear main oil seal, then bolt the retainer to the block (see Section 23).

23 Rear main oil seal - installation

Refer to illustrations 23.3 and 23.5

1 The crankshaft must be installed first and the main bearing caps bolted in place, then the new seal should be installed in the retainer and the retainer bolted to the block.

2 Check the seal contact surface on the crankshaft very carefully for scratches and nicks that could damage the new seal lip and cause oil leaks. If the crankshaft is damaged, the only alternative is a new or different crankshaft.

3 The old seal can be removed from the retainer by driving it out from the back side with a hammer and punch **(see illustration)**. Be sure to note how far it's recessed into the bore before removing it; the new seal will have to be recessed an equal amount. Be very careful not to scratch or otherwise damage the bore in the retainer or oil leaks could develop.

4 Make sure the retainer is clean, then apply a thin coat of engine oil to the outer edge of the new seal. The seal must be pressed squarely into the bore, so hammering it into place isn't recommended. If you don't have access to a press, sandwich the housing and seal between two smooth pieces of wood and press the seal into place with the jaws of a large vise. The pieces of wood must be thick enough to distribute the force evenly around the entire circumference of the seal. Work slowly and make sure the seal enters the bore squarely.

5 As a last resort, the seal can be tapped into the retainer with a hammer. Use a block of wood to distribute the force evenly and make sure the seal is driven in squarely **(see illustration)**.

6 The seal lips must be lubricated with clean engine oil or multi-purpose grease before the seal/retainer is slipped over the crankshaft and bolted to the block. Use a new gasket and make sure the dowel pins are in place before installing the retainer.

7 Tighten the bolts a little at a time until they're all at the torque listed in the Chapter 2 Part A Specifications.

24 Pistons/connecting rods - installation and rod bearing oil clearance check

Refer to illustrations 24.5, 24.9, 24.11, 24.13 and 24.17

1 Before installing the piston/connecting rod assemblies, the cylinder walls must be perfectly clean, the top edge of each cylinder must be chamfered, and the crankshaft must be in place.

2 Remove the cap from the end of the number one connecting rod (refer to the marks made during removal). Remove the original bearing inserts and wipe the bearing surfaces of the connecting rod and cap with a clean, lint-free cloth. They must be kept spotlessly clean.

Connecting rod bearing oil clearance check

3 Clean the back side of the new upper bearing insert, then lay it in place in the connecting rod. Make sure the tab on the bearing fits into the recess in the rod so the oil holes line up. Don't hammer the bearing insert into place and be very careful not to nick or gouge the bearing face. Don't lubricate the bearing at this time.

4 Clean the back side of the other bearing insert and install it in the rod cap. Again, make sure the tab on the bearing fits into the recess in the cap, and don't apply any lubricant. It's critically important that the mating surfaces of the bearing and connecting rod are perfectly clean and oil free when they're assembled.

5 Position the piston ring gaps at staggered intervals around the piston **(see illustration)**.

6 Slip a section of plastic or rubber hose over each connecting rod cap bolt.

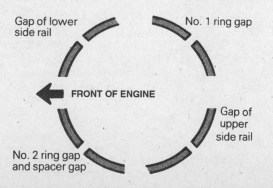

Gap of lower side rail

No. 1 ring gap

FRONT OF ENGINE

Gap of upper side rail

No. 2 ring gap and spacer gap

24.5 Ring end gap positions

2B

24.9 Turn the piston when installing it to make sure the mark/notch (arrow) in the piston faces the front (timing chain end) of the engine

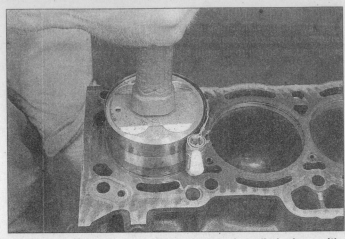

24.11 The piston can be driven gently into the cylinder bore with the end of a wooden or plastic hammer handle

7   Lubricate the piston and rings with clean engine oil and attach a piston ring compressor to the piston. Leave the skirt protruding about 1/4-inch to guide the piston into the cylinder. The rings must be compressed until they're flush with the piston.

8   Rotate the crankshaft until the number one connecting rod journal is at BDC (bottom dead center) and apply a coat of engine oil to the cylinder walls.

9   With the mark on top of the piston **(see illustration)** facing the front (timing chain end) of the engine, gently insert the piston/connecting rod assembly into the number one cylinder bore and rest the bottom edge of the ring compressor on the engine block.

10   Tap the top edge of the ring compressor to make sure it's contacting the block around its entire circumference.

11   Gently tap on the top of the piston with the end of a wooden or plastic hammer handle **(see illustration)** while guiding the end of the connecting rod into place on the crankshaft journal. The piston rings may try to pop out of the ring compressor just before

entering the cylinder bore, so keep some pressure on the ring compressor. Work slowly, and if any resistance is felt as the piston enters the cylinder, stop immediately. Find out what's hanging up and fix it before proceeding. Do not, for any reason, force the piston into the cylinder - you might break a ring and/or the piston. **Caution:** *Do not push the piston assembly further into the bore than is necessary. The rings can catch under the cylinder liner and prevent upward movement.*

12   Once the piston/connecting rod assembly is installed, the connecting rod bearing oil clearance must be checked before the rod cap is permanently bolted in place.

13   Cut a piece of the appropriate size Plastigage slightly shorter than the width of the connecting rod bearing and lay it in place on the number one connecting rod journal, parallel with the journal axis **(see illustration)**.

14   Clean the connecting rod cap bearing face, remove the protective hoses from the connecting rod bolts and install the rod cap. Make sure the mating mark on the cap is on the same side as the mark on the connecting rod. Check the cap to make sure the front

mark is facing the timing chain end of the engine.

15   Apply a light coat of oil to the under sides of the nuts, then install and tighten them to the torque listed in this Chapter's Specifications. Use a thin-wall socket to avoid erroneous torque readings that can result if the socket is wedged between the rod cap and nut. If the socket tends to wedge itself between the nut and the cap, lift up on it slightly until it no longer contacts the cap. Do not rotate the crankshaft at any time during this operation.

16   Remove the nuts and detach the rod cap, being very careful not to disturb the Plastigage.

17   Compare the width of the crushed Plastigage to the scale printed on the Plastigage envelope to obtain the oil clearance **(see illustration)**. Compare it to the Specifications to make sure the clearance is correct.

18   If the clearance is not as specified, the bearing inserts may be the wrong size (which means different ones will be required). Before deciding that different inserts are needed, make sure that no dirt or oil was between the

24.13 Lay the Plastigage strips on each rod bearing journal, parallel to the crankshaft centerline

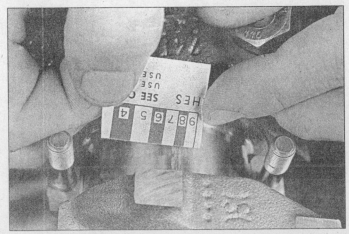

24.17 Measuring the width of the crushed Plastigage to determine the rod bearing oil clearance (be sure to use the correct scale - standard and metric ones are included)

bearing inserts and the connecting rod or cap when the clearance was measured. Also, recheck the journal diameter. If the Plastigage was wider at one end than the other, the journal may be tapered (see Section 18).

Final connecting rod installation

19 Carefully scrape all traces of the Plastigage material off the rod journal and/or bearing face. Be very careful not to scratch the bearing - use your fingernail or the edge of a credit card.

20 Make sure the bearing faces are perfectly clean, then apply a uniform layer of clean moly-base grease or engine assembly lube to both of them. You'll have to push the piston into the cylinder to expose the face of the bearing insert in the connecting rod - be sure to slip the protective hoses over the rod bolts first.

21 Slide the connecting rod back into place on the journal, remove the protective hoses from the rod cap bolts, install the rod cap and tighten the nuts to the torque listed in this Chapter's Specifications.

22 Repeat the entire procedure for the remaining pistons/connecting rods.

23 The important points to remember are:

a) Keep the back sides of the bearing inserts and the insides of the connecting rods and caps perfectly clean when assembling them.

b) Make sure you have the correct piston/rod assembly for each cylinder.

c) The dimple on the piston must face the front (timing chain end) of the engine.

d) Lubricate the cylinder walls with clean oil.

e) Lubricate the bearing faces when installing the rod caps after the oil clearance has been checked.

24 After all the piston/connecting rod assemblies have been properly installed, rotate the crankshaft a number of times by hand to check for any obvious binding.

25 As a final step, the connecting rod side clearance (endplay) must be checked (see Section 12).

26 Compare the measured side clearance to the Specifications to make sure it's correct. If it was correct before disassembly and the original crankshaft and rods were reinstalled, it should still be right. If new rods or a new crankshaft were installed, the side clearance may be inadequate. If so, the rods will have to be removed and taken to an automotive machine shop for re-sizing.

25 Initial start-up and break-in after overhaul

Warning: *Have a fire extinguisher handy when starting the engine for the first time.*

1 Once the engine has been installed in the vehicle, double-check the engine oil and coolant levels.

2 With the spark plugs out of the engine and the ignition system disabled (see Section 3), crank the engine until oil pressure registers on the gauge or the light goes out.

3 Install the spark plugs, hook up the plug wires and restore the ignition system functions (see Section 3).

4 Start the engine. It may take a few moments for the fuel system to build up pressure, but the engine should start without a great deal of effort.

5 After the engine starts, it should be allowed to warm up to normal operating temperature. While the engine is warming up, make a thorough check for fuel, oil and coolant leaks.

6 Shut the engine off and recheck the engine oil and coolant levels.

7 Drive the vehicle to an area with minimum traffic, accelerate at full throttle from 30 to 50 mph, then allow the vehicle to slow to 30 mph with the throttle closed. Repeat the procedure 10 or 12 times. This will load the piston rings and cause them to seat properly against the cylinder walls. Check again for oil and coolant leaks.

8 Drive the vehicle gently for the first 500 miles (no sustained high speeds) and keep a constant check on the oil level. It is not unusual for an engine to use oil during the break-in period.

9 At approximately 500 to 600 miles, change the oil and filter.

10 For the next few hundred miles, drive the vehicle normally. Do not pamper it or abuse it.

11 After 2000 miles, change the oil and filter again and consider the engine broken in.

2B

Notes

Chapter 3
Cooling, heating and air conditioning systems

Contents

Specifications

General

Expansion tank cap pressure rating	13 to 20 psi
Thermostat rating (opening temperature)	
Starts to open	182 to 190-degrees
Fully open	212-degrees
Cooling system capacity	See Chapter 1
Cooling system testing pressure	13 psi
Refrigerant type	
1994 and earlier	R-12
1995 and later	R-134a

Torque specifications

	Ft-lbs (unless otherwise indicated)
Expansion valve mounting bolts	89 in-lbs
Refrigerant line-to-expansion valve bolts	19 in-lbs
Refrigerant line-to-compressor bolts	19 in-lbs
Thermostat cover nuts	22
Water pump-to-engine block bolts	22

1 General information

Engine cooling system

All vehicles covered by this manual employ a pressurized engine cooling system with thermostatically controlled coolant circulation. An impeller type water pump mounted on the drivebelt end of the block pumps coolant through the engine. The coolant flows around each cylinder and toward the transaxle end of the engine. Cast-in coolant passages direct coolant around the intake and exhaust ports, near the spark plug areas and in close proximity to the exhaust valve guides.

A wax pellet type thermostat is located in a housing on the right front of the engine. During warm up, the closed thermostat prevents coolant from circulating through the radiator. As the engine nears normal operating temperature, the thermostat opens and allows hot coolant to travel through the radiator, where it's cooled before returning to the engine.

The cooling system is sealed by a pressure-type cap, which raises the boiling point of the coolant and increases the cooling efficiency of the radiator. The pressure cap is located on the translucent plastic expansion tank on the right side of the engine compartment. **Warning:** *Do not remove the pressure cap from the expansion tank until the engine has cooled completely and there's no pressure remaining in the cooling system.*

Heating system

The heating system consists of a blower fan and heater core located in the heater box, the hoses connecting the heater core to the engine cooling system and the heater/air conditioning control head on the dashboard. Hot engine coolant is circulated through the heater core. When the heater mode is activated, a flap opens to expose the heater box to the passenger compartment. A fan switch on the control head activates the blower motor, which forces air through the core, heating the air.

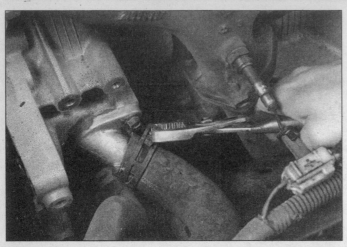

3.8 Loosen the hose clamp and detach the hose from the fitting

3.10 Remove the bolts (arrows) and detach the thermostat cover

Air conditioning system

The air conditioning system consists of a condenser mounted in front of the radiator, an evaporator mounted adjacent to the heater core, a compressor mounted on the engine, a receiver-drier which contains a high pressure relief valve and the plumbing connecting all of the above components.

A blower fan forces the warmer air of the passenger compartment through the evaporator core (sort of a radiator-in-reverse), transferring the heat from the air to the refrigerant. The liquid refrigerant boils off into low pressure vapor, taking the heat with it when it leaves the evaporator.

2   Antifreeze - general information

Warning: *Do not allow antifreeze to come in contact with your skin or painted surfaces of the vehicle. Rinse off spills immediately with plenty of water. Antifreeze is highly toxic if ingested. Never leave antifreeze lying around in an open container or in puddles on the floor; children and pets are attracted by it's sweet smell and may drink it. Check with local authorities about disposing of used antifreeze. Many communities have collection centers which will see that antifreeze is disposed of safely.*

The cooling system should be filled with a water/ethylene glycol based antifreeze solution, which will prevent freezing down to at least -20-degrees F, or lower if local climate requires it. It also provides protection against corrosion and increases the coolant boiling point.

The cooling system should be drained, flushed and refilled at the specified intervals (see Chapter 1). Old or contaminated antifreeze solutions are likely to cause damage and encourage the formation of corrosion and scale in the system. Use distilled water with the antifreeze.

Before adding antifreeze, check all hose connections, because antifreeze tends to leak through very minute openings. Engines don't normally consume coolant, so if the level goes down, find the cause and correct it.

The exact mixture of antifreeze-to-water which you should use depends on the relative weather conditions. The mixture should contain at least 50-percent antifreeze, but should never contain more than 70-percent antifreeze. Consult the mixture ratio chart on the antifreeze container before adding coolant. Hydrometers are available at most auto parts stores to test the coolant. Use antifreeze which meets the vehicle manufacturer's specifications.

3   Thermostat - check and replacement

Warning: *Do not remove the expansion tank cap, drain the coolant or replace the thermostat until the engine has cooled completely. Do not allow antifreeze to come in contact with your skin or painted surfaces of the vehicle. Rinse off spills immediately with plenty of water. Antifreeze is highly toxic if ingested. Never leave antifreeze lying around in an open container or in puddles on the floor; children and pets are attracted by it's sweet smell and may drink it. Check with local authorities about disposing of used antifreeze. Many communities have collection centers which will see that antifreeze is disposed of safely.*
Caution: *On models equipped with a theft-deterrent radio, make sure you have the correct activation code, or the theft deterrent system is turned off, before disconnecting the battery.*

Check

1   Before assuming the thermostat is to blame for a cooling system problem, check the coolant level, drivebelt tension (see Chapter 1) and temperature gauge operation.
2   If the engine seems to be taking a long time to warm up (based on heater output or temperature gauge operation), the thermostat is probably stuck open. Replace the thermostat with a new one.

3   If the engine runs hot, use your hand to check the temperature of the upper radiator hose. If the hose isn't hot, but the engine is, the thermostat is probably stuck closed, preventing the coolant inside the engine from escaping to the radiator. Replace the thermostat. **Caution:** *Don't drive the vehicle without a thermostat. The computer may stay in open loop, causing emissions and fuel economy to suffer.*
4   If the upper radiator hose is hot, it means that the coolant is flowing and the thermostat is open. Consult the *Troubleshooting* section at the front of this manual for cooling system diagnosis.

Replacement

Refer to illustrations 3.8, 3.10, 3.11 and 3.12
5   Disconnect the negative battery cable from the battery.
6   Drain the cooling system (see Chapter 1). If the coolant is relatively new or in good condition, save it and reuse it.
7   Follow the lower radiator hose (on the passenger side) to the engine to locate the thermostat housing.
8   Loosen the hose clamp and detach the hose from the fitting **(see illustration)**. If the hose is stuck, grasp it near the end with a pair of adjustable pliers and twist it to break the seal, then pull it off. If the hose is old or deteriorated, cut it off and install a new one.
9   If the outer surface of the large fitting that mates with the hose is deteriorated (corroded, pitted, etc.) it may be damaged further by hose removal. If it is, the thermostat housing cover will have to be replaced.
10   Remove the bolts and detach the thermostat cover **(see illustration)**. If the cover is stuck, tap it with a soft-face hammer to jar it loose. Be prepared for some coolant to spill as the gasket seal is broken.
11   Remove the thermostat and cover assembly **(see illustration)**.
12   Note how the thermostat element is installed in the cover. You must push down and turn the element to remove it from the cover **(see illustration)**.

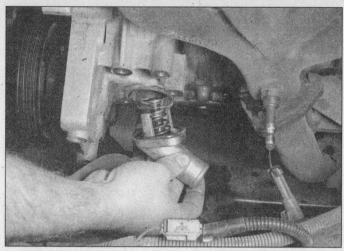

3.11 Remove the thermostat/cover assembly

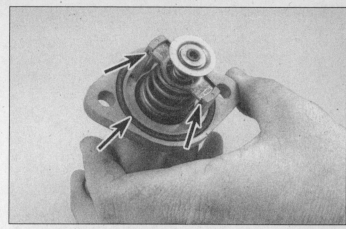

3.12 Note how the ears on the thermostat element engage with the slots in the cover - to remove the element, push it down and turn it (also, be sure to install a new O-ring if the old one is cracked or hardened)

13   Install the new thermostat in the cover.

14   The O-ring gasket may be re-used if it's in good condition. Be sure to check it carefully for nicks, hardening or signs of deterioration. If there is any doubt, install a new one.

15.   Install the thermostat cover and bolts. Tighten the bolts to the torque listed in this Chapter's Specifications.

16   Reattach the hose to the fitting and tighten the hose clamp securely. Reconnect the battery cable.

17   Refill the cooling system (see Chapter 1).

18   Start the engine and allow it to reach normal operating temperature, then check for leaks and proper thermostat operation (as described in Steps 3 and 4).

4   Engine cooling fan - check and replacement

Caution: *On models equipped with a theft-deterrent radio, make sure you have the correct activation code, or the theft deterrent system is turned off, before disconnecting the battery.*

1   The engine cooling fan is controlled by the PCM and activated through a relay which is mounted on the left side of the engine compartment in the relay center. When the coolant reaches a predetermined temperature or the air conditioning system is turned on, the PCM closes the relay, completing the circuit. The PCM will also turn the fan off when the vehicle reaches a predetermined speed.

Check

Refer to illustration 4.3

2   First, check the fuses (see Chapter 12).

3   To test the fan motor, unplug the electrical connector **(see illustration)** and use fused jumper wires to connect the fan directly to the battery. If the fan still does not work,

replace the motor.

4   If the motor tested okay, the fault lies in the relay, the coolant temperature sensor (see Chapter 6), the wiring harness or the PCM. Refer to Chapter 6 and check the PCM for stored trouble codes. Further diagnosis should be performed by a dealer service department or other repair shop.

Replacement

Refer to illustrations 4.8, 4.9, 4.11 and 4.12

5   Disconnect the negative battery cable from the battery.

6   Disconnect the air temperature sensor in the air intake duct.

7   Remove the air intake ducts that are mounted across the top of the radiator.

8   Remove the two upper radiator fan shroud bolts **(see illustration)**. The bottom two mounts just lock into place in the lower mounting brackets.

9   Remove the fan assembly from the vehi-

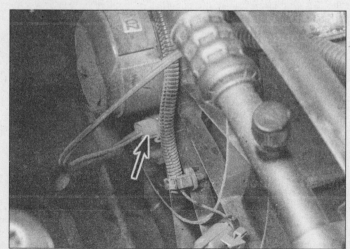

4.3 Disconnect the electrical connector from the fan motor (arrow) and, using jumper wires, connect the motor terminals directly to the battery - if the fan does not work, replace the motor

4.8 Remove the two upper fan shroud bolts (left side shown; there's one on the other side, too)

3

4.9 Remove the fan assembly from the vehicle by lifting the left (driver's) side up first - be careful not to damage the radiator fins)

4.11 Once the cooling fan has been removed the fan blade assembly can be separated from the motor by removing the nut (arrow) and lifting the fan blade assembly from the motor (the nut has left-hand threads; to unscrew, turn it clockwise)

4.12 Remove these three bolts to separate the motor from the shroud assembly

5.5a Squeeze the upper radiator hose clamp and slide it back on the hose . . .

cle **(see illustration)** by lifting the left (drivers) side up first and rotate the fan up and off its mounts.

10   Unplug the cooling fan electrical connector as the assembly is being removed from the vehicle.

11   Once the cooling fan has been removed the fan blade assembly can be separated from the electric fan motor by removing the nut **(see illustration)** and lifting the fan blade assembly from the motor. **Caution:** *This nut has left-hand threads - to remove it, turn it clockwise.*

12   The motor can be replaced by removing the three bolts and separating the motor from the shroud assembly **(see illustration)**.

13   Installation is the reverse of removal.

---

5   Radiator and coolant expansion tank - removal and installation

Warning: *Do not start this procedure until the*

engine is completely cool. Do not allow antifreeze to come in contact with your skin or painted surfaces of the vehicle. Rinse off spills immediately with plenty of water. Antifreeze is highly toxic if ingested. Never leave antifreeze lying around in an open container or in puddles on the floor; children and pets are attracted by it's sweet smell and may drink it. Check with local authorities about disposing of used antifreeze. Many communities have collection centers which will see that antifreeze is disposed of safely.

Caution: *On models equipped with a theft-deterrent radio, make sure you have the correct activation code, or the theft deterrent system is turned off, before disconnecting the battery.*

Radiator

Removal

Refer to illustrations 5.5a, 5.5b, 5.8 and 5.9

1   Disconnect the negative battery cable from the battery.

2   Raise the front of the vehicle and support it securely on jackstands. Remove the lower splash shields.

3   Drain the cooling system (see Chapter 1). If the coolant is relatively new or in good condition, save it and reuse it.

4   Remove the air intake duct that passes over the top of the radiator. On models equipped with MPFI, remove the air cleaner housing (see Chapter 4).

5   Loosen the upper and lower radiator hose clamps, then detach the radiator hoses from the fittings **(see illustrations)**. If they're stuck, grasp each hose near the end with a pair of adjustable pliers and twist it to break the seal, then pull it off - be careful not to damage the radiator fittings! If the hoses are old or deteriorated, cut them off and install new ones.

6   Disconnect the electrical connector from the cooling fan.

7   Remove the cooling fan (see Section 4).

8   If the vehicle is equipped with an automatic transaxle, remove the lower engine

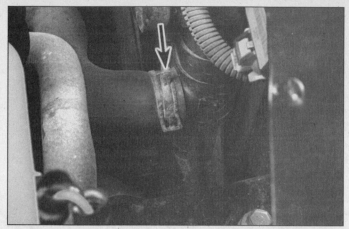

5.5b ... squeeze the lower radiator hose clamp and slide it back on the hose, then detach the radiator hoses from the fittings. If they're stuck, grasp each hose near the end with a pair of adjustable pliers and twist it to break the seal, then pull it off - be careful not to damage the radiator fittings!

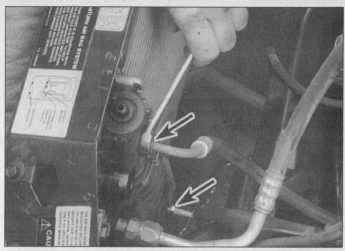

5.8 If the vehicle is equipped with an automatic transaxle, disconnect the transmission fluid cooler lines from the radiator

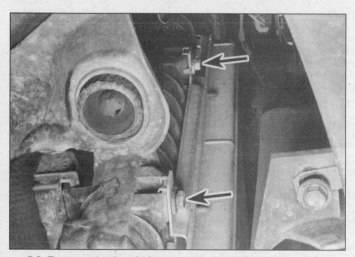

5.9 Remove the four bolts, two on each side of the radiator (arrows), that attach the radiator to the air conditioning condenser

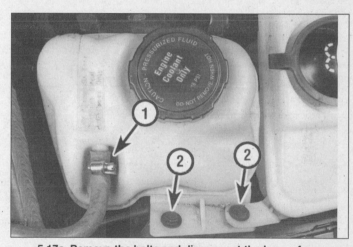

5.17a Remove the bolts and disconnect the hoses from the upper ...

1 Upper expansion tank hose

2 Expansion tank attaching bolts

splash shield and disconnect the transmission fluid cooler lines and plug the lines and fittings (see illustration). Be ready to catch some automatic transmission fluid when these are removed.

9   Unscrew the four bolts (two on each side of the radiator) that attach the radiator to the air conditioning condenser (see illustration).

10   Carefully tip the top of the radiator slightly towards the engine, remove the upper radiator seal and lift the radiator from the engine compartment.. Don't spill coolant on the vehicle or scratch the paint.

11   With the radiator removed, it can be inspected for leaks and damage. If it needs repair, have a radiator shop or dealer service department perform the work, as special tools and techniques are required.

12   Bugs and dirt can be removed from the radiator with a garden hose or a soft brush. Don't bend the cooling fins as this is done.

Installation

13   Installation is the reverse of the removal procedure. Be sure the rubber cushions are seated properly at the base of the radiator.

14   After installation, fill the cooling system with the proper mixture of antifreeze and water (see Chapter 1).

15   Start the engine and check for leaks. Allow the engine to reach normal operating temperature, indicated by the upper radiator hose becoming hot. Recheck the coolant level and add more if required.

16   If you're working on an automatic transaxle equipped vehicle, check and add fluid as needed (see Chapter 1).

Coolant expansion tank - removal and installation

Refer to illustrations 5.17a and 5.17b

17   Disconnect the hoses from the upper and lower expansion tank fittings (see illustrations) and inspect them for cracks or

5.17b ... and lower expansion tank fittings (arrow) and inspect them for cracks or signs of deterioration

7.5 Remove the three water pump pulley bolts (arrows)

7.6 Remove the water pump mounting bolts (arrows) and detach the water pump from the engine

signs of deterioration. Replace them at this time if necessary.

18   Remove the mounting bolts **(see illustration 5.17a)**.

19   Lift the coolant expansion tank up and disconnect the low coolant level sensor connection from underneath the tank. Remove the expansion tank from the vehicle.

20   Pour the coolant into a container. Wash out and inspect the reservoir for cracks and chafing. Replace it if it's damaged.

21   Installation is the reverse of removal.

---

6   Water pump - check

1   A failure in the water pump can cause serious engine damage due to overheating.

2   There are two ways to check the operation of the water pump while it's installed on the engine. If the pump is defective, it should be replaced with a new or rebuilt unit.

3   Loosen the tension on the serpentine drivebelt (see Chapter 1). Grasp the water pump pulley and try to rock it up-and-down. If any play is felt, the shaft bearings are worn out and the pump should be replaced.

4   Water pumps are equipped with weep or vent holes. If a failure occurs in the pump seal, coolant will leak from the hole. In most cases you'll need a flashlight to find the hole on the water pump from underneath to check for leaks.

5   If the water pump shaft bearings fail there may be a howling sound at the drivebelt end of the engine while it's running. Don't mistake drivebelt slippage, which causes a squealing sound, for water pump bearing failure.

---

7   Water pump - removal and installation

Warning: *Wait until the engine is completely cool before beginning this procedure. Do not allow antifreeze to come in contact with your skin or painted surfaces of the vehicle. Rinse*

off spills immediately with plenty of water. Antifreeze is highly toxic if ingested. Never leave antifreeze lying around in an open container or in puddles on the floor; children and pets are attracted by it's sweet smell and may drink it. Check with local authorities about disposing of used antifreeze. Many communities have collection centers which will see that antifreeze is disposed of safely.

Removal

Refer to illustrations 7.5 and 7.6

1   Disconnect the negative battery cable from the battery.

2   Drain the cooling system (see Chapter 1). If the coolant is relatively new or in good condition, save it and reuse it.

3   Remove the accessory drivebelt (see Chapter 1).

4   **Note:** *It is easier to remove the water pump through the wheel opening, but it is possible to perform the job from the engine compartment* **(see illustration)**, *without the need to jack the car up and remove the inner fender splash shield (space is limited, so you'll have to judge for yourself if you have the room to work). If you decide to work through the wheel well, remove the right front tire and the inner fender splash shield.*

5   Remove the three water pump pulley bolts **(see illustration). Note 1:** *You'll have to immobilize the pulley while unscrewing these bolts. This can be done with a strap wrench wrapped around the pulley or by bracing a screwdriver between two of the bolts and breaking the other bolt loose (if you use the latter method, be sure to loosen all three of the bolts before removing any of them).* **Note 2:** *On some vehicles, it is necessary to leave the pulley on the water pump after the pulley bolts are unscrewed about one inch. The pulley can then be pulled away from the water pump enough to remove the water pump bolts. The pulley and pump can then be removed as an assembly.*

6   Remove the water pump mounting bolts **(see illustration)** and detach the water pump

from the engine. If the water pump is stuck, gently tap it with a soft faced hammer to break the seal.

Installation

Refer to illustration 7.9

7   Clean the bolt threads and the threaded holes in the engine to remove corrosion and sealant. Remove all traces of old gasket material from the sealing surfaces.

8   Compare the new pump to the old one to make sure they're identical.

9   Apply a thin film of RTV sealant to the pump flange **(see illustration)** and place a new gasket on the pump.

10   Carefully mate the pump to the engine.

11   Install the bolts. Tighten them to the torque listed in this Chapter's Specifications. Tighten the bolts in a criss-cross pattern until the proper torque is reached.

12   Reinstall all parts removed for access to the pump.

13   Refill the cooling system (see Chapter 1). Run the engine and check for leaks.

7.9 Apply a thin film of RTV sealant to the pump flange (arrow) and place a new gasket on the pump

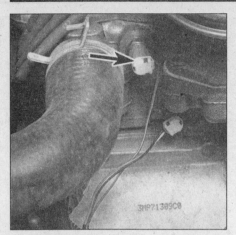

8.3 The coolant temperature sending unit is located on the left end of the cylinder head (arrow)

8   Coolant temperature sending unit - check and replacement

Refer to illustration 8.3
Warning: *The engine must be completely cool before removing the sending unit.*

Check

1   If the coolant temperature gauge is inoperative, check the fuses first (see Chapter 12).
2   If the temperature indicator shows excessive temperature after running awhile, see the *Troubleshooting* section in the front of the manual.
3   If the temperature gauge indicates Hot shortly after the engine is started cold, disconnect the wire at the coolant temperature sending unit - it's located on the left end (driver's side) of the cylinder head **(see illustration)**. If the gauge reading drops, replace the sending unit. If the reading remains high, the wire to the gauge may be shorted to ground or the gauge is faulty.

4   If the coolant temperature gauge fails to indicate after the engine has been warmed up (approximately 10 minutes) and the fuses checked out okay, shut off the engine. Disconnect the wire at the sending unit and, using a jumper wire, connect it to a clean ground on the engine. Turn on the ignition without starting the engine. If the gauge now indicates Hot, replace the sending unit.
5   If the gauge still does not work, the circuit may be open or the gauge may be faulty. See Chapter 12 for additional information.

Replacement

6   With the engine completely cool, remove the cap from the expansion tank to release any pressure, then reinstall the cap. This reduces coolant loss during sending unit replacement.
7   Disconnect the electrical connector from the sending unit.
8   Prepare the new sending unit for installation by wrapping the threads with Teflon tape.
9   Unscrew the sending unit from the engine and quickly install the new one to prevent coolant loss.
10   Tighten the sending unit securely and connect the electrical connector.
11   Check the coolant level and add, if necessary. Start the engine and check for leaks and proper gauge operation.

9   Heater and air conditioning blower motor circuit - check and component replacement

The blower motor resistor is connected in series to the blower motor and is located under the dash, which requires removal of the dash pad and ventilation system duct work. It controls the different speed selections for the blower motor.

The high blower is controlled by a relay which is in the relay panel located on the right side of the floor console behind a trim cover (see Chapter 12).

Check and replacement

1   If the blower motor speed does not correspond to the setting selected on the blower switch, or the blower motor does not operate at all, the problem could be a bad fuse, relay, switch, blower motor resistor, blower motor or blower motor circuit wiring.
2   Before checking the blower motor or circuit, always check the fuse and relay (if equipped) first (see Chapter 12).
3   With the ignition key in the ON position, turn the blower switch to each position and, using a test light or voltmeter, check the voltage at the motor. If the motor is receiving voltage but not operating, either the motor ground is bad, the motor itself is faulty or the fan is binding.
4   To check for a faulty blower motor, disconnect the electrical connector from the blower motor, connect a jumper wire between the ground wire terminal on the blower motor and a good ground, then connect a fused jumper wire between the battery positive terminal and the positive terminal on the blower motor. If the motor now operates properly, the problem probably lies with the blower motor resistor, the blower switch or related wirings. Proceed to Step 6. If the motor does not operate, either the motor is faulty or the fan is binding.
5   If you suspect the blower motor fan is binding, remove the blower motor (see Section 10) to check for free operation of the fan.

Blower motor resistor

Refer to illustrations 9.7 and 9.8
6   Remove the upper dash pad (see Chapter 11).
7   Remove the defroster duct **(see illustration)**.
8   Locate the electrical connector for the heater blower motor resistor **(see illustration)**, usually attached to a heating/air conditioning duct or blower motor case. Verify that the resistor is getting current from the blower switch.

9.7 Remove the defroster duct after the dash pad has been removed for access to the blower motor resistor

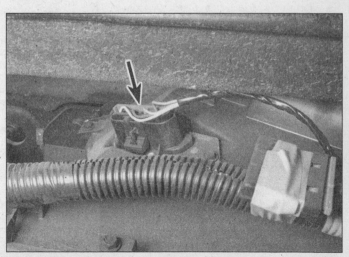

9.8 Locate the electrical connector for the heater blower motor resistor (arrow) and disconnect it

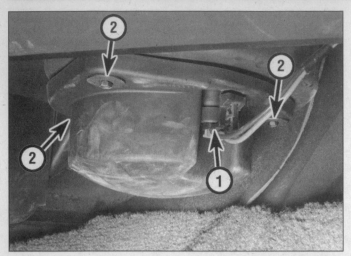

10.3 Disconnect the electrical connector at the blower motor

1 Electrical connector	*2 Mounting screws*

a) *If the resistor is not getting current, check the wires and the connectors between the resistor and the switch. Check for loose or corroded connections and damaged wires.*

b) *If the wires and connectors are good and the blower switch is getting current, verify current is flowing out of the blower switch to the resistor. If it is not, replace the switch.*

9   Unplug the electrical connector from the blower resistor and check the continuity across the resistor terminals with an ohmmeter. If the resistor does not have continuity between all the terminals, replace the resistor.

Blower relay

10   Remove the cover from the relay panel at the right side of the console, just under the dash (see Chapter 11).

11   Remove the 'High blower' relay and check for voltage and continuity at the relay connector.

12   If there is both voltage and continuity from the control head switch to the relay connector when the system is switched to the high speed position, and the blower motor doesn't operate, replace the 'High blower' relay.

10   Heater and air conditioning blower motor - removal and installation

Refer to illustrations 10.3 and 10.5

Warning: *On models equipped with a Supplemental Inflatable Restraint (SIR) system, always disable the airbag(s) before working in the vicinity of the steering wheel, instrument panel or SIR system components to avoid the possibility of accidental deployment of the airbag(s), which could cause personal injury (see Chapter 12).*

Caution: *On models equipped with a theft-deterrent radio, make sure you have the cor-*

10.5 On 1992 and later models, separate the cover from the blower motor assembly (arrows) by simply pulling the two apart

11.4 Remove the three screws (arrows) holding the control unit in the dash and pull the assembly from the dash

rect activation code, or the theft deterrent system is turned off, before disconnecting the battery.

1   Disconnect the negative cable from the battery.

2   Locate the blower motor on passenger side underneath where the carpeting meets the firewall.

3   Disconnect the electrical connector at the blower motor **(see illustration)**.

4   Remove the three blower motor retaining screws **(see illustration 10.3)** and lower the unit from the housing.

5   On 1992 and later models separate the insulated cover from the blower motor assembly **(see illustration)**. **Note:** *This noise reduction cover can be added to any earlier models that came without it.*

6   If you are replacing the motor, detach the fan and transfer it to the new motor. **Caution:** *The fan cage is a balanced assembly. Be careful to note the weights, if any, that are attached to the cage. They must remain in same location(s) or the fan will become unbalanced.*

7   Installation is the reverse of removal.

8   Run the blower and check for proper operation.

11   Heater and air conditioning control assembly - removal and installation

Refer to illustrations 11.4, 11.5a, 11.5b, 11.6 and 11.7

Warning: *On models equipped with a Supplemental Inflatable Restraint (SIR) system, always disable the airbag(s) before working in the vicinity of the steering wheel, instrument panel or SIR system components to avoid the possibility of accidental deployment of the airbag(s), which could cause personal injury (see Chapter 12).*

Caution: *On models equipped with a theft-deterrent radio, make sure you have the correct activation code, or the theft deterrent system is turned off, before disconnecting the battery.*

1   Disconnect the negative cable from the battery.

2   Remove the trim panel which surrounds the radio and heater control assembly (see Chapter 11).

3   Remove the radio (see Chapter 12).

4   Remove the three screws **(see illustration)** holding the control unit in the dash and

11.5a Disconnect the electrical connectors from the control assembly at the fan speed control and . . .

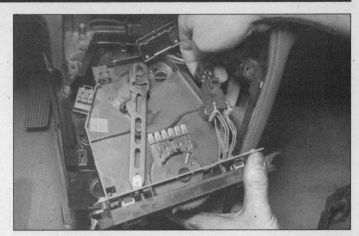

11.5b . . . at the air conditioning selector button

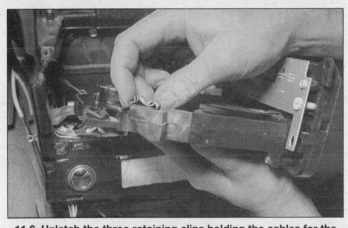

11.6 Unlatch the three retaining clips holding the cables for the temperature, mode control and fresh/recirculated air levers to the control assembly

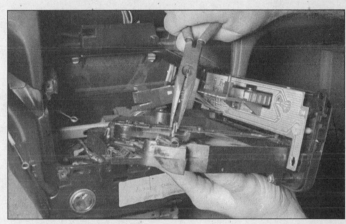

11.7 Squeeze the post that the cable loop connects with and slide the cable off the control lever

3

pull the assembly from the dash.

5   Disconnect the electrical connectors from the control assembly **(see illustrations)**.

6   Unlatch the three retaining clips holding the cables for the temperature, mode control and fresh/recirculated air levers **(see illustration)** to the control assembly.

7   Using needle-nose pliers, squeeze the post that the cable loop connects with and slide the cable off the control lever **(see illustration)**.

8   Remove the control assembly.

9   Adjustment for the temperature and mode control cables are as follows:

Mode control cable (black):

a) *Place the mode control lever all the way to the full vent position (left).*

b) *Place the mode cable loop over the post on the lever.*

c) *Snap the mode cable hold-down/adjustment clip over the cable.*

d) *Check the mode lever travel for spring back resistance.*

Temperature control cable (blue or white):

a) *Place the temperature control lever all the way to the left position (cold).*

b) *Make sure the lever on the air plenum, the housing for the air doors, is in the full rearward position.*

c) *Place the temperature cable loop over the post on the lever.*

d) *Snap the temperature cable hold-down/adjustment clip over the cable.*

e) *Check the temperature lever travel for spring back resistance.*

10   Make small fine tuning adjustments by unlocking the retaining clips and move the cable slightly in the direction necessary to adjust for proper operation.

11   The remainder of installation is the reverse of removal.

---

12 Heater core - removal and installation

Refer to illustrations 12.9a, 12.9b, 12.10, 12.12, 12.13 and 12.14

Warning: *On models equipped with a Supplemental Inflatable Restraint (SIR) system, always disable the airbag(s) before working in the vicinity of the steering wheel, instrument panel or SIR system components to avoid the possibility of accidental deployment of the* airbag(s), *which could cause personal injury* (see Chapter 12).

Caution: *On models equipped with a theft-deterrent radio, make sure you have the correct activation code, or the theft deterrent system is turned off, before disconnecting the battery.*

1   Disconnect the negative cable from the battery.

2   Drain the cooling system (see Chapter 1).

3   On MPFI models remove the air cleaner housing cover and the air induction hose at the intake manifold. On TBI models remove the air cleaner housing. **Note:** *In both situations this will give enough room to access the heater hoses for removal.*

4   Raise the vehicle and support it properly on jackstands. **Note:** *This done in order to get at the hose clamps. This may be possible from the top after the air cleaner housing has been removed, but the way the manufacturer places the clamps, access from above is very difficult.*

5   Loosen and move the heater hose clamps up the hose enough so the hoses can be removed from the heater core inlet and outlet tubes.

6   Lower the vehicle.

7   Working in the engine compartment,

12.9a Remove the screws on the lower heater duct (two screws on either side of the duct) . . .

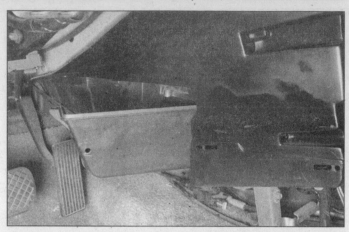

12.9b . . . and slide the duct out the drivers side

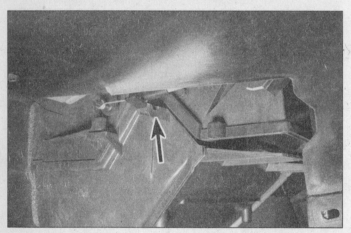

12.10 Disconnect the temperature control cable and the mode control cable at the heater case (arrow)

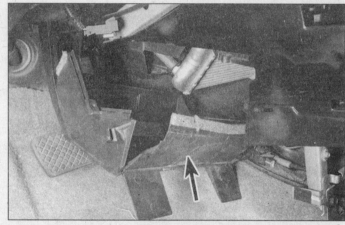

12.12 Remove the screws and reattach the lower heater core cover (arrow)

disconnect the heater hoses where they connect with the heater core at the firewall. **Note:** *If compressed air is available, blow through the opened heater core lines to remove coolant to avoid spilling any in the interior when the heater core is removed (be sure to place a drain pan under the vehicle, and wear eye protection). Jacking the back end of the car up slightly will also help in draining the heater core.*

8   Remove the left and right lower trim panel extensions by disconnecting the lower velcro fasteners and pulling out at the upper clip locations (see Chapter 11). Remove the instrument panel lower closeout panel by pulling out the top edge and then rotating the top down.

9   Remove the screws on the lower heater duct and slide the duct out the drivers side **(see illustrations)**.

10   Disconnect the temperature control cable and the mode control cable at the heater case **(see illustration)**.

11   Disconnect the electrical connector on the drivers side mounted near the temperature control cable.

12   Remove the screws and remove the lower heater core cover **(see illustration)**.

13   Remove the screws and take off the

12.13 Remove the screws and take off the heater core side cover (arrow) to expose the heater core pipe clamp

heater core side cover **(see illustration)**.

14   Remove the screw and heater core pipe clamp **(see illustration)**.

15   Remove the screw and the lower heater core retainer from under the heater core.

16   Remove the heater core, being careful not to spill any coolant.

12.14 Remove the screw and heater core pipe clamp (arrow)

17   Reassemble the heater unit and check the operation of the control flaps. If any parts bind, correct the problem before installation.

18   Reinstall the remaining parts in the reverse order of removal.

19   Refill the cooling system (see Chapter 1), reconnect the battery and run the engine. Check for leaks and proper system operation.

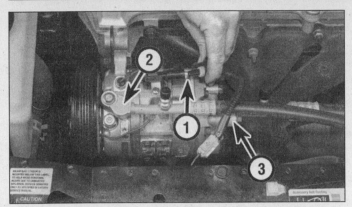

14.3 Locate the wires on top of the air conditioning compressor, follow them back and unplug the electrical connector

1	Electrical connector	3	Low pressure (suction)
2	High pressure (discharge) hose		hose

14.7 Remove the three rear compressor mounting bolts (arrows)

13 Air conditioning and heating system - check and maintenance

Warning: *The air conditioning system is under high pressure. Do not loosen any fittings or remove any components until after the system has been discharged. Air conditioning refrigerant should be properly discharged into an EPA-approved container at a dealer service department or an automotive air conditioning repair facility. Always wear eye protection when disconnecting air conditioning system fittings.*

Caution: *There are two types of refrigerants used on the models covered by this manual. 1994 and earlier models use R-12 refrigerant, while 1995 and later models use the new "environmentally friendly" R-134a refrigerant. The two refrigerants (and their appropriate refrigerant oils) are not compatible and must never be mixed or components will be damaged.*

1   The following maintenance checks should be performed on a regular basis to ensure that the air conditioner continues to operate at peak efficiency.

a) *Check the compressor drivebelt. If it's worn or deteriorated, replace it (see Chapter 1).*

b) *Check the drivebelt tension and, if necessary, replace the belt (see Chapter 1).*

c) *Check the system hoses. Look for cracks, bubbles, hard spots and deterioration. Inspect the hoses and all fittings for oil bubbles and seepage. If there's any evidence of wear, damage or leaks, replace the hose(s).*

d) *Inspect the condenser fins for leaves, bugs and other debris. Use a "fin comb" or compressed air to clean the condenser.*

e) *Make sure the system has the correct refrigerant charge.*

f) *Check the evaporator housing drain tube for blockage.*

Long term non-use can cause hardening, and

subsequent failure, of the seals.

2   Because of the complexity of the air conditioning system and the special equipment necessary to service it, in-depth troubleshooting and repairs are not included in this manual. However, simple checks and component replacement procedures are provided in this Chapter. For more complete information on the air conditioning system, refer to the *Haynes Automotive Heating and Air Conditioning Manual.*

3   The most common cause of poor cooling is simply a low system refrigerant charge. If a noticeable drop in cool air output occurs, one of the following quick checks will help you determine if the refrigerant level is low.

4   Warm the engine up to normal operating temperature.

5   Place the air conditioning temperature selector at the coldest setting and put the blower at the highest setting. Open the doors (to make sure the air conditioning system doesn't cycle off as soon as it cools the passenger compartment).

6   With the compressor engaged - the clutch will make an audible click and the center of the clutch will rotate - inspect the sight glass, if equipped. If the refrigerant looks foamy, it's low. Have the system charged by a dealer service department or other qualified repair shop.

7   If there's no sight glass, feel the inlet and outlet pipes at the compressor. One side should be much colder than the other. If there's no perceptible difference between the two pipes, there's something wrong with the compressor or the system. It might be a low charge - it might be something else. Take the vehicle to a dealer service department or other qualified repair shop.

14 Air conditioning compressor - removal and installation

Refer to illustrations 14.3, 14.7 and 14.8
Warning: *The air conditioning system is*

under high pressure. Do not loosen any fittings or remove any components until after the system has been discharged. Air conditioning refrigerant should be properly discharged into an EPA-approved container at a dealer service department or an automotive air conditioning repair facility. Always wear eye protection when disconnecting air conditioning system fittings.*

Caution 1: *There are two types of refrigerants used on the models covered by this manual. 1994 and earlier models use R-12 refrigerant, while 1995 and later models use the new "environmentally friendly" R-134a refrigerant. The two refrigerants (and their appropriate refrigerant oils) are not compatible and must never be mixed or components will be damaged.*

Caution 2: *On models equipped with a theft-deterrent radio, make sure you have the correct activation code, or the theft deterrent system is turned off, before disconnecting the battery.*

Removal

1   Have the refrigerant discharged at a dealer service department or an automotive air conditioning repair facility.

2   Disconnect the negative cable from the battery and remove the battery.

3   Disconnect the electrical connector(s) at the compressor **(see illustration)**.

4   On MPFI models, remove the air intake duct (see Chapter 4). **Note:** *Some models are equipped with the AIR air pump system. These models require the removal of the hoses, combination valve and air pipe. The combination valve need not be completely removed; simply position it away from the air conditioning compressor bracket.*

5   Remove the accessory drivebelt (see Chapter 1).

6   Detach the refrigerant lines from the compressor and immediately cap the open fittings to prevent the entry of dirt and moisture **(see illustration 14.3)**.

7   Remove the three rear mounting bolts **(see illustration)**.

3

14.8 Remove the three front compressor mounting bolts (arrows) - shown here with the compressor removed for clarity

8   Remove the three front mounting bolts **(see illustration)** and remove the compressor from its mounting bracket. **Note:** *Keep the compressor level during handling and storage. If the compressor seized or you find metal particles in the refrigerant lines, the system must be flushed out by an air conditioning technician and the receiver/drier must be replaced* (see Section 16).

Installation

9   Prior to installation, turn the center of the clutch six times to disperse any oil that has collected in the head.
10   Install the compressor in the reverse order of removal. Be sure to use new O-rings (designed specifically for air conditioning system use) on the refrigerant line fittings. Coat them with a little refrigerant oil before installation.
11   If you are installing a new compressor, refer to the manufacturer's instructions for adding refrigerant oil to the system. **Note:** *If you are installing the same compressor, drain and measure the amount of oil that is present in the compressor upon removal, then add the same amount of new refrigerant oil to the compressor on installation. New compressors come with 200 cc of refrigerant oil already installed.*
12   Have the system evacuated, charged and leak tested by the shop that discharged it.

15   Air conditioning condenser - removal and installation

Refer to illustration 15.3
Warning: *The air conditioning system is under high pressure. Do not loosen any fittings or remove any components until after the system has been discharged. Air conditioning refrigerant should be properly discharged into an EPA-approved container at a*

dealer service department or an automotive air conditioning repair facility. Always wear eye protection when disconnecting air conditioning system fittings.
Caution: *There are two types of refrigerants used on the models covered by this manual. 1994 and earlier models use R-12 refrigerant, while 1995 and later models use the new "environmentally friendly" R-134a refrigerant. The two refrigerants (and their appropriate refrigerant oils) are not compatible and must never be mixed or components will be damaged.*

Removal

1   Have the refrigerant discharged at a dealer service department or an automotive air conditioning repair facility.
2   Remove any components which interfere with the condenser fittings. On later models these may include the left headlamp assembly. On MPFI models, remove the air cleaner housing cover and the air induction hose at the intake manifold. On TBI models remove the air cleaner housing (see Chapter 4). On later models, remove the lower splash shield and access the lines from underneath. **Note:** *In each situation this will give enough room to access the refrigerant hose for disconnection.*
3   Remove the compressor discharge hose and the receiver/drier hose from the condenser **(see illustration)**. Be sure to use a backup wrench on the fittings so the condenser tubes aren't damaged. **Caution:** *Immediately cap all open refrigerant lines and connections to prevent moisture or debris from entering the system.*
4   Raise the vehicle and support it securely on jackstands.
5   Remove the lower splash shield.
6   From the front of the condenser, behind the grill, remove the four mounting bolts that attach the condenser to the radiator **(see illustration 15.3).**
7   From the front of the vehicle, pull the condenser towards you slightly to separate it from the mounting pads. Then carefully lower the condenser straight down. As it is moved

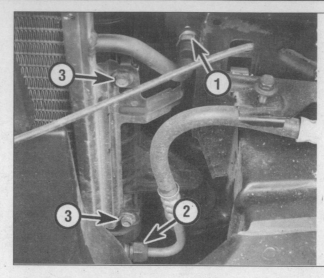

15.3 Remove the compressor discharge hose and the receiver/drier hose from the condenser

1   *Compressor high pressure (discharge) hose*
2   *Receiver/drier hose*
3   *Condenser mounting bolts (two per side)*

down rotate it so the inlet pipe is straight up.
8   Once the condenser is almost out the bottom of the vehicle, rotate the bottom of the condenser rearward. The condenser should now be parallel to the ground, and can be removed from under the car.

Installation

9   If a new condenser is to be installed, add one ounce of new refrigerant oil to the system.
10   Installation is the reverse of removal. Be sure to install new O-rings on the lines and lubricate them with clean refrigerant oil.
11   Have the system evacuated, charged and leak tested by the shop that discharged it.

16   Air conditioning receiver/drier - removal and installation

Refer to illustration 16.4
Warning: *The air conditioning system is under high pressure. Do not loosen any fittings or remove any components until after the system has been discharged. Air conditioning refrigerant should be properly discharged into an EPA-approved container at a dealer service department or an automotive air conditioning repair facility. Always wear eye protection when disconnecting air conditioning system fittings.*
Caution: *There are two types of refrigerants used on the models covered by this manual. 1994 and earlier models use R-12 refrigerant, while 1995 and later models use the new "environmentally friendly" R-134a refrigerant. The two refrigerants (and their appropriate refrigerant oils) are not compatible and must never be mixed or components will be damaged.*

Removal

1   Have the refrigerant discharged at a dealer service department or an automotive air conditioning repair facility.
2   Raise the vehicle and support it securely

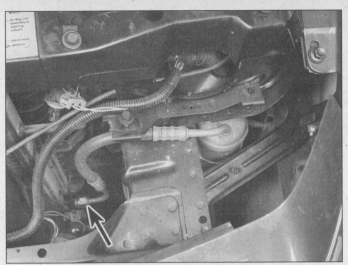

16.4 Disconnect the receiver-drier hose from the condenser outlet (arrow)

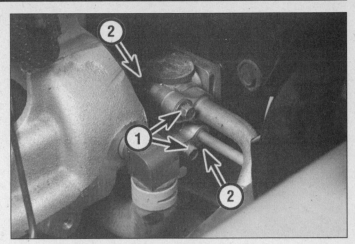

17.5 Disconnect the two refrigerant lines from the expansion valve

1   *Expansion valve-to-refrigerant lines connection*
2   *Expansion valve-to-evaporator mounting bolts*

on jackstands.

3   Remove the splash shield from under the car.

4   Disconnect the receiver-drier hose from the condenser outlet **(see illustration)**.

5   Disconnect the remaining refrigerant line from the top of the receiver-drier. **Caution:** *Immediately cap the open fittings to prevent the entry of dirt and moisture.*

6   Remove the receiver/drier mounting bolt, detach it from the condenser and lower it out through the bottom of the vehicle.

Installation

7   Install new O-rings on the lines and lubricate them with clean refrigerant oil.

8   Installation is the reverse of removal. **Note:** *Do not remove the sealing caps until you are ready to reconnect the lines.*

9   If a new receiver/drier is installed, add one fluid ounce of refrigerant oil to the system. **Caution:** *Due to the tight fit of the receiver-drier to the surrounding sheet metal and engine components, make sure that none of the refrigerant lines rub on anything or a refrigerant leak could develop.*

10   Have the system evacuated, charged and leak tested by the shop that discharged it.

---

17   Air conditioning evaporator core and expansion valve - removal and installation

Warning: *The air conditioning system is under high pressure. Do not loosen any fittings or remove any components until after the system has been discharged. Air conditioning refrigerant should be properly discharged into an EPA-approved container at a dealer service department or an automotive air conditioning repair facility. Always wear eye protection when disconnecting air condi-*

tioning system fittings.

Caution 1: *There are two types of refrigerants used on the models covered by this manual. 1994 and earlier models use R-12 refrigerant, while 1995 and later models use the new "environmentally friendly" R-134a refrigerant. The two refrigerants (and their appropriate refrigerant oils) are not compatible and must never be mixed or components will be damaged.*

Caution 2: *On models equipped with a theft-deterrent radio, make sure you have the correct activation code, or the theft deterrent system is turned off, before disconnecting the battery.*

Expansion valve

Refer to illustration 17.5

1   Have the refrigerant discharged at a dealer service department or an automotive air conditioning repair facility.

2   Disconnect the cable from the negative terminal of the battery.

3   On models with MPFI, remove the air cleaner housing cover and the air induction hose at the intake manifold.

4   On models with TBI, remove the air cleaner housing.

5   Disconnect the two refrigerant lines from the expansion valve **(see illustration)**.

6   Remove the two bolts that secure the expansion valve to the evaporator inlet fitting.

7   Remove the expansion valve from the evaporator inlet. **Caution:** *Immediately cap the open fittings to prevent the entry of dirt and moisture.*

8   Installation is the reverse of the removal. **Note:** *Always use new O-rings lubricated with clean refrigerant oil when reassembling air conditioning components.*

Evaporator core

Warning: *On models equipped with a Supplemental Inflatable Restraint (SIR) system, always disable the airbag(s) before working in*

the vicinity of the steering wheel, instrument panel or SIR system components to avoid the possibility of accidental deployment of the airbag(s), which could cause personal injury (see Chapter 12).

Note: *Throughout the evaporator core removal procedure there will be references to removal of interior components and trim. Many steps will be referenced to the specific Section in which it was removed. Information regarding interior trim will be found in Chapter 11 and information on chassis electrical components will found in Chapter 12.*

Removal of HVAC module

9   Have the refrigerant discharged at a dealer service department or an automotive air conditioning repair facility.

10   Disconnect the cable from the negative terminal of the battery.

11   Remove the expansion valve (see Steps 1 through 7).

12   Follow the heater core removal procedure (see Section 12) up to the point of disconnecting the heater core fasteners. **Note:** *There is no need to remove the heater core - it can be removed with the entire module.*

13   Remove the heater/air conditioning controls (see Section 11).

14   Remove the evaporator case retaining nuts from the firewall in the engine compartment.

15   Disconnect all electrical connectors from the air conditioning thermostatic switch and the blower motor resistor (see Section 9).

16   Unbolt the evaporator case assembly from the firewall/dash framework.

17   Disconnect the heater blower motor resistor connector and the blower (see Section 9) and all other wire harness connections that are located under the dash trim panels as they are removed. **Note:** *Label them to avoid reassembly problems later.*

18   Tilt the steering wheel down and pull the cluster trim panel rearward to disengage the

3

retainers.

19   Disconnect the electrical connectors from the instrument panel lighting rheostat and rear window defogger switches.

20   Remove the glove box and the instrument panel lower support.

21   Remove the cluster trim panel.

22   Remove the screws and ALDL (diagnostic) connector from the driver's side under the dash.

23   Remove the screws and the steering column filler panel.

24   Remove the hood release lever screw.

25   Remove the screws and pull the instrument panel cluster out far enough to reach in and disconnect the electrical connectors.

26   Remove the instrument cluster.

27   Apply the parking brake, if not already applied.

28   Remove the center console (see Chapter 11).

29   Remove the fuse block and any grounds or reinforcements from the lower reinforcement.

30   On vehicles equipped with an automatic transaxle, disconnect the instrument panel-to-body harness connector.

31   Remove the wiring harness and antenna lead clips and pull them out of the instrument panel retainer.

32   Remove the nuts and bolts, lift the lower reinforcement bracket off of the studs, and slide the bracket rearward.

33   Remove the screws and nuts from the dash retainer and reinforcement bracket.

34   Remove the bolts and lower the steering column assembly onto the front seat.

35   Carefully remove the instrument panel/dash assembly.

36   In the engine compartment, remove the fuel vapor return line and its clip from the HVAC module mounting stud in order to get access to the nut.

37   Remove three screws and two nuts that hold the HVAC module assembly to the firewall.

38   Inside the vehicle, remove the HVAC module assembly from under the dash and remove it from the vehicle.

Removal of evaporator core from the module

39   Remove the mode valve assembly from the evaporator case.

40   Separate the evaporator upper and lower cases by removing the screws and lifting the upper case half off.

41   Remove the evaporator pipe clamp and screw.

42   Lift the evaporator core straight up and out from the lower case half. **Note:** *Keep track of all the spacers and seals as they are removed.*

43   Check the evaporator fins for blockage; if they are dirty, clean them with compressed air - never use water for this purpose!

44   Check fittings for cracks and signs of wear; replace parts as necessary.

45   Inspect all the seals, spacers and water filter. Replace if necessary.

46   Installation is the reverse of the removal procedure. Be sure to replace all O-rings removed during disassembly with new ones.

47   If a new evaporator was installed, add 2.25 ounces of clean refrigerant oil to the system.

48   Have the system evacuated, charged and leak tested by the shop that discharged it.

49   The remainder of installation is in the reverse of removal.

Chapter 4
Fuel and exhaust systems

Contents

4

Specifications

Fuel pressure

Fuel system pressure (at idle)	
TBI models	26 to 31 psi
MPFI models	
1991 through 1997	
Vacuum hose attached................................	31 to 36 psi
Vacuum hose removed................................	37 to 45 psi
1998 through 1999..	40 to 55 psi
Fuel system pressure decay (maximum after 5 minutes)	
TBI models	6 psi
MPFI models	
1991 through 1997	3 psi
1998 through 1999..	8 psi
Fuel pump pressure	
1991 through 1997	46 to 94 psi
1998 through 1999..	65 to 94 psi
Fuel pump pressure decay (maximum)	
1991 through 1997	6 psi
1998 through 1999..	8 psi

Injector resistance

TBI models...	1.0 to 2.0 ohms
MPFI models	
1995 and earlier...	1.5 to 2.5 ohms
1996 and later	11.5 to 12.5 ohms

Torque specifications

	Ft-lbs
Throttle body-to-intake manifold bolts........................	24
Fuel rail mounting bolts	22

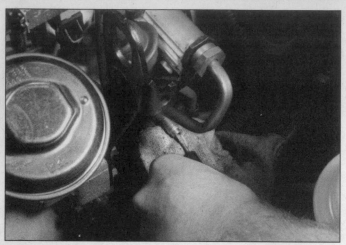

2.5 Wrap a shop towel around the Schrader valve on the fuel rail, then depress the valve to bleed any excess fuel from the rail

3.2a The fuel pump fuse (10 amp) is clearly labeled adjacent to the relay panel under the center console

1 General information

The fuel system consists of the fuel tank, an electric fuel pump (located in the fuel tank), an air cleaner assembly, the fuel injection system and the hoses and lines connecting the components. 1994 and earlier Single Overhead Camshaft (SOHC) engines are equipped with a Throttle Body Injection (TBI) system. 1995 and later SOHC and all Double Overhead Camshaft (DOHC) engines are equipped with a Multi Point Fuel Injection system.

Throttle Body Injection (TBI) system

The throttle body system utilizes a single injector, centrally mounted in a carburetor-like housing. The injector is an electrical solenoid, with fuel delivered to the injector at a constant pressure level. To maintain the fuel pressure at a constant level, excess fuel is returned to the fuel tank.

A signal from the Powertrain Control Module (PCM) opens the solenoid, allowing fuel to spray from the injector into the throttle body. The amount of time the injector is held open by the PCM determines the fuel/air mixture ratio.

Multi Point Fuel Injection (MPFI) system

Multi Point Fuel Injection uses timed impulses to inject the fuel directly into the intake port of each cylinder. The injectors are controlled by the Powertrain Control Module (PCM). The PCM monitors various engine parameters and delivers the exact amount of fuel required into the intake ports. The throttle body serves only to control the amount of air passing into the system. Because each cylinder is equipped with its own injector, much better control of the fuel/air mixture ratio is possible.

Fuel pump and lines

Fuel is circulated from the fuel tank to the fuel injection system, and back to the fuel tank, through a pair of metal lines running along the underside of the vehicle. An electric fuel pump is located inside the fuel tank. A vapor return system routes all vapors back to the fuel tank through a separate return line.

The fuel pump will operate as long as the engine is cranking or running and the PCM is receiving ignition reference pulses from the electronic ignition system. If there are no reference pulses, the fuel pump will shut off after two or three seconds.

Exhaust system

The exhaust system includes an exhaust manifold fitted with an exhaust oxygen sensor, a catalytic converter, an exhaust pipe, and a muffler.

The catalytic converter is an emission control device added to the exhaust system to reduce pollutants. A single-bed converter is used in conjunction with a three-way (reduction) catalyst. Refer to Chapter 6 for more information regarding the catalytic converter.

2 Fuel pressure relief procedure

Refer to illustration 2.5

Warning 1: *Gasoline is extremely flammable, so take extra precautions when you work on any part of the fuel system. Don't smoke or allow open flames or bare light bulbs near the work area, and don't work in a garage where a natural gas-type appliance (such as a water heater or clothes dryer) with a pilot light is present. If you spill any fuel on your skin, rinse it off immediately with soap and water. When you perform any kind of work on the fuel system, wear safety glasses and have a Class B type fire extinguisher on hand.*

Warning 2: *After the fuel pressure has been relieved, wrap shop towels around any fuel*

connection you'll be disconnecting. *They'll absorb the residual fuel that may leak out, reducing the risk of fire and preventing contact with your skin.*

1   Before servicing any fuel system component, you must relieve the fuel pressure to minimize the risk of fire or personal injury.

2   Remove the fuel filler cap - this will relieve any pressure built up in the tank.

3   Remove the fuel pump relay from the relay panel located under the center console **(see illustration 3.2a** for the location of the relay).

4   Start the engine and wait for the engine to stall, then turn off the ignition key.

5   Position a small screwdriver in the fuel pressure test port and carefully depress the valve to allow any excess fuel to bleed out of the fuel line **(see illustration)**. The fuel system is now depressurized. **Warning:** *Place a rag around the fuel line before removing any fuel line fitting to prevent any residual fuel from spilling onto the engine or your skin.*

3 Fuel pump/fuel pressure – check

Warning: *Gasoline is extremely flammable, so take extra precautions when you work on any part of the fuel system. Don't smoke or allow open flames or bare light bulbs near the work area, and don't work in a garage where a natural gas-type appliance (such as a water heater or clothes dryer) with a pilot light is present. If you spill any fuel on your skin, rinse it off immediately with soap and water. When you perform any kind of work on the fuel system, wear safety glasses and have a Class B type fire extinguisher on hand.*

Note 1: *To perform the fuel pressure test, you will need to obtain a fuel pressure gauge compatible with the Schrader valve test port.*

Note 2: *The fuel pump will operate as long as the engine is cranking or running and the PCM is receiving ignition reference pulses from the electronic ignition system. If there*

3.2b With the ignition key ON (engine not running), unplug the fuel pump relay and connect a jumper wire from the bottom terminal (battery voltage) to the upper terminal to activate the fuel pump (the relay panel is located under the center console)

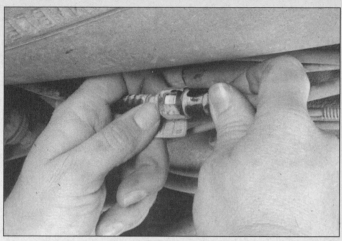

3.15 First pinch the fuel line tabs and pull back to release the fitting, then attach a fuel gauge to the line from the fuel tank using a special connector or adapter made for Saturn type fuel lines

are no reference pulses, the fuel pump will shut off after two or three seconds.

Preliminary inspection

Refer to illustrations 3.2a and 3.2b

1   Should the fuel system fail to deliver the proper amount of fuel, or any fuel at all, inspect it as follows. Remove the fuel filler cap. Have an assistant turn the ignition key to the On position (engine not running) while you listen at the fuel filler opening. You should hear a whirring sound that lasts for a couple of seconds.

2   If you don't hear anything, check the fuel pump fuse **(see illustration)**. If the fuse is blown, replace it and see if it blows again. If it does, trace the fuel pump circuit for a short. If it isn't blown, remove the fuel pump relay and install a jumper wire into the fuel pump relay terminals that power the fuel pump **(see illustration)**. Listen at the fuel filler opening again – if you now hear the whirring sound, the fuel pump relay or its control circuit is faulty. If there still is no whirring sound, there is a problem in the fuel pump circuit from the relay panel to the fuel pump, or a defective fuel pump.

3   Check for battery voltage to the fuel pump relay connector **(see illustration 3.2b)**. If there is battery voltage present, remove one of the other relays in the panel and install it in the fuel pump relay position (the relays are interchangeable). If the fuel pump now runs, the original relay was bad. Replace it with a new one.

4   If there is no voltage present, check the fuse(s) and the wiring circuit for the fuel pump relay (see Chapter 12).

Operating pressure check

5   Relieve the fuel system pressure (see Section 2).

6   Detach the cable from the negative battery terminal.

7   Remove the cap from the fuel pressure test port **(see illustration 2.5)** and attach a

fuel pressure gauge.

8   Attach the cable to the negative battery terminal, then start the engine.

9   Note the fuel pressure and compare it with the pressure listed in this Chapter's Specifications.

10   If the system fuel pressure is less than specified:

  a) *Inspect the system for a fuel leak. Repair any leaks and recheck the fuel pressure.*

  b) *If the fuel pressure is still low, replace the fuel filter (it may be clogged) and recheck the pressure.*

  c) *If the pressure is still low, check the fuel pump output pressure (see below) and the fuel pressure regulator (see Section 14 [TBI models] or Section 16 [MPFI models]).*

11   If the pressure is higher than specified:

  a) *Check the fuel return line for an obstruction.*

  b) *Check the fuel pressure regulator (see Section 16).*

12   Turn the ignition switch to Off, wait five minutes then check the pressure on the gauge. Compare the reading with the hold pressure listed in this Chapter's Specifications. If the hold pressure is less than specified:

  a) *The fuel lines may be leaking.*

  b) *The fuel pressure regulator may be allowing the fuel pressure to bleed through to the fuel return line (see Section 16).*

  c) *A fuel injector (or injectors) may be leaking.*

Fuel pump output pressure check

Refer to illustration 3.15

Warning: *For this test it is necessary to use a fuel pressure gauge with a bleeder valve in order to relieve the fuel pressure after the test is completed (the normal procedure for pressure relief will not work because the gauge is*

connected directly to the the fuel pump).

13   Relieve the system fuel pressure (see Section 2).

14   Detach the cable from the negative battery terminal.

15   Attach a fuel pressure gauge directly to the fuel feed line at the fuel tank **(see illustration)**.

16   Attach the cable to the negative battery terminal.

17   Remove the fuel pump relay and, using a jumper wire, bridge the terminals on the fuel pump relay connector **(see illustration 3.2b)**.

18   Note the pressure reading on the gauge and compare the reading to the value listed in this Chapter's Specifications.

19   If the indicated pressure is less than specified, inspect the fuel line for leaks between the pump and gauge. If no leaks are found, replace the fuel pump.

20   Turn the ignition key to Off and wait five minutes. Note the reading on the gauge and compare it to the hold pressure listed in this Chapter's Specifications. If the hold pressure is less than specified, check the fuel line between the pump and gauge for leaks. If no leaks are found, replace the fuel pump.

21   Remove the jumper wire and reinstall the fuel pump relay.

22   Open the bleeder valve on the gauge and allow the pressurized fuel drain into an approved fuel container. Remove the gauge and reconnect the fuel line.

---

4   Fuel lines and fittings – inspection and replacement

Warning: *Gasoline is extremely flammable, so take extra precautions when you work on any part of the fuel system. Don't smoke or allow open flames or bare light bulbs near the work area, and don't work in a garage where a natural gas-type appliance (such as a water heater or clothes dryer) with a pilot light is present. If you spill any fuel on your skin, rinse*

5.8 Disconnect the clamps from the filler neck and vent tube and separate them from the rubber hoses

5.11a Remove the bolts (arrow) from the fuel tank straps and allow the straps to swing down

it off immediately with soap and water. When you perform any kind of work on the fuel system, wear safety glasses and have a Class B type fire extinguisher on hand.

Inspection

1   Once in a while, you will have to raise the vehicle to service or replace some component (an exhaust pipe hanger, for example). Whenever you work under the vehicle, always inspect fuel lines and all fittings and connections for damage or deterioration.
2   Check all hoses and pipes for cracks, kinks, deformation or obstructions.
3   Make sure all hoses and pipe clips attach their associated hoses or pipes securely to the underside of the vehicle.
4   Verify all hose clamps attaching rubber hoses to metal fuel lines or pipes are snug enough to assure a tight fit between the hoses and pipes.

Replacement

5   If you must replace any damaged sections, use original equipment replacement hoses or pipes constructed from exactly the same material as the section you are replacing. Do not install substitutes constructed from inferior or inappropriate material or you could cause a fuel leak or a fire.
6   Always, before detaching or disassembling any part of the fuel system, note the routing of all hoses and pipes and the orientation of all clamps and clips to assure that replacement sections are installed in exactly the same manner.
7   Before detaching any part of the fuel system, be sure to relieve the fuel system pressure (see Section 2). Also cover the fitting being disconnected with a rag to absorb any fuel that may spray or leak out.

5   Fuel tank – removal and installation

Refer to illustrations 5.8, 5.11a and 5.711b
Warning: Gasoline is extremely flammable, so take extra precautions when you work on

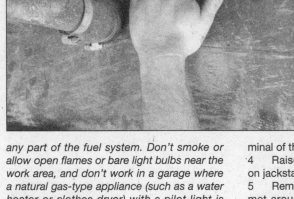

5.11b Slowly lower the fuel tank until there is enough clearance to reach the top of the tank, then disconnect the electrical connector

any part of the fuel system. Don't smoke or allow open flames or bare light bulbs near the work area, and don't work in a garage where a natural gas-type appliance (such as a water heater or clothes dryer) with a pilot light is present. If you spill any fuel on your skin, rinse it off immediately with soap and water. When you perform any kind of work on the fuel system, wear safety glasses and have a Class B type fire extinguisher on hand.
Caution: On models equipped with a theft-deterrent radio, make sure you have the correct activation code, or the theft deterrent system is turned off, before disconnecting the battery.
Note: Don't begin this procedure until the fuel gauge indicates the tank is empty or nearly empty. If the tank must be removed when it's full (for example, if the fuel pump malfunctions), siphon any remaining fuel from the tank prior to removal.
1   Unless the vehicle has been driven far enough to completely empty the tank, it's a good idea to siphon the residual fuel out before removing the tank from the vehicle.
Warning: DO NOT start the siphoning action by mouth! Use a siphoning kit (available at most auto parts stores).
2   Relieve the fuel pressure (see Section 2).
3   Detach the cable from the negative ter-

minal of the battery.
4   Raise the vehicle and place it securely on jackstands.
5   Remove the filler cap, the rubber grommet around the top of the fill pipe and the screw that is under the grommet.
6   Remove the wheel housing liner only if it interferes with fuel tank removal on your particular model.
7   Disconnect the wiring from the fuel tank sensor and the fuel pump unit. Note: It may be necessary to lower the tank slightly on some models to access the components on top of the tank.
8   Disconnect the fill pipe and the vent tube from the fuel tank (see illustration). Remove the fill pipe.
9   On later models with the fuel filter/pressure regulator located near the fuel tank, disconnect the lines from the filter/regulator assembly and slide it out of the way.
10   Support the fuel tank with a block of wood and a floor jack.
11   Remove the two strap retaining bolts and carefully lower the tank (see illustrations).
12   Installation is the reverse of removal.
Caution: Replace the plastic retainers at the fuel filter/pressure regulator assembly. Make certain that all quick-connect fittings are firmly connected.

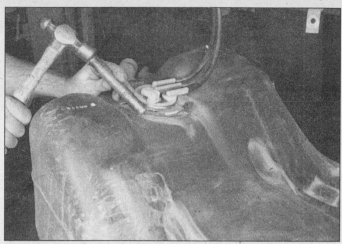

7.4 Use a brass punch and carefully tap the locking ring counter-clockwise until the ring slots are disengaged

7.5 Angle the fuel pump/sending unit assembly to allow the float to be removed without any damage

6 Fuel tank cleaning and repair – general information

1   The fuel tanks installed in the vehicles covered by this manual are made of plastic and are not repairable.

2   If the fuel tank is removed from the vehicle, it should not be placed in an area where sparks or open flames could ignite the fumes coming out of the tank. Be especially careful inside garages where a natural gas-type appliance is located, because the pilot light could cause an explosion.

7 Fuel pump – removal and installation

Refer to illustrations 7.4 and 7.5

Warning: *Gasoline is extremely flammable, so take extra precautions when you work on any part of the fuel system. Don't smoke or allow open flames or bare light bulbs near the work area, and don't work in a garage where a natural gas-type appliance (such as a water heater or clothes dryer) with a pilot light is*
present. If you spill any fuel on your skin, rinse it off immediately with soap and water. When you perform any kind of work on the fuel system, wear safety glasses and have a Class B type fire extinguisher on hand.
Caution: *On models equipped with a theft-deterrent radio, make sure you have the correct activation code, or the theft deterrent system is turned off, before disconnecting the battery.*

Removal

1   Relieve the fuel pressure (see Section 2).

2   Remove the cable from the negative battery terminal.

3   Remove the fuel tank (see Section 5).

4   The fuel pump/fuel level sending unit assembly is located inside the fuel tank. It's held in place by a lock ring. Using a brass punch and hammer, carefully tap the lock ring counterclockwise and separate the cover from the tank **(see illustration). Warning:** *Do not use a steel punch - a spark could cause an explosion.*

5   Pull the fuel pump assembly out of the tank **(see illustration). Caution:** *Be slow and cautious as you manipulate the module out of*
the fuel tank. It will be necessary to angle it in various positions as you extract it.

6   Remove the O-ring from the opening.

7   Inspect the filter on the lower end of the fuel pump. If it's dirty, clean it with solvent and blow it out with compressed air. If it's too dirty to be cleaned, replace it.

8   The fuel pump assembly cannot be disassembled except for removal of the fuel level sending unit (see Section 8). In the event of a defective fuel pump, replace the entire assembly as a single unit or module.

Installation

9   Position a new O-ring around the opening of the fuel tank and guide the fuel pump assembly into the tank.

10   Install and tighten the lock ring by reversing the removal procedure.

11   Install the fuel tank (see Section 5).

8 Fuel level sending unit – check and replacement

Refer to illustrations 8.2, 8.9, 8.11a and 8.11b

Check

1   Raise the vehicle and support it securely on jackstands.

2   Disconnect the electrical connector for the fuel level sending unit **(see illustration)** located in the trunk area behind the carpet.

3   Position the probes of an ohmmeter into the electrical connector and check the resistance. Use the 200-ohm scale on the ohmmeter.

4   With the fuel tank completely full, the resistance should be about 35.5 ohms. With the fuel tank nearly empty, the resistance of the sending unit should be about 236 ohms.

5   If the readings are incorrect, replace the sending unit. **Note:** *A more accurate check of the sending unit can be made by removing it from the fuel tank and checking its resistance while manually operating the float arm.*

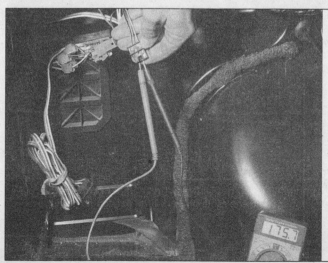

8.2 Remove the trim panel from the left side of the trunk and unplug the electrical connector for the fuel level sending unit

4

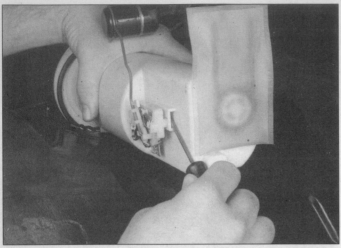

8.9 Use a small screwdriver to lift the plastic tabs and separate the fuel level sending unit from the module on early models

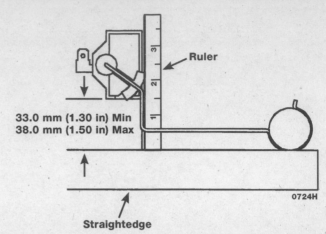

8.11a Use a straightedge or ruler and measure the distance from the bottom of the float to the bottom edge of the sending unit on early models

9.7a Remove the bolts (arrows) that retain the air cleaner assembly (MPFI model shown)

9.7b Remove the bolts (arrows) that retain the air cleaner assembly (TBI model shown)

Replacement

6   Remove the fuel pump/sending unit assembly from the vehicle (see Section 7).

7   Carefully angle the sending unit out of the opening without damaging the fuel level float located at the bottom of the assembly.

8   Remove the screws that secure the electrical connectors to the fuel pump module.

9   Use a small screwdriver and lift the tabs that retain the sending unit to the fuel pump module (**see illustration**). On 1997 through 1999 vehicles, the sending unit can be removed by simply sliding it upwards.

10   Connect the new sending unit to the module, making sure the tabs engage completely.

11   If the float arm and float wire loop were damaged or bent during removal and installation, adjust the angle and travel distance as necessary (**see illustrations**).

12   Install the the fuel pump/sending unit assembly (see Section 7).

9   Air cleaner housing – removal and installation

Refer to illustrations 9.3, 9.7a and 9.7b
Caution: *On models equipped with a theft-deterrent radio, make sure you have the correct activation code, or the theft deterrent system is turned off, before disconnecting the battery.*

1   Disconnect the negative battery cable. Position it so that it cannot contact the positive cable.

2   On early models where the cruise control assembly interferes, disconnect and remove it along with its bracket.

3   Disconnect the fresh air hose from the valve cover if it interferes.

4   Disconnect the wiring from the IAT sensor, if so equipped.

5   Remove the air intake duct from the throttle body and the filter housing.

6   Remove the air filter (see Chapter 1) and the plastic resonator (if so equipped).

7   Remove the bolts from the housing (**see illustrations**) and lift it from the engine compartment.

10   Accelerator cable – removal, installation and adjustment

Removal

Refer to illustrations 10.2, 10.3 and 10.4

1   Remove the air cleaner assembly on TBI models (see Section 9) or the air intake duct on MPFI models from the throttle body. On later models, disconnect the PCV valve and pull it out of the way.

2   Detach the accelerator cable from the throttle lever (**see illustration**).

3   On MPFI models, using a pair of needle-nose pliers, pinch the tabs on the cable housing (**see illustration**) and push the assembly through the hole in the bracket.

4   On TBI models, remove the nuts from the accelerator cable bracket assembly (**see illustration**).

10.2 Rotate the throttle lever and remove the cable end from the slotted portion of the lever

10.3 Use a pair of needle-nose pliers and squeeze the plastic tabs while pushing the cable assembly through the bracket opening

10.4 On SOHC models, remove the two nuts from the bracket and lift the cable assembly from the engine

5   Working under the dash, pull the cable end back and pass it through the slot in the end of the pedal arm.

6   Still working under the dash, squeeze the tabs on the cable housing and push the cable through the firewall into the engine compartment.

7   On later models, remove the coolant expansion tank mounting screws, then lift it up and disconnect the low coolant sensor wiring connector. Move the windshield washer bottle aside.

8   On later models, disconnect the cable from the top of the suspension strut bolt.

Installation

9   Installation is the reverse of removal. Be sure the cable is routed correctly, and be sure the cable housing is securely connected on each end.

10   If necessary, at the engine compartment side of the firewall, apply sealant around the accelerator cable to prevent water and air from entering the passenger compartment.

Adjustment

11   It is not necessary or possible to adjust the accelerator cable. Be sure the accelerator cable is the correct length and part number when purchasing it from the parts department.

11   Fuel Injection system – general information

Two types of fuel injection are used on these vehicles. Throttle Body Injection (TBI) on 1994 and earlier SOHC engines and Multi-Port Fuel Injection (MPFI) on 1995 and later SOHC and all DOHC models. Fuel injection provides optimum mixture ratios at all stages of combustion. Combined with its immediate response characteristics, electronic fuel injection (EFI) provides optimum fuel/air mixture ratios at all stages of combustion. Fuel injection permits the engine to run on the leanest possible air/fuel mixture, which greatly reduces exhaust gas emissions.

The fuel injection system is controlled directly by the vehicle's Powertrain Control Module (PCM), which automatically adjusts the air/fuel mixture in accordance with engine load and performance.

Throttle Body Injection (TBI)

The main component of the TBI system is the Throttle Body Injection (TBI) unit, which is mounted on the intake manifold just like a carburetor. The TBI unit is made up of two major assemblies: the throttle body and the fuel metering assembly.

The throttle body contains a single throttle valve, controlled by the accelerator pedal, similar to a carburetor. Attached to the exterior of the body are the Throttle Position Sensor (TPS), which sends throttle position information to the PCM, and the Idle Air Control (IAC) assembly, which is used by the PCM to maintain a constant idle speed during normal engine operation.

The fuel metering assembly contains the fuel pressure regulator and the single fuel injector. The regulator dampens the pulsations of the fuel pump and maintains a steady pressure at the injector. The fuel injector is controlled by the PCM through an electrically operated solenoid. The amount of fuel injected into the intake manifold is varied by the length of time the injector plunger is held open.

Multi-Port Fuel Injection (MPFI)

Multi-Port Fuel Injection (MPFI) consists of an air intake manifold, the throttle body, the injectors, the fuel rail assembly, an electric fuel pump and attendant plumbing.

Air is drawn through the air cleaner and throttle body. A Mass Air Flow (MAF) sensor mounted between the air cleaner and the throttle body measures the mass (weight) of air passing through the manifold and compensates for temperature and pressure variations.

While the engine is running, the fuel constantly circulates through the fuel rail, which removes vapors and keeps the fuel cool while maintaining sufficient pressure to the injectors under all running conditions.

4

12.7 Use a stethoscope or screwdriver to determine if the injectors are working properly - they should make a steady clicking sound that rises and falls with engine speed changes

12.8 Install the "noid" light into the fuel injector electrical connector and confirm that it blinks when the engine is running

As with TBI, the operation of the MPFI injection system is controlled by the PCM so that it works in conjunction with the rest of the vehicle functions to provide optimum driveability and emissions control.

Because the MPFI system meters fuel and air precisely, it is important to the proper operation of the vehicle that the fuel and air filters be changed at the specified intervals.

Both systems

The PCM controlling both types of fuel injection systems has a learning capability for certain performance conditions. If the battery is disconnected, part of the PCM memory is erased, which makes it necessary to "re-teach" the computer. This is done by thoroughly warming up the engine and operating the vehicle at part throttle, stop and go and idle conditions.

A fuel pump relay is used to control fuel pump operation during start-up. When the ignition is turned on, the fuel pump relay immediately supplies current to the fuel pump to pressurize the system. After two seconds, it will shut off. The fuel pump is kept running during normal engine operation by the PCM as it receives pulses from the ignition system or the oil pressure sensor.

The throttle stop screw, used to regulate the minimum idle speed, is adjusted at the factory and sealed with a plug to discourage unnecessary adjustment.

---

12   Fuel injection system – check

Warning: *Gasoline is extremely flammable, so take extra precautions when you work on any part of the fuel system. Don't smoke or allow open flames or bare light bulbs near the work area, and don't work in a garage where a natural gas-type appliance (such as a water heater or clothes dryer) with a pilot light is present. If you spill any fuel on your skin, rinse*

it off immediately with soap and water. When you perform any kind of work on the fuel system, wear safety glasses and have a Class B type fire extinguisher on hand.
Note: *the following procedure is based on the assumption that the fuel pump is working and the fuel pressure is adequate (see Section 3).*

Preliminary checks

1   Check all electrical connectors that are related to the system. Loose electrical connectors and poor grounds can cause many problems that resemble more serious malfunctions.

2   Check to see that the battery is fully charged, as the control unit and sensors depend on an accurate supply voltage in order to properly meter the fuel.

3   Check the air filter element - a dirty or partially blocked filter will severely impede performance and economy (see Chapter 1).

4   If a blown fuse is found **(see illustration 3.2a)**, replace it and see if it blows again. If it does, search for a grounded wire in the harness to the fuel pump.

Multi-Port Fuel Injection (MPFI) systems

Refer to illustrations 12.7, 12.8 and 12.9

5   Check the air passage from the Manifold Absolute Pressure (MAP) sensor to the intake manifold for leaks, which will result in an excessively lean mixture (see Chapter 6). Also check the condition of the vacuum hoses connected to the intake manifold.

6   Remove the air intake duct from the throttle body and check for dirt, carbon or other residue build-up in the throttle body, particularly around the throttle plate. If it's dirty, clean it with carburetor cleaner and a toothbrush.

7   With the engine running, place a stethoscope against each injector, one at a time,

and listen for a clicking sound, indicating operation **(see illustration)**. **Note:** *If you don't have a stethoscope, place the tip of a screwdriver against the injector and listen through the handle.*

8   If an injector isn't functioning (not clicking), purchase a special injector test light (sometimes called a "noid" light) and install it into the injector electrical connector **(see illustration)**. Start the engine and check to see if the noid light flashes. If it does, the injector is receiving proper voltage. If it doesn't flash, further diagnosis should be performed by a dealer service department or other repair shop.

9   With the engine OFF and the fuel injector electrical connectors disconnected, measure the resistance of each injector **(see illustration)**. Check the Specifications listed in the front of this Chapter for the correct resistance values.

10   The remainder of the system checks can be found in Section 17 and Chapter 6.

Throttle Body Injection (TBI) systems

Refer to illustration 12.13

11   Set the parking brake, remove the air cleaner top plate and, with the engine idling in Park, observe the operating fuel injector. The spray pattern should be even and conical in shape. The spray should touch the throttle body bore.

   a) If the spray is weak or uneven, the injector is clogged or faulty. Gasoline additives designed to clean fuel injectors can sometimes clear a clogged injector. If not, a dealer service department or other qualified shop has more effective cleaning equipment.

   b) If the injector is not operating at all, check its electrical connector. If the connection is good and the injector is receiving voltage, but the injector still doesn't work, the injector is faulty.

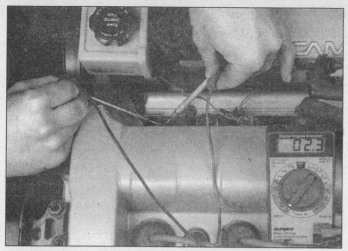

12.9 Measure the resistance of each injector - and compare it to this Chapter's Specifications (MPFI models)

12.13 Measure the resistance of the injector at the top of the TBI unit - and compare it to this Chapter's Specifications (TBI models)

12   Turn the engine off and observe the injector. There shouldn't be any leakage or dripping. If the injector does drip, either the injector seals are faulty or the injector itself is defective. Usually a problem like this will result in a hard-starting condition and/or a puff of smoke as the engine is started.

13   With the engine OFF and the fuel injector electrical connector disconnected, measure the resistance of the injector **(see illustration)**. Check the Specifications listed in the front of this Chapter for the resistance values.

13   Throttle body assembly and minimum idle speed adjustment – check, removal and installation

Caution: *On models equipped with a theft-deterrent radio, make sure you have the correct activation code, or the theft deterrent system is turned off, before disconnecting the battery.*

Check

Refer to illustration 13.2

1   Remove the air cleaner assembly on TBI models (see Section 9) or the air intake duct on MPFI models from the throttle body.

2   Have an assistant depress the throttle pedal while you watch the throttle valve. Check that the throttle valve moves smoothly when the throttle is moved from closed (idle position) to fully open (wide open throttle). **Note:** *Spray carburetor cleaner into the throttle body, especially around the shaft area, to free-up any binding caused by the accumulation of carbon deposits or sludge buildup* **(see illustration).**

3   Wiggle the throttle lever while watching the throttle shaft inside the bore. If it appears worn (loose), replace the throttle body unit.

Removal and installation

Refer to illustration 13.10

4   Disconnect the cable from the negative terminal of the battery.

5   Detach the air cleaner assembly on

SOHC models or the air intake duct on MPFI models (see Section 9).

6   Unplug the Idle Air Control (IAC) valve, the Throttle Position Sensor (TPS) electrical connectors and on TBI models, the fuel injector electrical connector.

7   Mark and disconnect any vacuum hoses connected to the throttle body.

8   Disconnect the accelerator cable from the throttle lever, then detach the cable housing from its bracket on MPFI models or detach the cable with the bracket assembly on TBI models (see Section 10).

9   Remove the breather hose.

10   Remove the throttle body bolts and detach the throttle body **(see illustration)**. Clean off all traces of old gasket material.

11   Install the throttle body and a new gasket and tighten the bolts to the torque listed in this Chapter's Specifications.

12   Installation is the reverse of removal.

Minimum idle speed adjustment

Note: *The minimum idle speed is pre-set at*

4

13.2 Clean the throttle body with carburetor cleaner to remove sludge deposits

13.10 Remove the throttle body mounting bolts (arrows) (MPFI model shown)

14.5 Pry the injector out of the fuel meter body using one screwdriver as a fulcrum and another as a lever

14.18 The IAC valve pintle must not extend more than 1-1/8 inch - also, replace the O-ring if it is brittle

A   *Distance of pintle extension*
C   *Gasket/O-ring*
B   *Pintle*

the factory and should not require adjustment under normal operating conditions; however if the throttle body has been replaced or you suspect the minimum idle speed has been tampered with (for example, if the idle stop screw was turned), use the following procedure to check the minimum idle speed.

13   Make sure the IAC valve pintle is extended. With the engine warmed up to operating temperature, turn the ignition OFF and remove the IAC valve from the throttle body (see Section 14 for TBI models or Section 15 for MPFI models).

14   Observe that the valve is completely extended and not in the seated position. Install the IAC valve back into the throttle body and do NOT connect the electrical connector to the IAC valve.

15   Remove the idle stop screw plug from the throttle body by piercing it with a sharp punch or awl.

16   Insert the IAC air plug into the throttle body.

17   Adjust the idle stop screw to obtain the specified idle speed in Park. The idle speed should be approximately 500 to 600 rpm. **Note:** *The special idle air plug and handle will remain in the throttle body while the engine is running. Do not assemble the air cleaner housing (TBI models) or air intake duct (MPFI models) until the procedure is completed.*

18   Turn the ignition off and reconnect the IAC valve electrical connector.

19   Remove the special tool, unplug any plugged vacuum ports and reconnect the hoses.

20   Install the air cleaner housing.

21   Disconnect the cable from the negative terminal of the battery for at least ten seconds. This will erase any stored trouble codes that may have been set by unplugging the IAC valve and running the engine.

---

14   Throttle Body Injection (TBI) unit - component check and replacement

Warning: *Gasoline is extremely flammable,*

so take extra precautions when you work on any part of the fuel system. Do not smoke or allow open flames or bare light bulbs near the work area, and don't work in a garage where a natural gas-type appliance (such as a water heater or clothes dryer) with a pilot light is present. If you spill any fuel on your skin, rinse it off immediately with soap and water. When you perform any kind of work on the fuel system, wear safety glasses and have a Class B type fire extinguisher on hand. **Caution:** *On models equipped with a theft-deterrent radio, make sure you have the correct activation code, or the theft deterrent system is turned off, before disconnecting the battery.*

Fuel injector

Check

1   Refer to Section 12 for the fuel injector checking procedure. Also check for stored trouble codes in the PCM (see Chapter 6).

Replacement

Refer to illustration 14.5

2   Disconnect the negative battery cable.

3   Unplug the electrical connector from the fuel injector.

4   Remove the injector retainer screw and the retainer

5   Using one screwdriver as a fulcrum on the fuel meter body, place another screwdriver tip under the ridge on the fuel injector opposite the electrical connector end and gently pry the injector out **(see illustration)**.

6   If the injector is to be reused, replace the upper and lower O-rings on the injector and in the fuel injector cavity. Install the upper O-ring in the groove on the injector and the lower O-ring flush against the filter element.

7   Install the injector assembly in the fuel meter body by pushing it straight down. Make sure the connector end is facing in the direction of the opening in the fuel meter body for the wire harness grommet.

8   Install the injector retainer and screw. Use a thread locking compound on the retainer screw.

9   Reconnect the negative battery cable. Pressurize the fuel system by turning the ignition key to the On position, then inspect the area around the injector for leaks.

10   Plug the electrical connector into the injector and start the engine to check for correct operation.

Pressure regulator assembly

Check

11   Refer to Section 3 and perform the fuel pressure checks, which will diagnose a malfunctioning fuel pressure regulator.

Replacement

12   Underneath the pressure regulator cover assembly is a large spring which is highly compressed. Repairs to this component should be performed by a dealer service department or other repair shop due to the possibility of personal injury. Also, the tension on this spring affects fuel pressure and is set at the factory - any tampering with this component would be in violation of Federal law.

Idle Air Control valve

Check

13   Refer to Chapter 6 and check for trouble codes stored in the PCM. If the IAC valve is malfunctioning, a trouble code indicating this condition would most likely have been set.

Replacement

Refer to illustration 14.18

14   Disconnect the negative battery cable.

15   Remove the air cleaner and unplug the electrical connector from the IAC valve.

16   Remove the two valve retaining screws and pull the valve out of the throttle body.

17   If the same valve is to be reinstalled, be sure to use a new O-ring.

18   Before installing the valve, measure the distance from the end of the pintle to the mounting flange **(see illustration)**. If the distance exceeds 1-1/8 inch, the IAC valve must be retracted by applying firm hand pressure to the pintle. A slight side-to-side motion may help.

14.30 The TPS mounts to the side of the throttle body and is not adjustable

15.2 Working on the harness side of the IAC electrical connector, probe the green/white wire and the green/black wire with a voltmeter and with the ignition key ON, confirm that the voltage fluctuates between 1.5 and 10.5 volts

19 Position the valve on the throttle body and install the screws. Plug in the electrical connector to the valve.

20 No adjustment of the IAC valve is necessary, as it is automatically reset by the PCM.

Minimum idle speed adjustment

21 Refer to Section 13, Steps 13 through 21 for this adjustment.

Throttle Position Sensor (TPS)

General information and check

22 The Throttle Position Sensor (TPS) is connected to the throttle shaft on the TBI unit. As the throttle valve angle is changed (as the accelerator pedal is moved), the output of the TPS also changes. At a closed throttle position, the output of the TPS is below 1.25-volts. As the throttle valve opens, the output increases so that, at wide-open throttle, the output voltage is approximately 5-volts.

23 A broken or loose TPS can cause intermittent bursts of fuel from the injector and an unstable idle, because the PCM thinks the throttle is moving. If a problem with the TPS sensor or circuit develops, a trouble code most likely will be set (see Chapter 6).

24 Connect a digital voltmeter from the TPS electrical connector terminal "C" (dark blue wire) to terminal "B" (black wire) (you'll have to fabricate jumper wires for terminal access).

25 With the ignition on and the engine off, TPS voltage should be less than 1.25 volts. If it's more than specified, check and, if necessary, adjust the minimum idle speed before condemning the TPS (see Section 13).

26 The TPS is not adjustable. If the TPS malfunctions, it must be replaced as a unit.

Replacement

Refer to illustration 14.30

27 Disconnect the cable from the negative battery terminal.

28 Remove the air cleaner housing.

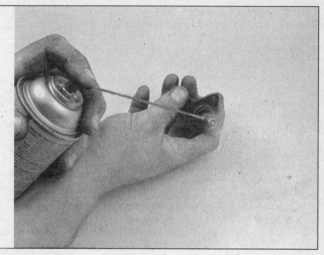

15.3 Clean the IAC valve pintle with carburetor cleaner to remove carbon deposits

29 Unplug the electrical connector from the throttle position sensor.

30 Remove the two sensor mounting screws and pull the sensor from the throttle body **(see illustration)**.

31 To install the TPS, align the slot in the rear of the sensor with the throttle shaft and insert the sensor into the throttle body. Install the mounting screws and tighten them securely. This style TPS is not adjustable.

32 The remainder of installation is the reverse of the removal procedure.

---

15 Idle Air Control (IAC) valve (MPFI models) – check, removal and installation

---

Refer to illustrations 15.2, 15.3 and 15.6

Note: *The minimum idle speed is pre-set at the factory and should not require adjustment under normal operating conditions; however if the throttle body has been replaced or you suspect the minimum idle speed has been tampered with (for example, if the idle stop screw was removed) follow the procedure in Section 13.*

Check

1 The idle air control valve (IAC) controls the engine idle speed. This output actuator is mounted on the throttle body and is controlled by voltage pulses sent from the PCM (computer). The IAC valve pintle moves in or out allowing more or less intake air into the system according to the engine conditions. To increase idle speed, the PCM retracts the IAC valve pintle away from the seat and allows more air to bypass the throttle bore. To decrease idle speed, the PCM extends the IAC valve pintle towards the seat, reducing the air flow.

2 To check the system, first check for the voltage signal from the PCM. Turn the ignition key On (engine not running) and with a voltmeter, probe the green/white and green/black wires of the terminals of the IAC valve electrical connector. It should fluctuate between 1.5 volts and 10.5 volts **(see illustration)**. This indicates that the IAC valve is receiving the proper signal from the PCM.

3 Next, remove the valve (see Step 4) and inspect it. Check the pintle for excessive carbon deposits. If necessary, clean it with carburetor cleaner spray. Also clean the IAC

4

15.6 With the throttle body removed from the engine, remove the two screws (lower arrows) from the IAC valve (the upper arrows point to the TPS screws)

16.6a Carefully watch the fuel pressure gauge with vacuum applied. The fuel pressure should remain steady (refer to the Specifications for the correct fuel pressure)

16.6b Now release the vacuum and confirm that fuel pressure increases as vacuum decreases

valve housing to remove any deposits **(see illustration)**.

Removal

4   Unplug the electrical connector from the IAC valve.

5   Remove the throttle body if necessary (see Section 13).

6   Remove the two IAC valve attaching screws and withdraw the assembly **(see illustration)**.

7   Check the condition of the rubber O-ring. If it's hardened or deteriorated, replace it.

8   Clean the sealing surface and the bore of the idle air/vacuum signal housing assembly to ensure a good seal. **Caution:** *The IAC valve itself is an electrical component and must not be soaked in any liquid cleaner, as damage may result.*

9   Before installing the ISC valve, the position of the pintle must be checked. If the pintle is extended too far, damage to the assembly may occur (see Section 14, Step 18).

Installation

10   Position the new O-ring on the IAC valve. Lubricate the O-ring with a light film of engine oil.

11   Install the IAC valve and tighten the screws securely.

12   Plug in the electrical connector at the IAC valve assembly.

16   Fuel pressure regulator (MPFI models) – check and replacement

Warning: *Gasoline is extremely flammable, so take extra precautions when you work on any part of the fuel system. Don't smoke or allow open flames or bare light bulbs near the work area, and don't work in a garage where a natural gas-type appliance (such as a water*

heater or clothes dryer) with a pilot light is present. If you spill any fuel on your skin, rinse it off immediately with soap and water. When you perform any kind of work on the fuel system, wear safety glasses and have a Class B type fire extinguisher on hand.*
Caution: *On models equipped with a theft-deterrent radio, make sure you have the correct activation code, or the theft deterrent system is turned off, before disconnecting the battery.*

Check

Refer to illustrations 16.6a, 16.6b and 16.10
Note: *This procedure assumes the fuel filter is in good condition.*

1   Relieve the fuel system pressure (see Section 2).

2   Detach the cable from the negative battery terminal.

3   Connect a fuel pressure gauge to the fuel rail (see Section 3). Reconnect the battery cable.

4   Start the engine and check for leakage around the gauge connection.

5   Disconnect the vacuum hose from the fuel pressure regulator and hook up a hand-

held vacuum pump to the port on the fuel pressure regulator.

6   Read the fuel pressure gauge with vacuum applied to the pressure regulator and also with no vacuum applied **(see illustrations)**. The fuel pressure should decrease as vacuum increases (and increase as vacuum decreases). Compare your readings with the values listed in this Chapter's Specifications.

7   Reconnect the vacuum hose to the regulator and check the fuel pressure at idle, comparing your reading with the value listed in this Chapter's Specifications. Disconnect the vacuum hose and watch the gauge – the pressure should jump up considerably as soon as the hose is disconnected. If it doesn't, proceed to Step 10.

8   If the fuel pressure is low, pinch the fuel return line shut and watch the gauge. If the pressure doesn't rise, the fuel pump is defective or there is a restriction in the fuel feed line. If the pressure rises sharply, replace the pressure regulator.

9   If the fuel pressure is too high, turn the engine off. Disconnect the fuel return line and blow through it to check for a blockage. If there is no blockage, replace the fuel pres-

16.10 Make sure there is vacuum to the fuel pressure regulator hose at idle

16.14a First, remove the bolt (arrow) that retains the return line to the intake manifold . . .

16.14b . . . then use an open end wrench to remove the fuel return line from the fuel pressure regulator (fuel feed line removed for clarity)

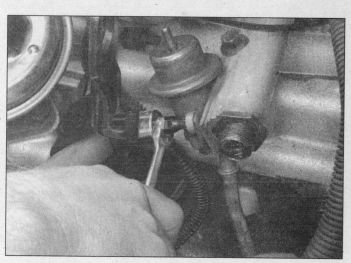

16.15 Use a Torx driver to remove the bolt from the fuel pressure regulator

sure regulator.

10   Connect a vacuum gauge to the vacuum hose to the pressure regulator. Start the engine and check for vacuum (**see illustration**). If there isn't vacuum present, check for a clogged hose or vacuum port. If the amount of vacuum is adequate, replace the fuel pressure regulator.

Replacement

Note: *Early models through 1997 use a fuel pressure regulator which is mounted on the fuel rail in the engine compartment. 1998 through 1999 models use a combination fuel filter/ pressure regulator assembly which is mounted under the vehicle near the fuel tank. Refer to illustrations 16.14a, 16.14b and 16.15*

11   Relieve the fuel pressure from the system (see Section 2). Disconnect the cable from the negative terminal of the battery.

12   Clean any dirt from around the fuel pressure regulator.

Vehicles through 1997

13   Detach the vacuum hose from the fuel pressure regulator.

14   Detach the fuel return line from the pressure regulator (**see illustrations**).

15   Remove the bolt that retains the fuel pressure regulator (**see illustration**). Pull the regulator from the fuel rail.

16   Install new O-rings and lubricate them with a light coat of oil.

17   Installation is the reverse of removal. Tighten the pressure regulator mounting bolt and the fuel return line securely.

1998 through 1999 vehicles

18   Raise the vehicle and support it securely on jackstands.

19   Unbolt the fuel filter/pressure regulator assembly from beneath the vehicle.

20   Unclip the fuel line bundle form its retainer. Take care to avoid damaging the retainer.

21   Disconnect the lines from the unit as

you slide it from its bracket. Note: The plastic fuel line retainers must be replaced upon reassembly.

22   Installation is the reverse of removal.

---

17   Fuel rail and injectors (MPFI models) – removal and installation

---

Warning: *Gasoline is extremely flammable, so take extra precautions when you work on any part of the fuel system. Don't smoke or allow open flames or bare light bulbs near the work area, and don't work in a garage where a natural gas-type appliance (such as a water heater or clothes dryer) with a pilot light is present. If you spill any fuel on your skin, rinse it off immediately with soap and water. When you perform any kind of work on the fuel system, wear safety glasses and have a Class B type fire extinguisher on hand.*

Caution: *On models equipped with a theft-*

4

17.8 Remove the bolts (arrows) that hold the fuel rail to the intake manifold

17.9 Carefully pry the fuel rail away from the intake manifold using a prybar or large screwdriver

17.10 The injector is retained in the fuel rail by a clip - slide the retaining clip along the rail to release the injector

17.11a Be sure to replace the O-rings with new ones

deterrent radio, make sure you have the correct activation code, or the theft deterrent system is turned off, before disconnecting the battery.

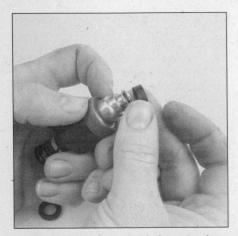

17.11b Be sure to check the ceramic collar on the injector for wear or cracks before replacing the O-ring

Fuel rail and related components

Refer to illustrations 17.8 and 17.9

Caution: *To prevent dirt from entering the engine, the area around the injectors should be cleaned before servicing.*

1   Relieve the fuel system pressure (see Section 2).

2   Detach the negative battery cable from the battery.

3   Remove the air intake and fresh air tubes if they interfere.

4   Disconnect the fuel supply and return lines from the fuel rail or the pressure regulator. Replace the plastic retainers upon reassembly. Disconnect the vacuum line from the regulator, if so equipped.

5   Remove the PCV hose if it interferes.

6   Disconnect the throttle cable. Remove the throttle cable bracket and move it aside.

7   Label and then disconnect the injector electrical connectors.

8   Remove the fuel rail bolts **(see illustration)**.

9   Pull the fuel rail upwards to remove the

injectors from their bores (see illustration). Carefully manipulate the assembly out of the engine compartment. Caution: Use care when handling the fuel rail assembly to avoid damaging the injectors.

Fuel injectors

Refer to illustrations 17.10, 17.11a and 17.11b

8   To remove the fuel injectors, first, slide the injector retaining clip from the fuel rail **(see illustration)** and then carefully wiggle the end of the injector from the housing and remove it from the fuel rail.

9   Inspect the injector O-ring seals. These should be replaced whenever the fuel rail is removed. Also inspect the ceramic collars on the injectors **(see illustrations)**.

10   Install the new O-ring(s) on the injector(s) and lubricate them with a light film of clean engine oil.

11   Install the injectors on the fuel rail.

12   Secure the injectors with the retaining clips.

13   Installation is the reverse of removal.

18 Exhaust system servicing – general information

Warning: *Inspection and repair of exhaust system components should be done only after enough time has elapsed after driving the vehicle to allow the system components to cool completely. Also, when working under the vehicle, make sure it is securely supported on jackstands.*

1   The exhaust system consists of the exhaust manifold, the catalytic converter, the muffler, the tailpipe and all connecting pipes, brackets, hangers and clamps. The exhaust system is attached to the body with mounting brackets and rubber hangers. If any of the parts are improperly installed, excessive noise and vibration will be transmitted to the body.

2   Conduct regular inspections of the exhaust system to keep it safe and quiet.

Look for any damaged or bent parts, open seams, holes, loose connections, excessive corrosion or other defects which could allow exhaust fumes to enter the vehicle. Deteriorated exhaust system components should not be repaired; they should be replaced with new parts.

3   If the exhaust system components are extremely corroded or rusted together, welding equipment will probably be required to remove them. The convenient way to accomplish this is to have a muffler repair shop remove the corroded sections with a cutting torch. If, however, you want to save money by doing it yourself (and you don't have a welding outfit with a cutting torch), simply cut off the old components with a hacksaw. If you have compressed air, special pneumatic cutting chisels can also be used. If you do decide to tackle the job at home, be sure to wear safety goggles to protect your eyes from metal chips and work gloves to protect your hands.

4   Here are some simple guidelines to follow when repairing the exhaust system:

a) *Work from the back to the front when removing exhaust system components.*

b) *Apply penetrating oil to the exhaust system component fasteners to make them easier to remove.*

c) *Use new gaskets, hangers and clamps when installing exhaust systems components.*

d) *Apply anti-seize compound to the threads of all exhaust system fasteners during reassembly.*

e) *Be sure to allow sufficient clearance between newly installed parts and all points on the underbody to avoid overheating the floor pan and possibly damaging the interior carpet and insulation. Pay particularly close attention to the catalytic converter and heat shield.*

4

Notes

Chapter 5
Engine electrical systems

Contents

Specifications

Coil

Primary resistance	700 to 900 ohms
Secondary resistance	
1991 through 1996 vehicles	7,000 to 10,000 ohms
1997 through 1999 vehicles	8,000 to 15,000 ohms

1   General information

Warning: *Because of the very high voltage generated by the ignition system, extreme care should be taken whenever an operation involving ignition components is performed. This not only includes the coil(s), module and spark plug wires, but related items that are* connected to the systems as well, such as plug connections, tachometer and testing equipment.

Ignition systems

The engines covered by this manual are equipped with a Direct (distributorless) Ignition System (DIS). It is a "waste-spark" method of spark distribution. Each cylinder is paired with an opposing cylinder in the firing order (1-4 and 2-3) so that one cylinder on the compression fires simultaneously with its opposing cylinder on the exhaust stroke. Since the cylinder on exhaust requires very little of the available voltage to fire its plug, most of the voltage is used to fire the cylinder on compression. The DIS system includes a coil pack, an ignition module, crankshaft sensor and the PCM.

2.1 Using a box-end wrench, first remove the negative cable from the battery, followed by the positive battery cable

2.2 Remove the lower bracket bolt (left arrow) and then the upper bracket bolt (right arrow) from the protective cover

Charging system

The charging system consists of a belt-driven alternator with an internal voltage regulator and the battery. These components work together to supply electrical power for the ignition system, the lights and all accessories.

All Saturn models are equipped with sealed alternator units. In the event of voltage regulator or alternator brush failure, it will be necessary to replace the alternator as a complete unit. All types use a conventional pulley and fan.

Starting system

The starting circuit consists of the battery, starter motor, ignition switch and other related electrical wiring. The starter motor is equipped with a solenoid mounted directly to the assembly.

The starter/solenoid assembly is a non-serviceable component. In the event of failure, replace the assembly as a single unit with a new or rebuilt part from a dealer parts department or an auto parts store.

2   Battery - removal and installation

Refer to illustrations 2.1 and 2.2
Warning: *Hydrogen gas is produced by the battery, so keep open flames and lighted cigarettes away from it at all times. Always wear eye protection when working around the battery. Rinse off spilled electrolyte immediately with large amounts of water.*
Caution: *On models equipped with a theft-deterrent radio, make sure you have the correct activation code, or the theft deterrent system is turned off, before disconnecting the battery.*

1   Disconnect the cables from the negative and positive terminals of the battery **(see illustration)**. **Caution:** *To prevent arcing, dis-connect the negative (-) cable first, then remove the positive cable (+).*
2   Remove the upper and lower bolts from the battery cover, if so equipped **(see illustration)**.
3   Remove the bolts or nuts which secure the battery brace or hold-down bracket. Remove the brace or bracket.
4   Remove the battery cover from the engine compartment.
5   Lift out the battery. Be careful - it's heavy.
6   While the battery is out, inspect the carrier (tray) for corrosion.
7   If you are replacing the battery, make sure that you get one that's identical, with the same dimensions, amperage rating, cold cranking rating, etc.
8   Installation is the reverse of removal.

3   Battery - emergency jump starting

Refer to the *Booster battery (jump) starting* procedure at the front of this manual.

4   Battery cables - check and replacement

1   Periodically inspect the entire length of each battery cable for damage, cracked or burned insulation and corrosion. Poor battery cable connections can cause starting problems and decreased engine performance.
2   Check the cable-to-terminal connections at the ends of the cables for cracks, loose wire strands and corrosion. The presence of white, fluffy deposits under the insulation at the cable terminal connection is a sign that the cable is corroded and should be replaced. Check the terminals for distortion, missing mounting bolts and corrosion.

3   When removing the cables, always disconnect the negative cable first and hook it up last or the battery may be shorted by the tool used to loosen the cable clamps. Even if only the positive cable is being replaced, be sure to disconnect the negative cable from the battery first (see Chapter 1 for further information regarding battery cable removal).
4   Disconnect the old cables from the battery, then trace each of them to their opposite ends and detach them from the starter solenoid and ground terminals. Note the routing of each cable to ensure correct installation.
5   If you are replacing either or both of the old cables, take them with you when buying new cables. It is vitally important that you replace the cables with identical parts. Cables have characteristics that make them easy to identify: positive cables are usually red, larger in cross-section and have a larger diameter battery post clamp; ground cables are usually black, smaller in cross-section and have a slightly smaller diameter clamp for the negative post.
6   Clean the threads of the solenoid or ground connection with a wire brush to remove rust and corrosion. Apply a light coat of battery terminal corrosion inhibitor, or petroleum jelly, to the threads to prevent future corrosion.
7   Attach the cable to the solenoid or ground connection and tighten the mounting bolt securely.
8   Before connecting a new cable to the battery, make sure that it reaches the battery post without having to be stretched.
9   Connect the positive cable first, followed by the negative cable.

5   Ignition system - general information

The distributorless ignition system (DIS)

6.3 To use a calibrated ignition tester, simply disconnect a spark plug wire, clip the tester to a convenient ground (like a valve cover bolt) and operate the starter - if there is enough power to fire the plug, sparks will be visible between the electrode tip and the tester body

6.4 Measure the resistance of each plug wire. It should be less than 12,000 ohms

is controlled by the PCM (Powertrain Control Module). This system offers no moving parts, less maintenance, more coil cool-down between spark plug firing and the elimination of the mechanical timing adjustments.

The DIS ignition systems use a "waste spark" method of spark distribution. Each cylinder is paired with its opposing cylinder in the firing order (1-4, 2-3) so one cylinder under compression fires simultaneously with its opposing cylinder, where the piston is on the exhaust stroke. Since the cylinder on exhaust requires very little of the available voltage to fire its plug, most of the voltage is used to fire the cylinder on compression.

The DIS system includes coil packs, an ignition module, a crankshaft position sensor, spark plug wires, and the PCM (Powertrain Control Module).

Conventional ignition coils have one end of the secondary winding connected to the engine ground. On DIS, neither end of the secondary winding is grounded; instead, each end of the coil's secondary winding is directly attached to the spark plug and its companion.

The crankshaft sensor (CKP) is mounted at the rear of the engine block near the knock sensor (see Chapter 6). This sensor will disrupt the signal voltage causing an ON-OFF-ON-OFF etc. reaction much the same as the opening and closing of the ignition points on the older ignition systems. There are several different systems contained within the DIS system and here is a list of the systems particular to the ignition module:

1) The BYPASS system signals whether the system is in ignition control (IC) mode (5-volt signal) or Bypass mode (0 volts).
2) The reference (REF) output signal is usually approximately 5 volts and it is used by the PCM to determine crankshaft position and engine rpm.
3) The 6X output signal is a 5-volt sine wave that initially starts high but

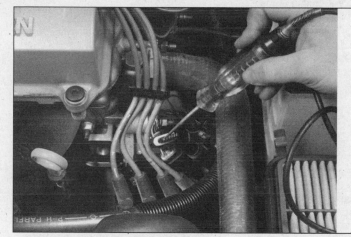

6.5 It is better to use a voltmeter, but a test light is adequate to check for battery voltage at the pink/black wire on the 6-pin terminal

switches low at each of the 60-degree marks of crankshaft revolution.
4) The Electronic Spark Timing (EST) system uses a five-volt square signal wave that is supplied by the PCM. This ignition control signal defines the desired spark timing.

6 Ignition system - check

Warning: *Because of the very high voltage generated by the ignition system, extreme care should be taken whenever an operation is performed involving ignition components. This not only includes the coils, control module and spark plug wires, but related items connected to the system as well, such as the plug connections, tachometer and any test equipment.*
Note: *The following test procedures apply to 1995 and earlier vehicles. They can be performed on 1996 and later vehicles, however the OBD-II system used on these later models requires that a dealership or other properly equipped service facility deal with the OBD-II system and any trouble codes it may generate (refer to Chapter 6 if necessary).*

General checks

Refer to illustration 6.3 and 6.4

1   With the ignition switch turned to the "ON" position, the "SERVICE ENGINE SOON" light should be on. This means there's ignition and battery supply to the PCM.
2   Check all ignition wiring connections for tightness, cuts, corrosion or any other signs of a bad connection. A faulty or poor connection at a spark plug could also result in a misfire. Also check for carbon deposits inside the spark plug boots.
3   Use a calibrated ignition tester to verify adequate available secondary voltage (25,000 volts) at each spark plug **(see illustration)**.
4   Using an ohmmeter, check the resistance of the spark plug wires. Each wire should measure approximately 12,000 ohms or less **(see illustration)**. If the resistance is over 12,000 ohms, replace the wire(s).

If there's no spark at any wire

Refer to illustration 6.5 and 6.7

5   The following tests are necessary if there is no spark. Disconnect the 6-pin connector from the ignition module, and, with the

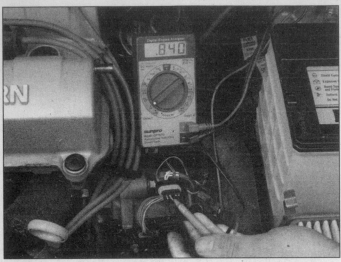

6.7 Check the resistance of the crankshaft sensor by probing the yellow and purple wires on the 5-pin connector. It should be 700 to 900 ohms

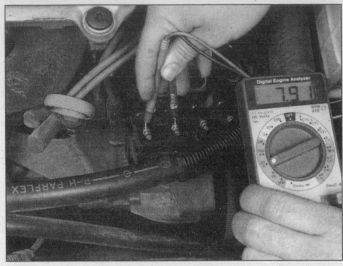

7.2 Measure the secondary resistance of each coil pack

ignition key ON (engine not running), measure the voltage on the pink/black wire **(see illustration)**. There should be battery voltage present. If there is no voltage available, check the electronic ignition fuse (7.5 amp) in that circuit for an open or short.

6   If battery voltage is available, disconnect the 5-pin connector from the ignition module and, using an ohmmeter, check the resistance of the ground wire (black/white wire). The resistance should be 200 ohms. If not, check for an open or shorted circuit.

7   Disconnect the 5-pin connector from the ignition module and, working on the harness side, measure the resistance between the yellow and the purple wire **(see illustration)**. The resistance should be 700 to 900 ohms. If the resistance is incorrect, remove the crankshaft sensor (see Chapter 6) and measure the resistance. It should be 700 to 900 ohms. If not, replace it with a new part.

8   If the resistance is as specified, check the voltage signal. Set a digital voltmeter on the A/C (alternating current) scale and, while cranking the engine, measure the voltage between the yellow and the purple wire. It should be over 200 millivolts. If it is not, replace the crankshaft sensor (see Chapter 6).

9   If all the tests are correct, check the PCM and all the connector pins for bends, breaks or corrosion that might cause intermittent shorts or grounds.

If there's a misfire at idle or under load

10   If the engine misfires at idle, continue to next step.

11   **Test #2:** Use a spark tester to verify adequate available secondary voltage (25,000 volts) at the spark plug.

12   **Test #3:** If the spark jumps the test gap after grounding the opposite plug wire, it indicates excessive resistance in the plug which

was bypassed. A faulty or poor connection at that plug could also result in a miss condition. Also check for carbon deposits inside the spark plug boot.

13   **Test #4:** If carbon tracking is evident, replace the coil and be sure that the plug wires relating to that coil are clean and tight. Excessive wire resistance or faulty connections could cause damage to the coil.

14   **Test #5:** If a no spark condition disappears when the coil is switched with another coil, the original coil is faulty. If not, the ignition module is the cause of the no spark condition. This test can also be performed by substituting a known good coil for the one causing the no spark condition.

15   **Test #6:** If the engine misfires under a load;use a spark tester to verify adequate available secondary voltage (25,000 volts) at the spark plug. The spark should jump the test gap on all four leads. This simulates a "load" condition.

16   **Test #7:** If the spark jumps the tester gap after grounding the opposite plug wire, it indicates excessive resistance in the plug which was bypassed. A faulty or poor connection at that plug could also result in the miss condition. Also check for carbon deposits inside the spark plug boot.

17   **Test #8:** If carbon tracking is evident, replace the coil and be sure that the plug wires attached to that coil are clean and tight. Excessive wire resistance or faulty connections could damage the coil.

18   **Test #9:** If the no spark condition vanishes when the suspected coil is replaced by one of the other coils, that coil is faulty. If not, the ignition module is the reason there is no spark. This test could also be performed by substituting a known good coil for the one causing the no spark condition.

If the engine cranks but won't run

19   With the ignition ON, if the "SERVICE

ENGINE SOON" light is blinking, check for ignition and battery supply to the PCM.

20   Install an injector test light (sometimes called a "noid" light) (available at most auto parts stores) and check to see if the PCM is controlling each fuel injector. A blinking test while cranking the engine indicates that the PCM is controlling the injector and that ignition reference signal to the PCM is good.

21   Install the calibrated ignition tester and check to see if the problem is fuel or ignition related **(see illustration 6.3)**.

22   Check to see if the fuel pump and relay are operating correctly (see Chapter 4). The fuel pump should run for only two seconds after the ignition is turned on.

23   Check the ignition module to see if the PCM is receiving a reference signal from the ignition system (see Section 8).

7   Ignition coil - check and replacement

Caution: *On models equipped with a theft-deterrent radio, make sure you have the correct activation code, or the theft deterrent system is turned off, before disconnecting the battery.*

Check

Refer to illustration 7.2

1   Refer to Section 6 and perform the ignition system checks there first.

2   Use an ohmmeter and check coil secondary resistance of each coil pack **(see illustration)**. Refer to the Specifications listed in this Chapter for the correct amount of resistance.

Replacement

3   Detach the cable from the negative terminal of the battery.

8.3 With the ignition key on (engine not running), check for voltage at the white wire. It should be less than 1 volt

4   Unplug the electrical connectors from the module.
5   If the plug wires are not numbered, label them and detach the plug wires at the coil assembly.
6   Remove the module/coil assembly mounting bolts and lift the assembly from the vehicle.
7   Remove the bolts attaching the coils to the module and separate them.
8   Installation is the reverse of removal.
9   When installing the coils, make sure they are connected properly.

8   Ignition module - check and replacement

Note: *The following diagnostic procedure applies to 1995 and earlier models only.*

Check

Refer to illustration 8.3
1   First, perform the ignition system checks detailed in Section 6.
2   To check the ignition module, first check for an ignition control (IC) signal at the 6 pin electrical connector that connects to the ignition module.
3   With the ignition key ON (engine not running) measure the voltage at the white wire (terminal C) **(see illustration)**. It should be 1 volt or less.
4   If the voltage reading is correct, turn the ignition key OFF, disconnect the 6-pin electrical connector and measure the resistance of the ground circuit on the same wire (white wire). It should be above 200 ohms.
5   If all the tests are correct, the circuits are operating correctly and the ignition module is receiving the proper voltage signals. If a no-start condition persists, replace the ignition module with a new one.

Replacement

6   Disconnect the ignition module connec-

tor from the module.
7   Remove the bolts, then unplug the two individual coil packs from the top of the ignition module.
8   Remove the bolts that retain the ignition module to the engine.
9   Installation is the reverse of removal.

9   Charging system - general information and precautions

The charging system consists of a belt-driven alternator with an integral voltage regulator and the battery. These components work together to supply electrical power for the ignition system, the lights and all accessories.

All Saturn models are equipped with sealed alternator units. In the event of voltage regulator or alternator brush failure, it will be necessary to replace the alternator as a complete unit. All types use a conventional pulley and fan.

The purpose of the voltage regulator is to limit the alternator's voltage to a preset value. This prevents power surges, circuit overloads, etc., during peak voltage output. On all models with which this manual is concerned, the voltage regulator is contained within the alternator housing.

The charging system does not ordinarily require periodic maintenance. The drivebelts, electrical wiring and connections should, however, be inspected at the intervals suggested in Chapter 1.

Take extreme care when making circuit connections to a vehicle equipped with an alternator and note the following. When making connections to the alternator from a battery, always match correct polarity. Before using arc welding equipment to repair any part of the vehicle, disconnect the wires from the alternator and the battery terminal. Never start the engine with a battery charger connected. Always disconnect both battery leads

before using a battery charger.
The charging indicator lamp on the dash lights when the ignition switch is turned on and goes out when the engine starts. If the lamp stays on or comes on once the engine is running, a charging system problem has occurred. See Section 10 for the proper diagnosis.

10   Charging system - check

1   If a malfunction occurs in the charging circuit, do not immediately assume that the alternator is causing the problem. First check the following items:
   a) *The battery cables where they connect to the battery. Make sure the connections are clean and tight.*
   b) *The battery state of charge. If it is low, charge the battery (see Chapter 1).*
   c) *Check the external alternator wiring and connections. They must be in good condition.*
   d) *Check the drivebelt condition and tension (see Chapter 1).*
   e) *Make sure the alternator mounting bolts are tight.*
   f) *Run the engine and check the alternator for abnormal noise (may be caused by a loose drive pulley, loose mounting bolts, worn or dirty bearings, defective diode or defective stator).*

2   Using a voltmeter, check the battery voltage with the engine off. It should be approximately 12 volts.
3   Start the engine and check the battery voltage again. It should now be approximately 14 to 15 volts.
4   If the alternator (which includes the internal regulator) is not functioning properly, it must be replaced (see the next Section). These alternators are sealed units that cannot be rebuilt by the home mechanic.

11   Alternator - removal and installation

Refer to illustrations 11.4, 11.7 and 11.8
Caution: *On models equipped with a theft-deterrent, radio, make sure you have the correct activation code, or the theft deterrent system is turned off, before disconnecting the battery.*
Note: *The alternator is a sealed unit that the home mechanic cannot rebuild. Remove the alternator and replace it with a rebuilt unit from an auto parts store.*
1   Detach the cable from the negative terminal of the battery.
2   Raise the vehicle and support it securely it on jackstands.
3   Remove the serpentine drivebelt (see Chapter 1).
4   Working underneath the engine compartment, remove the mounting bolts **(see**

5

11.4 Remove these bolts (arrows) and separate the brace from the engine compartment (if equipped)

11.7 Remove the right front wheel for easier access to the alternator lower bolt (arrow)

illustration) from the alternator brace (if equipped) and separate the brace from the engine.

5   Remove the alternator protective cover and then detach the electrical connector(s) from the back of the alternator.

6   Also remove the nut and detach the battery positive cable from the back of the alternator.

7   Remove the alternator mounting bolts **(see illustration)**.

8   Detach any wiring harness routing clips and remove the alternator from the vehicle **(see illustration)**.

9   Installation is the reverse of the removal procedure.

12   Starting system - general information

The function of the starting system is to crank the engine. The starting system is composed of a starting motor, solenoid and battery. The battery supplies the electrical energy to the solenoid, which then completes

11.8 Remove the alternator from underneath the engine compartment

the circuit to the starting motor, which does the actual work of cranking the engine.

The solenoid and starting motor are mounted together at the lower front side of the engine. No periodic lubrication or maintenance is required.

The electrical circuitry of the vehicle is arranged so that the starter motor can only be operated when the clutch pedal is depressed (manual transmission) or the transmission selector lever is in Park or Neutral (automatic transmission).

Never operate the starter motor for more than 10 seconds at a time without pausing to allow it to cool for at least two minutes. Excessive cranking can cause overheating, which can seriously damage the starter.

13   Starter motor - testing in vehicle

1   If the starter motor does not turn at all when the switch is operated, make sure that the shift lever is in Neutral or Park (automatic transmission) or that the clutch pedal is depressed (manual transmission).

2   Make sure that the battery is charged and that all cables, both at the battery and starter solenoid terminals, are secure.

3   If the starter motor spins but the engine is not cranking, the overrunning clutch in the starter motor is slipping and the motor must be removed from the engine for replacement.

4   If, when the switch is actuated, the starter motor does not operate at all, but the solenoid clicks, then the problem lies with either the battery, the main solenoid contacts or the starter motor itself. **Note:** *Before diagnosing starter problems, make sure that the battery is fully charged.*

5   If the solenoid plunger cannot be heard when the switch is actuated, the solenoid itself is defective or the solenoid circuit is open.

6   To check the solenoid, connect a jumper lead between the battery (+) and the "S" terminal on the solenoid. If the starter motor now operates, the solenoid is OK and the problem is in the ignition switch, neutral start switch or in the wiring.

7   If the starter motor still does not operate, remove the starter/solenoid assembly for testing and replacement.

8   If the starter motor cranks the engine at an abnormally slow speed, first make sure that the battery is charged and that all terminal connections are tight. If the engine is partially seized, or has the wrong viscosity oil in it, it will crank slowly.

9   Run the engine until normal operating temperature is reached, then disconnect the primary electrical connector from the coil pack.

10   Connect a voltmeter positive lead to the starter motor terminal of the solenoid and then connect the negative lead to ground.

11   Crank the engine and take the voltmeter readings as soon as a steady figure is indicated. Do not allow the starter motor to turn for more than 10 seconds at a time. A reading of 9 volts or more, with the starter motor turning at normal cranking speed, is normal. If the reading is 9 volts or more but the cranking speed is slow, the motor is faulty.

14.3 Remove the nuts and bolts (arrows) from the brace and separate the brace from the starter

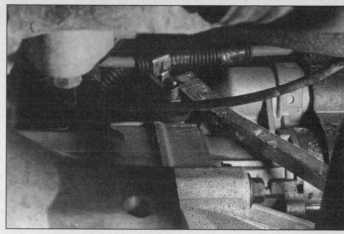

14.4 Use a prybar or a long screwdriver to separate the wire harness from the bellhousing bracket

14 Starter motor - removal and installation

Refer to illustrations 14.3 and 14.4

Caution: *On models equipped with a theft-deterrent radio, make sure you have the correct activation code, or the theft deterrent system is turned off, before disconnecting the battery.*

Note: *The starter is a sealed unit that the home mechanic cannot rebuild. Remove the starter and replace it with a rebuilt unit from an auto parts store.*

1   Detach the cable from the negative terminal of the battery.

2   Raise the vehicle and support it on jackstands.

3   Working under the vehicle, remove the starter brace **(see illustration)**.

4   Remove the wire harness from the brace on the transaxle bellhousing **(see illustration)**.

5   Remove the two bolts holding the starter to the engine. The top bolt must be removed from the engine compartment working above, near the intake manifold.

6   Partially lower the starter to gain access to the wires on the starter solenoid. Disconnect the electrical connectors from the solenoid terminals.

7   Remove the starter from the engine.

8   Installation is the reverse of removal.

5

Notes

Chapter 6
Emissions and engine control systems

Contents

Specifications

Crankshaft position sensor

Resistance .... 700 to 900 ohms

Coolant Temperature sensor and Intake Air Temperature sensor resistance

-20-degrees F	30K to 53k ohms
0-degrees F	21K to 27k ohms
60-degrees F	3.9K to 4.5K ohms
120-degrees F	1.0K to 1.1K ohms
160-degrees F	430 to 480 ohms
200-degrees F	215 to 235 ohms

1   General information

Refer to illustration 1.7

To prevent pollution of the atmosphere from incompletely burned and evaporating gases, and to maintain good driveability and fuel economy, a number of emission control devices are incorporated.
They include the:

Air pump system
Exhaust Gas Recirculation (EGR) system
Evaporative Emission Control System (EECS)
Positive Crankcase Ventilation (PCV) system
Transmission Converter Clutch (TCC)
Catalytic converter (CAT)
Air conditioning (A/C) control

All of these systems are linked, directly or indirectly, to the Computer Command Control (CCC or C3) system.

The Sections in this Chapter include general descriptions, checking procedures within the scope of the home mechanic and component replacement procedures (when possible) for each of the systems listed above.

Before assuming that an emissions control system is malfunctioning, check the fuel and ignition systems carefully. The diagnosis of some emission control devices requires specialized tools, equipment and training. If checking and servicing become too difficult or if a procedure is beyond the scope of your skills, consult your dealer service department.

This doesn't mean, however, that emission control systems are particularly difficult to maintain and repair. You can quickly and easily perform many checks and do most (if not all) of the regular maintenance at home with common tune-up and hand tools. **Note:** *The most frequent cause of emissions problems is simply a loose or broken vacuum hose or wiring connection, so always check the hose and wiring connections first.*

Pay close attention to any special precautions outlined in this Chapter. It should be noted that the illustrations of the various systems may not exactly match the system installed on your vehicle because of changes made by the manufacturer during production or from year-to-year.

A Vehicle Emissions Control Information (VECI) label is located in the engine compartment **(see illustration)**. This label contains important emissions specifications as well as

6

1.7 The Vehicle Emission Control Information (VECI) label is located on the radiator support and contains information of the emission devices on your vehicle, vacuum line routing, etc.

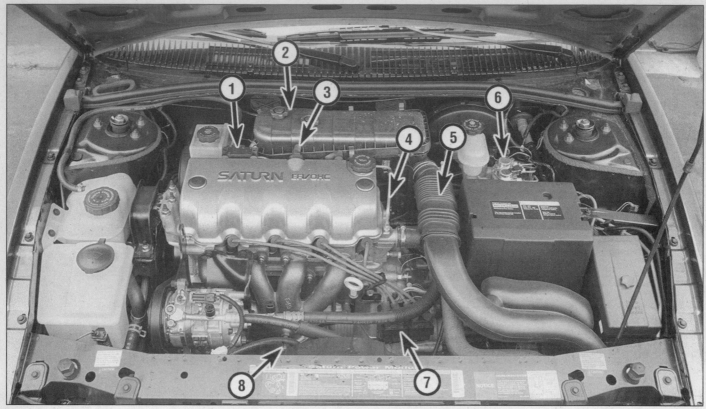

2.1a Emission and engine control components on the TBI model

1 MAP sensor	5 Air intake duct
2 Throttle body unit (under air cleaner housing)	6 ABS unit
3 PCV valve	7 DIS coil packs
4 EGR valve (near cylinder head)	8 Oxygen sensor

a vacuum hose schematic and emissions components identification guide. When servicing the engine or emissions systems, the VECI label in your particular vehicle should always be checked for up-to-date information.

2   Computer Command Control (CCC) system and trouble codes

Refer to illustrations 2.1a, 2.1b and 2.5
Note: *The following procedure ;for extracting trouble codes and the trouble code charts, apply to 1995 and earlier models only. On 1996 and later models, the On-Board Diagnostic (OBD-II) system and trouble codes can only be accessed using a hand-held SCAN tool. DO NOT attempt to jump the new 16-pin ALDL connector, trouble codes will not flash on the SERVICE ENGINE SOON light.*

The Computer Command Control (CCC) system consists of a Powertrain Control Module (PCM) and information sensors which monitor various functions of the engine and send data back to the PCM **(see illustrations).** The PCM is also equipped with an Erasable Programmable Read Only Memory (EEPROM). The calibrations are stored in the PCM within the EEPROM. If the PCM must be

replaced, it is necessary to have the EEPROM programmed with a special scanning tool called the "TECH #1," available only at dealer service departments and some auto repair shops. **Note:** *The EEPROM is not replaceable on these vehicles. In the event of any malfunction with the EEPROM (Code 51), the vehicle must be taken to a dealership service department for diagnosis and reprogramming.*

The PCM controls the following systems:

Fuel control
Electronic spark timing
Exhaust gas recirculation
Canister purge
Engine cooling fan
Idle Air Control (IAC)
Transmission converter clutch
Air conditioning clutch control

The CCC system is analogous to the central nervous system in the human body. The sensors (nerve endings) constantly relay information to the PCM (brain), which processes the data and, if necessary, sends out a command to change the operating parameters of the engine (body).

Here's a specific example of how one portion of this system operates: An oxygen sensor, located in the exhaust manifold, constantly monitors the oxygen content of the

exhaust gas. If the percentage of oxygen in the exhaust gas is incorrect, an electrical signal is sent to the PCM. The PCM takes this information, processes it and then sends a command to the fuel injection system, telling it to change the air/fuel mixture. This happens in a fraction of a second and it goes on continuously when the engine is running. The end result is an air/fuel mixture ratio which is constantly maintained at a predetermined ratio, regardless of driving conditions.

One might think that a system which uses an on-board computer and electrical sensors would be difficult to diagnose. This is not necessarily the case. The CCC system has a built-in diagnostic feature which indicates a problem by flashing a SERVICE ENGINE SOON light on the instrument panel. When this light comes on during normal vehicle operation, a fault in one of the information sensor circuits or the PCM itself has been detected. More importantly, a trouble code is stored in the PCM's memory.

To retrieve this information from the PCM memory, you must use a short jumper wire to ground a diagnostic terminal. This terminal is part of an electrical connector known as the Assembly Line Data Link (ALDL) **(see illustration).** The ALDL is located underneath the dashboard, just below the instrument

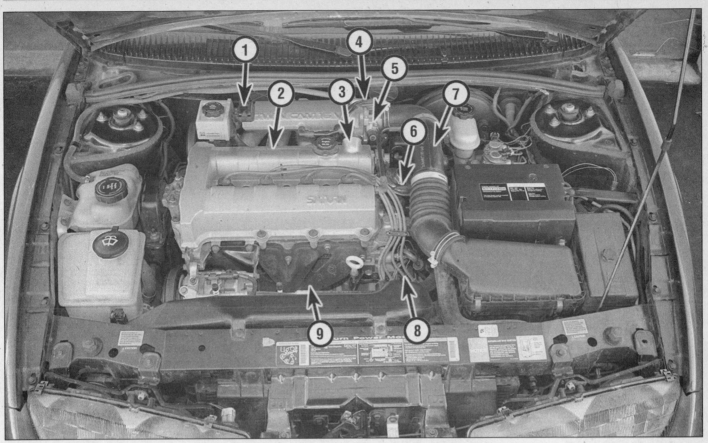

2.1b Emission and engine control components on the MPFI model

1	MAP sensor	4	TPS
2	Fuel injector	5	Throttle body unit
3	PCV valve	6	EGR valve

7	Air intake duct	9	Oxygen sensor (under
8	DIS coil packs		exhaust manifold)

panel and to the left of the driver's foot area. To use the ALDL, remove the plastic cover (if equipped) by sliding it toward you. With the electrical connector exposed to view, push one end of the jumper wire into the diagnostic terminal (B) and the other end into the ground terminal (A).

When the diagnostic terminal is grounded with the ignition on and the engine stopped, the system will enter the Diagnostic Mode. **Caution:** *Don't start or crank the engine with the diagnostic terminal grounded*. In this mode the PCM will display a "Code 12" by flashing the SERVICE ENGINE SOON light, indicating that the system is operating. A code 12 is simply one flash, followed by a brief pause, then two flashes in quick succession. This code will be flashed three times. If no other codes are stored, Code 12 will continue to flash until the diagnostic terminal ground is removed.

After flashing Code 12 three times, the PCM will display any stored trouble codes. Each code will be flashed three times, then Code 12 will be flashed again, indicating that the display of any stored trouble codes has been completed. If more than one trouble code is stored, the diagnostic trouble codes will be displayed from lowest to highest in number with the exception of number 11 -

transaxle trouble code indication. This trouble code will flash last, followed by codes flashed on another dash light (SHIFT TO D2 on 1991 and 1992 models or on the HOT light on 1993 and later models). Refer to the transaxle trouble code chart for the specific code and its probable cause.

When the PCM sets an engine trouble code, the SERVICE ENGINE SOON light will come on and a trouble code will be stored in memory. If the problem is intermittent, the

light will go out after 10 seconds, when the fault goes away. However, the trouble code will stay in the PCM memory until the battery voltage to the PCM is interrupted. Removing battery voltage for 10 seconds will clear all stored trouble codes. Trouble codes should always be cleared after repairs have been completed. **Caution:** *To prevent damage to the PCM, the ignition switch must be OFF when disconnecting or connecting power to the PCM.*

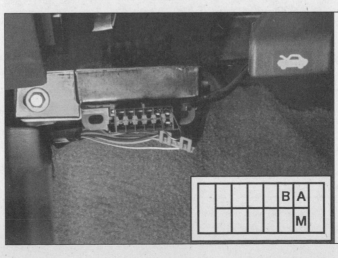

2.5 The assembly Line Data Link (ALDL) is located under the driver's side dashboard near the kick panel. To activate the diagnostic codes, jump terminals B and A (1995 and earlier)

Diagnostic tool information
Refer to illustrations 2.9, 210 and 2.4

A digital multimeter is a necessary tool for checking fuel injection and emission related components **(see illustration)**. A digital volt-ohmmeter is preferred over the older style analog multimeter for several reasons. The analog multimeter cannot display the volts-ohms or amps measurement in hundredths and thousandths increments. When working with electronic circuits which are often very low voltage, this accurate reading is most important. Another good reason for the digital multimeter is the high impedance circuit. The digital multimeter is equipped with a high resistance internal circuitry (10 million ohms). Because a voltmeter is hooked up in parallel with the circuit when testing, it is vital that none of the voltage being measured should be allowed to travel the parallel path set up by the meter itself. This dilemma does not show itself when measuring larger amounts of voltage (9 to 12 volt circuits) but if you are measuring a low voltage circuit such as the oxygen sensor signal voltage, a fraction of a volt may be a significant amount when diagnosing a problem.

Hand-held scanners are the most powerful and versatile tools for analyzing engine management systems used on later model vehicles **(see illustration)**. Early model scanners handle codes and some diagnostics for many OBD I systems. Each brand scan tool must be examined carefully to match the year, make and model of the vehicle you are working on. Often interchangeable cartridges are available to access the particular manufacturer; Ford, GM, Chrysler, etc.). Some manufacturers will specify by continent; Asia, Europe, USA, etc. Seek the advice of your local parts retailer.

With the arrival of the federally mandated emission control system (OBD II), a specially designed scanner has been developed. At this time, several manufacturers have released OBD II scan tools for the home mechanic. Ask the parts salesperson at a local auto parts store for additional information concerning these tools. **Note:** *Although OBD II codes cannot be accessed without a Scan tool, follow the simple component checks in Section 4.*

Another type of code reader and less expensive is available at parts stores **(see illustration)**. These tools simplify the procedure for extracting codes from the engine management computer by simply "plugging in" to the diagnostic connector on the vehicle wiring harness.

Following is a list of the typical trouble codes for the engine and transaxle which may be encountered while diagnosing the Computer Command Control System. Also included are simplified troubleshooting procedures. If the problem persists after these checks have been made, more detailed service procedures will have to be done by a dealer service department.

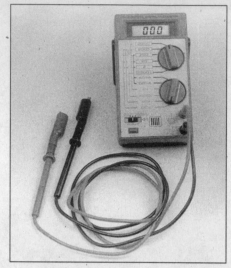

2.9 Digital multimeters can be used for testing all types of circuits; because of their high impedance, they are much more accurate than analog meters for measuring millivolts in low-voltage computer circuits

2.10 Scanners like the Actron Scantool and the AutoXray XP240 are powerful diagnostic aids - programmed with comprehensive diagnostic information, they can tell you just about anything you want to know about your engine management system

2.12 Trouble code tools simplify the task of extracting the trouble codes

Engine code chart (1995 and earlier)

Trouble code	Circuit or system	Probable cause
Code 11 (1 flash, pause 1 flash)	Transaxle codes present	This indicates that there are trouble codes for the transaxle unit stored in the PCM. Read the codes after the engine code sequence on the SHIFT TO D2 light (1991 and 1992 models) or HOT light (1993 and later models).
Code 12 (1 flash, pause, 2 flashes)	Diagnostic check only	Indicates the system is ready (ALDL grounded) and ready to flash the engine codes.
Code 13 (1 flash, pause, 3 flashes)	Oxygen sensor circuit	Possible oxygen sensor ground loose. Check the wiring and connectors from the oxygen sensor. Replace the oxygen sensor (see Section 4).

Trouble code	Circuit or system	Probable cause
Code 14 (1 flash, pause, 4 flashes)	Coolant sensor/high temperature	If the engine is experiencing cooling system problems, the problem must be rectified before continuing. Check all the wiring and the connectors associated with the coolant temperature sensor (see Section 4). Replace if necessary.*
Code 15 (1 flash, pause, 5 flashes)	Coolant sensor/low temperature	See above then check the wiring harness connector at the PCM for damage (see Section 4).
Code 17 (1 flash, pause, 7 flashes)	PCM fault - Pull-up resistor	Faulty PCM resistor in PCM. Replace PCM (see Section 3).
Code 19 1 flash, pause, 9 flashes	6X signal fault (1992 and later models only)	PCM and/or ignition module may be defective. Check all connections and grounds (see Chapter 5).
Code 21 (2 flashes, pause, 1 flash)	Throttle position sensor voltage high	Check for a sticking or misadjusted TPS plunger. Check all the wiring and connections between the TPS and the PCM. Adjust or replace the TPS (see Section 4).*
Code 22 (2 flashes, pause, 2 flashes)	Throttle position sensor voltage low	Check the TPS adjustment. Check the PCM connector. Replace the TPS* (see Section 4).
Code 23) (2 flashes, pause, 3 flashes)	IAT circuit low	Intake air temperature sensor and/or circuit may be faulty. Check sensor and replace if necessary (see Section 4).*
Code 24 (2 flashes, pause, 4 flashes)	VSS circuit - no signal	A fault in this circuit should be indicated only when the vehicle is in motion. Disregard code 24 if it is set while the drive wheels are not turning (test situation). Check TPS and PCM.
Code 25 (2 flashes, pause, 5 flashes)	IAT circuit - temperature out of high range	Temperature range excessive causing a misreading by the PCM. Check IAT sensor (see Section 4).*
Code 26 (2 flashes, pause, 6 flashes)	Quad driver output fault	The PCM detects an improper voltage level on the circuit that is connected to the Quad Driver Module.
Code 32 (3 flashes, pause, 2 flashes)*	EGR system fault	Vacuum switch shorted to ground on start-up, switch not closed after the PCM has commanded the EGR for a specified period of time or the EGR solenoid circuit is open for a specified amount of time. Replace the EGR valve.*
Code 33 (3 flashes, pause, 3 flashes)	MAP circuit - voltage out of range high	Check the vacuum hoses from the MAP sensor. Check the electrical connections at the PCM. Replace the MAP sensor (see Section 4).*
Code 34 (3 flashes, pause, 4 flashes)	MAP circuit - voltage out of range low	Signal voltage from MAP sensor too low. Check MAP sensor circuit also TPS circuit (see Section 4).
Code 35 (3 flashes, pause, 5 flashes)	Idle air control (IAC) - rpm out of range	IAC motor possibly defective. Idle control is high or low. Possible PCM problem. Have the system diagnosed by a dealer service department (see Section 4).
Code 41 (4 flashes, pause, 1 flash)	Ignition control circuit - open or shorted	Possible defective ignition module. Also check circuit to PCM from ignition module (see Chapter 5).
Code 42 (4 flashes, pause, 2 flashes)	Bypass circuit - open or shorted	Bypass circuit from ignition module to PCM possibly open or shorted. (see Chapter 5).
Code 41 and 42 (light will flash each separately)	IC control circuit grounded/bypass open	Bypass circuit and/or ignition control circuit shorted causing no feedback pulses for the ignition cycle (see Chapter 5).

6

Engine code chart (1995 and earlier)

Trouble code	Circuit or system	Probable cause
Code 43 (4 flashes, pause, 3 flashes)	Knock sensor circuit open or shorted	Possible loose or defective knock sensor (see Section 4). Also check the knock sensor circuit.
Code 44 (4 flashes, pause, 4 flashes)	Oxygen sensor indicates lean exhaust	Check for vacuum leaks near the throttle body gasket, vacuum hoses or the intake manifold gasket. Also check for loose connections on PCM, oxygen sensor etc. Replace the oxygen sensor if necessary (see Section 4).
Code 45 (4 flashes, pause, 5 flashes)	Oxygen sensor indicates rich exhaust	Possibly rich or leaking injector. high fuel pressure or faulty TPS or MAP sensor. Also, check the charcoal canister and its components for the presence of fuel. Replace the oxygen sensor if necessary (see Section 4).*
Code 46 (4 flashes, pause, 6 flashes)	Power steering pressure circuit (1991 models only) - open or shorted	Possible defective power steering pressure switch (see Section 4). Also check the circuit to the switch (see Chapter 12).
Code 49 (4 flashes, pause, 9 flashes)	High idle indicates vacuum leak	Check all hoses to MAP sensor, PCV valve, brake booster, fuel pressure regulator, throttle body, intake manifold gasket and any other vacuum line.
Code 51 (5 flashes, pause, 1 flash)	PCM memory error	Possible defective EEPROM, RAM or EPROM. Have the vehicle diagnosed by a dealer service department.
Code 55	A/D error	Defective PCM. Have the vehicle diagnosed by a dealer service department.
Code 81 (8 flashes, pause, 1 flash)	ABS message fault (1993 vehicles only)	Defective ABS controller. Have the vehicle diagnosed by a dealer service department.
Code 82 (8 flashes, pause, 2 flashes)	PCM internal communication fault	Defective PCM. Have the vehicle diagnosed by a dealer service department.

* Component replacement may not cure the problem in all cases. For this reason, you may want to seek professional advice before purchasing replacement parts.

Transaxle trouble code chart (1995 and earlier)

Trouble code	Circuit or system	Probable cause
Code 13 (1 flash, pause, 3 flashes)	Line pressure high (1992 and later models only)	Possible stuck regulator valve in the valve body
Code 14 (1 flash, pause, 4 flashes)	Line pressure low (1993 and later models only)	Possible pump problem or a restriction in the sump suction filter
Code 16 (1 flash, pause, 6 flashes)	No first gear	Possible damaged first gear hub seal rings
Code 18 (1 flash, pause, 8 flashes)	No gears available (1993 and later models)	Possible misadjusted shift cable
Code 21 (2 flashes, pause, 1 flash)	Second gear stuck ON	Possible internal hydraulic leak from seals, gaskets or porosity
Code 22 (2 flashes, pause 2 flashes)	No second gear	Possible servo piston installed incorrectly

Trouble code	Circuit or system	Probable cause
Code 23 (2 flashes, pause, 3 flashes)	No third gear	Possible cross leak to another clutch pack
Code 24 (2 flashes, pause, 4 flashes)	No fourth gear	Possible hydraulic leak in fourth or failed clutch pack
Code 25 (2 flashes, pause, 5 flashes)	No TCC (torque converter clutch)	Possible poor electrical connection on the PCM/TC, actuator connector plate or the UHJB terminals
Code 26 (2 flashes, pause, 6 flashes)	TCC (torque converter clutch) stuck ON	TCC release circuit may be leaking at the seal ring on the input shaft
Code 31 (3 flashes, pause, 1 flash)	Transaxle temperature circuit open (cold)	Possible defective sensor or circuit
Code 32 (3 flashes, pause, 2 flashes)	Transaxle temperature circuit grounded (hot)	Possible wire harness problem
Code 34 (3 flashes, pause 4 flashes)	ECM communications failure (1992 and later models)	Possible PCM/TC failure
Code 35 (3 flashes, pause, 5 flashes)	No turbine speed signal	Possible defective turbine speed sensor and/or circuit
Code 36 (3 flashes, pause, 6 flashes)	Turbine speed signal noise	Possible defective turbine speed sensor
Code 41 (4 flashes, pause, 1 flash)	No vehicle speed sensor signal	Possible defective vehicle speed sensor or circuit
Code 42 (4 flashes, pause, 2 flashes)	Vehicle speed signal noise	Possible loose connection on PCM/TC, open circuit to VSS or defective VSS
Code 43 (4 flashes, pause, 3 flashes)	Master enable relay grounded/open (1993 and later models only)	Possible defective master enable relay or its circuit
Code 44 (4 flashes, pause, 4 flashes)	Master enable relay coil shorted to voltage (1993 and later models only)	Possible defective master enable relay or its circuit
Code 45 (4 flashes, pause, 5 flashes)	Selector switch - no data	Possible disconnected gear selector switch
Code 46 (4 flashes, pause, 6 flashes)	Park/Neutral switch - invalid data	Possible misadjusted gear selector switch
Code 48 (4 flashes, pause, 8 flashes)	Hold mode voltage low	Possible defective PCM/TC
Code 49 (4 flashes, pause, 9 flashes)	Park/Neutral switch - invalid data (1992 and later models only)	Possible misadjusted gear selector switch
Code 52 (5 flashes, pause, 2 flashes)	Hold mode stuck on (1991 and 1992 models only)	Possible defective PCM/TC

6

Transaxle trouble code chart (1995 and earlier) (continued)

Trouble code	Circuit or system	Probable cause
Code 53 (5 flashes, pause, 3 flashes)	Hold mode stuck off (1991 and 1992 models only)	Possible defective PCM/TC
Code 54 (5 flashes, pause, 4 flashes)	A/D error	Possible defective PCM/TC
Code 55 (5 flashes, pause, 5 flashes)	Transaxle temperature resistor failure	Possible defective PCM/TC
Code 56 (5 flashes, pause, 6 flashes)	Generic field - effect transistor driver failure	Possible defective PCM/TC
Code 75 (7 flashes, pause, 5 flashes)	Third gear stuck on (1993 and later models only)	Possible cross leaks from third gear clutch oil passages
Code 78 (7 flashes, pause, 8 flashes)	Fourth gear stuck on (1993 and later models only)	Possible internal cross leaks from third gear into another clutch pack

OBD II Trouble Codes (1996 and later)

Code	Code Definition	Location
P0101	Mass Air Flow (MAF) sensor error	See Section 4
P0107	Manifold Absolute Pressure (MAP) sensor circuit low input	See Section 4
P0108	Manifold Absolute Pressure (MAP) sensor circuit high input	See Section 4
P0112	Intake Air Temperature (IAT) sensor circuit low input	See Section 4
P0113	Intake Air Temperature (IAT) sensor circuit high input	See Section 4
P0117	Electronic Coolant Temperature (ECT) sensor circuit low input	See Section 4
P0118	Electronic Coolant Temperature (ECT) sensor circuit high input	See Section 4
P0121	Throttle Position Sensor (TPS) range/performance fault	See Section 4
P0122	Throttle Position Sensor (TPS) circuit low input	See Section 4
P0123	Throttle Position Sensor (TPS) circuit high input	See Section 4
P0131	Upstream heated O_2 sensor circuit low voltage (Bank 1, Sensor 1)	See Section 4
P0133	Upstream heated O_2 sensor circuit high voltage (Bank 1, Sensor 1)	See Section 4

Code	Code Definition	Location
P0137	Downstream heated O2 sensor circuit low voltage (Bank 1, Sensor 2)	See Section 4
P0138	Downstream heated O2 sensor circuit high voltage (Bank 1, Sensor 2)	See Section 4
P0141 (Bank 1, Sensor 2)	O2 sensor heater circuit fault	See Section 4
P0171	System Adaptive fuel too lean	See Chapter 4
P0172	System Adaptive fuel too rich	See Chapter 4
P0191	Injector Pressure sensor system performance	See Chapter 4
P0192	Injector Pressure sensor circuit low input	See Chapter 4
P0193	Injector Pressure sensor circuit high input	See Chapter 4
P0301	Cylinder number 1 misfire detected	See Chapter 5
P0302	Cylinder number 2 misfire detected	See Chapter 5
P0303	Cylinder number 3 misfire detected	See Chapter 5
P0304	Cylinder number 4 misfire detected	See Chapter 5
P0325	Knock sensor circuit 1 fault	See Section 4
P0326	Knock sensor circuit performance	See Section 4
P0351	COP ignition coil 1 primary circuit fault	See Chapter 5
P0352	COP ignition coil 2 primary circuit fault	See Chapter 5
P0353	COP ignition coil 3 primary circuit fault	See Chapter 5
P0354	COP ignition coil 4 primary circuit fault	See Chapter 5
P0400	EGR flow fault	See Section 6
P0401	EGR insufficient flow detected	See Section 6
P0402	EGR excessive flow detected	See Section 6
P0420	Catalyst system efficiency below threshold (Bank 1)	See Section 8
P0421	Catalyst system efficiency below threshold (Bank 1)	See Section 8
P0430	Catalyst system efficiency below threshold (Bank 2)	See Section 8

6

OBD II Trouble Codes

Code	Code Definition	Location
P0431	Catalyst system efficiency below threshold (Bank 2)	See Section 8
P0441	EVAP incorrect purge flow	See Section 5
P0443	EVAP VMV circuit fault	See Section 5
P0452	EVAP fuel tank pressure sensor low input	See Section 5
P0453	EVAP fuel tank pressure sensor high input	See Section 5
P0502	VSS circuit low input	See Section 4
P0503	VSS circuit range performance	See Section 4
P0506	IAC system rpm lower than expected	See Chapter 4
P0507	IAC system rpm higher than expected	See Chapter 4
P0602	PCM control module programming error	See Section 4
P0705	Transaxle Range sensor circuit malfunction	See Section 4
P1171	Inadequate fuel flow	See Chapter 4

Component replacement may not cure the problem in all cases. For this reason, you may want to seek professional advice before purchasing replacement parts.

3   Powertrain Control Module (PCM)

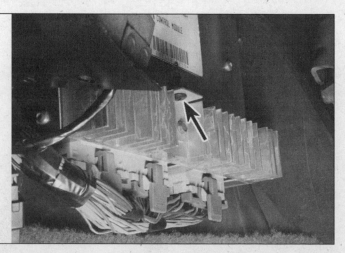

3.4 Remove the center nut (arrow) and carefully lower the PCM from the bracket assembly located under the dash in the corner of the passenger compartment

Refer to illustration 3.4

Warning: *On models equipped with a Supplemental Inflatable Restraint (SIR) system, always disable the airbag(s) before working in the vicinity of the steering wheel, instrument panel or SIR system components to avoid the possibility of accidental deployment of the airbag(s), which could cause personal injury (see Chapter 12).*

Caution: *On models equipped with a theft-deterrent radio, make sure you have the correct activation code, or the theft deterrent system is turned off, before disconnecting the battery.*

1   The Powertrain Control Module (PCM) is located inside the passenger compartment under the dashboard tucked into the corner near the ABS module on the driver's side. The PCM is easily distinguished by the aluminum fins attached to the bottom of the module.

2   Disconnect the negative battery cable from the battery. On 1996 and later models, remove the instrument panel pad (see Chapter 11).

3   Unplug all three electrical connectors from the PCM. **Caution:** *The ignition switch must be turned OFF when pulling out or plugging in the electrical connectors to prevent damage to the PCM.*

4   Remove the retaining nut **(see illustration)** from the PCM bracket.

5   On 1995 and earlier models, carefully slide the PCM down far enough to clear the kick panel. On 1996 and later models, lift the PCM up, towards the windshield. **Note:** *Avoid any static electricity damage to the computer by using gloves and a special anti-static pad to set the PCM on once it is removed.*

4.2 Check the resistance of the coolant temperature sensor with the engine completely cold and then with the engine at operating temperature

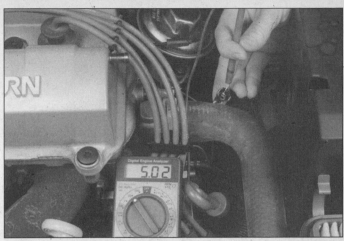

4.3 Working on the harness side, check the voltage from the PCM to the coolant temperature sensor with the ignition key ON and the engine not running. It should be approximately 5.0 volts

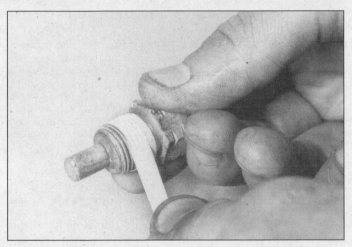

4.4 To prevent leakage, wrap the threads of the coolant temperature sensor with teflon tape before installing it

4.5 To remove the engine coolant temperature sensor (arrow), pry open the locking tab with a small screwdriver and unplug the electrical connector, then carefully unscrew the sensor with a deep socket or wrench

4   Information sensors

Note: *See component location illustrations in Section 2 for additional information on the location of the following information sensors.*

Engine coolant temperature sensor

Refer to illustrations 4.2, 4.3, 4.4 and 4.5

General description

1   The coolant sensor is a thermistor (a resistor which varies the value of its voltage output in accordance with temperature changes). The change in the resistance values will directly affect the voltage signal from the coolant sensor. As the sensor temperature DECREASES, the resistance values will INCREASE. As the sensor temperature INCREASES, the resistance values will DECREASE. A failure in the coolant sensor circuit should set a Code 14 or 15 on 1995 and earlier vehicles. This code indicates a failure in the coolant temperature circuit, so in

most cases the appropriate solution to the problem will be either repair of a wire or replacement of the sensor.

Check

2   To check the sensor, check the resistance value of the coolant temperature sensor while it is completely cold (50 to 80-degrees F = 5,600 to 2,400 ohms). Next, start the engine and warm it up until it reaches operating temperature **(see illustration)**. The resistance should be lower (180 to 200-degrees F = 300 to 200 ohms). **Note:** *Access to the coolant temperature sensor makes it difficult to position electrical probes on the terminals. If necessary, remove the sensor and perform the tests in a pan of heated water to simulate the conditions.*

3   If the resistance values on the sensor are correct, check the signal voltage to the sensor from the PCM **(see illustration)**. It should be approximately 5.0 volts.

Replacement

Warning: *Wait until the engine is completely cool before beginning this procedure.*

Remove the coolant expansion cap to relieve any residual pressure in the cooling system, then reinstall the cap.

4   Before removing the old sensor, wrap the threads of the new sensor with Teflon sealing tape to prevent leakage and thread corrosion **(see illustration)**.

5   To remove the sensor, release the locking tab, unplug the electrical connector **(see illustration)**, then carefully unscrew the sensor. **Caution:** *Handle the coolant sensor with care. Damage to this sensor will affect the operation of the entire fuel injection system.* Installation is the reverse of removal. Check the coolant level and add some, if necessary, to bring to the desired level (see Chapter 1).

Manifold Absolute Pressure (MAP) sensor

General description

6   The Manifold Absolute Pressure (MAP) sensor monitors the intake manifold pressure changes resulting from changes in engine load and speed and converts the information into a

6

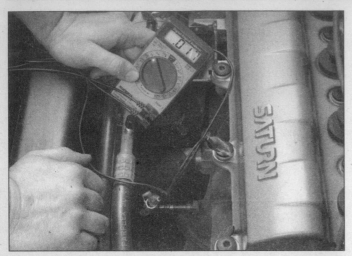

4.11 The oxygen sensor creates a very small voltage signal when the sensor is warmed up. A quick oxygen sensor check is to disconnect the sensor connector when it is warmed up and probe the sensor side of the connector for a voltage reading

4.12 Working on the harness side, probe the electrical connector and check for reference voltage. It should be approximately 0.5 volts with the ignition key on and the engine not running

voltage output. The PCM uses the MAP sensor to control fuel delivery and ignition timing.

7   The PCM will receive information as a voltage signal that will vary from 1.0 to 1.5 volts at closed throttle (high vacuum) and 4.0 to 5.0 volts at wide open throttle (low vacuum).

Check

8   A failure in the MAP sensor circuit should set a Code 33 or 34.

Oxygen sensor

Refer to illustration 4.11, 4.12 and 4.19

General description and check

9   The oxygen sensor, which is located in the exhaust manifold, monitors the oxygen content of the exhaust gas stream. **Note:** *1996 and later vehicles use two oxygen sensors. The front oxygen sensor is located in the exhaust manifold. The rear sensor is located aft of the catalytic converter. Its purpose is to monitor the performance of the catalytic converter.* The oxygen content in the exhaust reacts with the oxygen sensor to produce a voltage output which varies from 0.1-volt (high oxygen, lean mixture) to 0.9-volts (low oxygen, rich mixture). The PCM constantly monitors this variable voltage output to determine the ratio of oxygen to fuel in the mixture. The PCM alters the air/fuel mixture ratio by controlling the pulse width (open time) of the fuel injectors. A mixture ratio of 14.7 parts air to 1 part fuel is the ideal mixture ratio for minimizing exhaust emissions, thus allowing the catalytic converter to operate at maximum efficiency. It is this ratio of 14.7 to 1 which the PCM and the oxygen sensor attempt to maintain at all times.

10   The oxygen sensor produces no voltage when it is below its normal operating temperature of about 600-degrees F. During this initial period before warm-up, the PCM operates in open loop mode.

11   If the engine reaches normal operating temperature and/or has been running for two or more minutes, and if the oxygen sensor is producing a steady signal voltage between 0.35 and 0.55-volts **(see illustration)**, even though the TPS indicates that the engine is not at idle, the PCM will set a Code 13.

12   A delay of two minutes or more between engine start-up and normal operation of the sensor, followed by a low voltage signal or a short in the sensor circuit, will cause the PCM to set a Code 44 on 1995 and earlier vehicles. If a high voltage signal occurs, the PCM will set a Code 45 on 1995 and earlier vehicles **(see illustration)**.

13   When any of the above codes occur, the PCM operates in the open loop mode - that is, it controls fuel delivery in accordance with a programmed default value instead of feedback information from the oxygen sensor.

14   The proper operation of the oxygen sensor depends on four conditions:

a) **Electrical** - *The low voltages generated by the sensor depend upon good, clean connections which should be checked whenever a malfunction of the sensor is suspected or indicated.*

b) **Outside air supply** - *The sensor is designed to allow air circulation to the internal portion of the sensor. Whenever the sensor is removed and installed or replaced, make sure the air passages are not restricted.*

c) **Proper operating temperature** - *The PCM will not react to the sensor signal until the sensor reaches approximately 600-degrees F. This factor must be taken into consideration when evaluating the performance of the sensor.*

d) **Unleaded fuel** - *The use of unleaded fuel is essential for proper operation of the sensor. Make sure the fuel you are using is of this type.*

15   In addition to observing the above conditions, special care must be taken whenever the sensor is serviced.

a) *The oxygen sensor has a permanently attached pigtail and electrical connector which should not be removed from the sensor. Damage or removal of the pigtail or electrical connector can adversely affect operation of the sensor.*

b) *Grease, dirt and other contaminants should be kept away from the electrical connector and the louvered end of the sensor.*

c) *Do not use cleaning solvents of any kind on the oxygen sensor.*

d) *Do not drop or roughly handle the sensor.*

e) *The silicone boot must be installed in the correct position to prevent the boot from being melted and to allow the sensor to operate properly.*

Replacement

Note: *Because it is installed in the exhaust manifold or pipe, which contracts when cool, the oxygen sensor may be very difficult to loosen when the engine is cold. Rather than risk damage to the sensor (assuming you are planning to reuse it in another manifold or pipe), start and run the engine for a minute or two, then shut it off. Be careful not to burn yourself during the following procedure.* **Caution:** *On models equipped with a theft-deterrent radio, make sure you have the correct activation code, or the theft deterrent system is turned off, before disconnecting the battery.*

16   Disconnect the cable from the negative terminal of the battery.

17   Raise the vehicle and place it securely on jackstands.

18   Carefully disconnect the electrical connector from the sensor.

19   Carefully unscrew the sensor from the exhaust manifold **(see illustration)**. **Caution:** *Excessive force may damage the threads.*

20   Anti-seize compound must be used on the threads of the sensor to facilitate future

4.19 Use a long-handled wrench to slide over the wire loom and remove the oxygen sensor from the exhaust manifold

4.26a Use sharp-tipped electrical probes to pierce the wires on the backside of the TPS electrical harness in order to read signal voltage

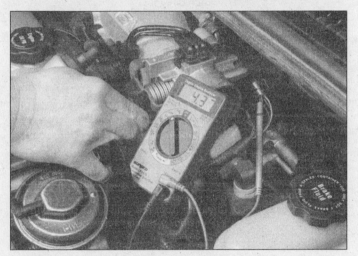

4.26b The voltage should increase as the throttle is opened

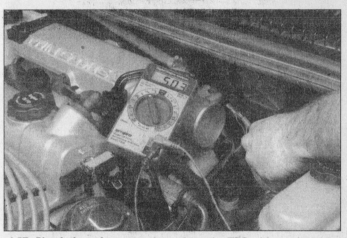

4.27 Check the reference voltage from the TPS with a voltmeter. Backprobe terminal A (gray wire) with the positive (+) probe of the voltmeter and make sure the reference voltage is approximately 5.0 volts

6

removal. The threads of new sensors will already be coated with this compound, but if an old sensor is removed and reinstalled, recoat the threads.

21   Install the sensor and tighten it securely.

22   Reconnect the electrical connector of the pigtail lead to the main engine wiring harness.

23   Lower the vehicle and reconnect the cable to the negative terminal of the battery.

Throttle Position Sensor (TPS)

General description

Refer to illustrations 4.26a, 4.26b and 4.27

24   The Throttle Position Sensor (TPS) is located on the end of the throttle shaft on the throttle body. By monitoring the output voltage from the TPS, the PCM can determine fuel delivery based on throttle valve angle (driver demand). A broken or loose TPS can cause intermittent bursts of fuel from the injector and an unstable idle because the PCM thinks the throttle is moving. Any problems in the TPS or circuit will set a code 21 or 22.

Check

25   To check the TPS, turn the ignition switch to ON (engine not running) and connect the probes of the voltmeter into the ground wire (terminal B) and signal wire (terminal C) on the backside of the electrical connector. This test checks for the proper signal voltage from the TPS. **Note:** *The signal wire is a dark blue wire. The ground wire is a black wire on all models.*

26   The sensor should read 0.25 to 0.98-volts at idle. Have an assistant depress the accelerator pedal to simulate full throttle and the sensor should increase voltage to 4.0 to 4.8-volts **(see illustrations)**. If the TPS voltage readings are incorrect, replace it with a new unit.

27   Also, check the TPS reference voltage. With the ignition key ON (engine not running), install the positive (+) probe of the ohmmeter **(see illustration)** onto the A terminal (gray wire). There should be approximately 5.0 volts sent from the PCM to the TPS.

28   A problem in any of the TPS circuits will

set a Code 21. Once a trouble code is set, the PCM will use an artificial default value for TPS and some vehicle performance will return.

Replacement

29   To replace the TPS, refer to Chapter 4, Section 14 for throttle body injection models. The TPS for multiport injection models requires the removal of the throttle body first, refer to Chapter 4, Section 13.

Park/Neutral Position switch

General description

30   The Park/Neutral Position (PNP) switch, located on the transaxle indicates to the PCM when the transmission is in Park or Neutral. This information is used for starting, Transmission Converter Clutch (TCC), Exhaust Gas Recirculation (EGR) and Idle Air Control (IAC) valve operation. For example, if the signal wire(s) become grounded, it may be difficult to start the engine in Park or Neutral. A problem with the Park/Neutral switch will flash a

4.38a An easy check to find out if the A/C compressor clutch is working is to disconnect the electrical connector to the compressor clutch (arrow) and install a jumper wire from the battery (+). The clutch should activate

4.38b If the relay is receiving battery voltage, check the A/C clutch by jumping the harness and applying voltage directly to the clutch. You should hear a loud "clunk" when the compressor clutch engages.

code 46 or 49 on the transaxle codes (see Section 2) on 1995 and earlier vehicles. **Caution:** *The vehicle should not be driven with the Park/Neutral switch disconnected because idle quality will be adversely affected and a false Code 24 (failure in the Vehicle Speed Sensor circuit).*

31   In the event there is a problem with the Park/Neutral switch, first check the terminal connectors for proper attachment.

32   Use a voltmeter and with the ignition key ON (engine not running), check for power to each of the signal wires (A through D) of the switch. There should be voltage present.

33   Check the adjustment of the switch (see Chapter 7B). If the switch is out of adjustment, perform the procedure and clear the codes. Recheck the system for any other problems.

34   Any further diagnostics of the Park/Neutral switch must be performed by a dealer service department or other repair shop because this system requires a special SCAN tool to access the working parameters from the PCM.

Adjustment

35   To adjust the Park/Neutral Position (PNP) switch or to replace the switch, refer to Chapter 7B.

Air conditioning clutch control

Refer to illustrations 4.38a, and 4.38b

Note: *Refer to Chapter 12 for additional information on the location of the relays.*

36   During air conditioning operation, the PCM controls the application of the air conditioning compressor clutch. The PCM controls the air conditioning clutch control relay to delay clutch engagement after the air conditioning is turned ON to allow the IAC valve to adjust the idle speed of the engine to compensate for the additional load.

37   The PCM also controls the relay to disengage the clutch in the event of an exces-

sively high or low pressure within the system or an overheating problem.

38   In most cases, if the air conditioning does not function, the problem is probably related to the air conditioning system relays and switches and not the PCM **(see illustrations)**.

39   If the air conditioning is operating properly and idle is too low when the air conditioning compressor turns on or is too high when the air conditioning compressor turns off, check for an open circuit between the air conditioning control relay and the PCM **(see illustration 4.37c)**.

Vehicle Speed Sensor (VSS)

Refer to illustration 4.40

General description

40   The Vehicle Speed Sensor (VSS) is located in the differential area of the transaxle **(see illustration)**. This sensor is a permanent magnetic variable reluctance sensor that produces a pulsing voltage whenever vehicle speed is over 3 mph. These pulses are translated by the PCM and provided for other systems for fuel and transaxle shift control. The VSS is part of the Transmission Converter Clutch (TCC) system. Any problems with the VSS will usually set a code 24 on 1995 and earlier vehicles.

Check

41   To check the vehicle speed sensor, remove the electrical connector in the wiring harness near the sensor. Using a voltmeter, check for signal voltage to the sensor. The signal wire should have 10 volts or more available. If there is no voltage available, have the PCM diagnosed by a dealer service department or other repair shop.

Replacement

42   To replace the VSS, disconnect the electrical connector from the VSS.

4.40 Location of the Vehicle Speed Sensor on the transaxle

43   Remove the retaining bolt and lift the VSS from the transaxle or simply unscrew the VSS.

44   Installation is the reverse of removal.

Intake Air Temperature (IAT) Sensor

Refer to illustrations 4.47 and 4.49

General description

45   The Intake Air Temperature (IAT) sensor is located inside the air intake duct. This sensor acts as a resistor which changes value according to the temperature of the air entering the engine. Low temperatures produce a high resistance value (for example, at 40-degrees F the resistance is 6.6K to 8.4K ohms) while high temperatures produce low resistance values (at 200-degrees F the resistance is 215 to 235 ohms. The PCM supplies approximately 5-volts (reference voltage) to the IAT sensor. The voltage will change according to the temperature of the incoming

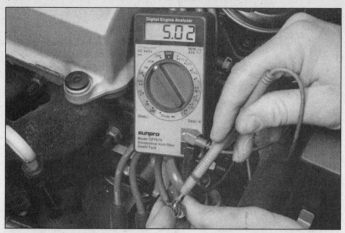

4.47 Unplug the electrical connector from the IAT sensor (located in air intake duct) and working on the harness side, check for reference voltage. It should be approximately 5.0 volts

4.49 Check the resistance of the IAT sensor cold and warm and compare to the specifications listed in this Chapter. It may be necessary to use simulated conditions to obtain accurate results (a hairdryer can be used to heat the sensor)

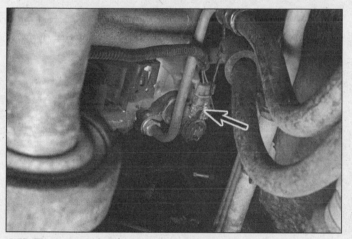

4.50 The power steering pressure switch (arrow) is located under the power steering pump

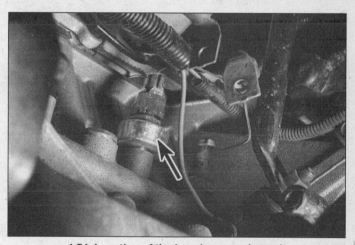

4.54 Location of the knock sensor (arrow)

6

air. The voltage will be high when the air temperature is cold and low when the air temperature is warm. Any problems with the IAT sensor will usually set a code 25 on 1995 and earlier vehicles.

Check

46   To check the IAT sensor, disconnect the two prong electrical connector and turn the ignition key ON but do not start the engine.

47   Measure the voltage (reference voltage) on terminal A. The VOM should read approximately 5-volts **(see illustration)**.

48   If the voltage signal is not correct, have the PCM diagnosed by a dealer service department or other repair shop.

49   Measure the resistance across the sensor terminals **(see illustration)**. The resistance should be HIGH when the air temperature is LOW. Next, start the engine and let it idle (cold). Wait awhile and let the engine reach operating temperature. Turn the ignition OFF, disconnect the IAT sensor and measure the resistance across the terminals. The resistance should be LOW when the air temperature is HIGH. If the sensor does not

exhibit this change in resistance, replace it with a new part.

Power steering pressure switch

Refer to illustration 4.50

50   Turning the steering wheel increases power steering fluid pressure and engine load. The pressure switch **(see illustration)** will close before the load can cause an idle problem. A problem in the power steering pressure switch circuit will set a code 46 (1991 models only).

51   The power steering switch is normally open to ground and the light blue/orange wire (power wire) will be battery voltage. Closing the switch will cause the circuit voltage to drop to 1 volt or less.

52   A pressure switch that will not open or an open circuit from the PCM will cause timing to retard at idle and this will affect idle quality.

53   A pressure switch that will not close or an open circuit may cause the engine to die when the power steering system is used heavily.

Knock sensor

General description

Refer to illustration 4.54

54   The knock sensor **(see illustration)** is located in the engine block directly behind the oil filter. The sensor plays a major role in the Electronic Spark Control (ESC) system. The knock sensor detects abnormal vibration in the engine. The sensor produces an AC output voltage which increases with the severity of the knock.

55   The signal is fed into the PCM and the timing is retarded up to 10-degrees to compensate for the severe "rattling" or "heavy" detonation. Any problems with the knock sensor will set a code 43 on 1995 and earlier vehicles.

Check

56   With the ignition key ON (engine not running) backprobe the sensor electrical connector and check for voltage. It should be between 1.5 and 5.0 volts. If the voltage is higher or lower, there is a short or open on the sensor circuit. Also check for a loose

4.60 Location of the crankshaft position sensor (arrow)

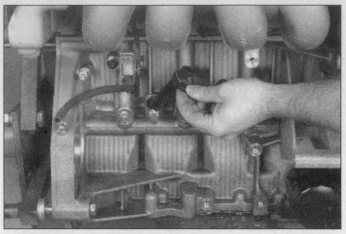

4.62 Remove the retaining bolt and lift the crankshaft position sensor from the engine block

knock sensor. Tighten with an open end wrench if necessary.

57   The operation of the sensor can be checked with the engine running. Reconnect the electrical connector and connect a timing light in accordance with the tool manufacturer's instructions. Start the engine and point the timing light at the timing marks on the crankshaft pulley. Have an assistant sharply tap the backside of the air intake plenum with a hammer. The ignition timing should momentarily retard a few degrees. If it does, the system is working properly. If the harness tests are correct but the system is not working properly, the sensor is probably faulty. Replace the knock sensor with a new part.

Handwheel sensor (1991 and 1992 coupes only)

58   The handwheel sensor is used to determine the rate at which the steering wheel is being turned. The PCM calculates the rate of change in the sensor signal which in turn controls the amount of power assist that is applied to the steering gear. The handwheel sensor is located on the lower portion of the steering column.

59   Any problems with the handwheel sensor should be repaired by the dealer service department.

Crankshaft position sensor

Refer to illustration 4.60

General information

60   The crankshaft position sensor **(see illustration)** defines the engine position to the PCM. A crankshaft position pulse occurs at each TDC, at 60-degrees after (A) TDC and 120-degrees ATDC. A seventh pulse known as the SYNC pulse occurs 70-degrees ATDC on number one cylinder only.

Check

61   Disconnect the wiring from the crankshaft position sensor and set it aside. Connect the leads of an ohmmeter to the termi-

5.10 Apply vacuum to the control vacuum signal tube and make sure the canister holds vacuum

nals of the sensor. Compare the reading with the value listed in the Specifications of this Chapter. If it is not within the limits, replace it.

Replacement

Refer to illustration 4.62

62   Remove the electrical connector and the retaining bolt and lift the assembly from the engine block **(see illustration)**.

63   Installation is the reverse of removal.

---

5   Evaporative Emission Control System (EECS)

Refer to illustration 5.10

General description

Note: *1991 through 1997 vehicles use a conventional EVAP system* **(see illustration 5.2a)**. *1998 through 1999 vehicles use an enhanced system which includes a canister vent solenoid and a fuel tank pressure sensor* **(see illustration 5.2b)**. *Both of these added components are connected to the PCM. The canister for the enhanced system is not located in the engine compartment. It is mounted on top of the fuel tank at the left*

rear, next to the canister vent solenoid.

1   This system is designed to trap and store fuel vapors that evaporate from the fuel tank, throttle body and intake manifold.

2   The Evaporative Emission Control System (EECS) consists of a charcoal-filled canister and the lines connecting the canister to the fuel tank, ported vacuum and intake manifold vacuum **(see illustration)**.

3   Fuel vapors are transferred from the fuel tank, throttle body and intake manifold to a canister where they are stored when the engine is not operating. When the engine is running, the fuel vapors are purged from the canister by a purge control solenoid which is PCM controlled and consumed in the normal combustion process.

Check

4   Poor idle, stalling and poor driveability can be caused by an inoperative purge control solenoid, a damaged canister, split or cracked hoses or hoses connected to the wrong tubes.

5   Evidence of fuel loss or fuel odor can be caused by fuel leaking from fuel lines or the TBI unit, a cracked or damaged canister, an inoperative bowl vent valve, an inoperative

6.7a Apply vacuum to the EGR valve and check with the tip of the finger for movement of the diaphragm. It should move smoothly without any binding with vacuum applied

6.7b With the engine running, check for vacuum to the EGR valve

6.7c With battery voltage applied to the EGR solenoid, there should be no vacuum to the EGR valve

purge valve, disconnected, misrouted, kinked, deteriorated or damaged vapor or control hoses or an improperly seated air cleaner or air cleaner gasket.

6   Inspect each hose attached to the canister for kinks, leaks and breaks along its entire length. Repair or replace as necessary.

7   Inspect the canister. If it is cracked or damaged, replace it.

8   Look for fuel leaking from the bottom of the canister. If fuel is leaking, replace the canister and check the hoses and hose routing.

9   Apply a short length of hose to the lower tube of the purge valve assembly and attempt to blow through it. Little or no air should pass into the canister (a small amount of air will pass because the canister has a constant purge hole).

10   With a hand-held vacuum pump, apply vacuum through the control vacuum signal tube near the throttle body to the purge control solenoid **(see illustration)** diaphragm.

11   If the purge control solenoid does not hold vacuum for at least 20 seconds, the purge control solenoid is leaking and must be replaced.

12   If the diaphragm holds vacuum, apply battery voltage to the purge control solenoid and observe that vacuum (vapors) are allowed to pass through to the intake system. The purge control solenoid is located on the rear side of the engine block next to the crankshaft position sensor and knock sensor **(see illustration 4.60)**.

Canister replacement

13   Clearly label, then detach, all vacuum lines from the canister.

14   Loosen the canister mounting clamp bolt and pull the canister out.

15   Installation is the reverse of removal.

6   Exhaust Gas Recirculation (EGR) system

Refer to illustrations 6.7a, 6.7b, 6.7c and 6.14

General description

1   The EGR system is used to lower NOx (oxides of nitrogen) emission levels caused by high combustion temperatures. It does

this by decreasing combustion temperatures.

2   The EGR system consists of a negative or positive backpressure EGR valve, a ported manifold vacuum source tube and an electronic vacuum regulator valve or more commonly called the EGR solenoid which controls this vacuum source.

Check

3   Too much EGR flow tends to weaken combustion, causing the engine to run rough or stop. When EGR flow is excessive, the engine can stop after a cold start or at idle after deceleration, the vehicle can surge at cruising speeds or the idle may be rough. If the EGR valve remains constantly open, the engine may not idle at all.

4   Too little or no EGR flow allows combustion temperatures to get too high during acceleration and load conditions. This can cause spark knock (detonation), engine overheating or emission test failure.

5   The following checks will help you pinpoint problems in the EGR system. Where the procedure says to lift up on the EGR valve diaphragm, it's a good idea to wear a heat-resistant glove to prevent burns.

EGR valve

6   The EGR valve is controlled by a normally open solenoid which allows vacuum to pass when energized. The PCM energizes the solenoid to turn on the EGR. The PCM controls the EGR when three conditions are present: engine coolant is above 113-degrees F, the TPS is at part throttle and the MAP sensor is in its mid-range. Code 32 will detect a faulty vacuum solenoid, temperature switch or vacuum supply on 1995 and earlier vehicles.

7   The accompanying photos **(see illustrations)** outline the testing procedures for the EGR valve.

EGR control system

8   If a Code 32 is displayed by the SER-

6

VICE ENGINE SOON light on 1995 and earlier vehicles, there are several possibilities for EGR failure. Engine coolant temperature sensor, TPS, MAP sensor, TCC system and the engine rpm govern the parameters the EGR system use for distinguishing the correct ON time.

9   The EGR solenoid uses an PCM controlled pulse width to modulate the EGR solenoid. The valve is normally open (engine at operating temperature) and the vacuum source is a ported signal. The PCM will turn the EGR ON and OFF (the "duty cycle") by grounding circuit 435 (gray wire). The duty cycle should be zero percent (no EGR) when in Park or Neutral, when the TPS input is below the specified value or when Wide Open Throttle (WOT) is indicated.

Component replacement

Caution: *On models equipped with a theft-deterrent radio, make sure you have the correct activation code, or the theft deterrent system is turned off, before disconnecting the battery.*

EGR valve

10   When buying a new EGR valve, make sure that you get the right valve. Use the stamped code located on the top of the EGR valve.
11   Detach the cable from the negative terminal of the battery.
12   Remove the air cleaner housing (see Chapter 4).
13   Detach the vacuum line from the EGR valve.
14   Remove the EGR valve mounting bolts **(see illustration)**.
15   Remove the EGR valve and gasket from the manifold. Discard the gasket.
16   With a wire wheel, buff the exhaust deposits from the EGR valve mounting surface on the manifold and, if you plan to reuse the same valve, the mounting surface of the valve itself. Look for exhaust deposits in the valve outlet. Remove deposit build-up with a screwdriver. **Caution:** *Never wash the valve in solvents or degreaser - both agents will permanently damage the diaphragm. Sand blasting is also not recommended because it will affect the operation of the valve.*
17   If the EGR passage contains an excessive build-up of deposits, clean it out with a wire wheel. Make sure that all loose particles are completely removed to prevent them from clogging the EGR valve or from being

6.14 Remove the two bolts (arrows) from the base of the EGR valve

ingested into the engine.
18   Installation is the reverse of removal.

EGR solenoid

19   Detach the cable from the negative terminal of the battery.
20   Remove the air intake duct from the air cleaner assembly (see Chapter 4).
21   Unplug the electrical connector from the solenoid.
22   Clearly label, then detach, both vacuum hoses.
23   Remove the solenoid mounting screw and remove the solenoid.
24   Installation is the reverse of removal.

---

7   Positive Crankcase Ventilation (PCV) system

---

1   The Positive Crankcase Ventilation (PCV) system reduces hydrocarbon emissions by scavenging crankcase vapors. It does this by circulating fresh air from the air cleaner through the crankcase, where it mixes with blow-by gases and is then rerouted through a PCV valve to the intake manifold.
2   The main components of the PCV system are the PCV valve, a fresh air filtered inlet and the vacuum hoses connecting these two components with the engine and the EECS system.
3   To maintain idle quality, the PCV valve restricts the flow when the intake manifold

vacuum is high. If abnormal operating conditions arise, the system is designed to allow excessive amounts of blow-by gases to flow back through the crankcase vent tube into the air cleaner to be consumed by normal combustion.

---

8   Catalytic converter

General description

1   The catalytic converter is an emission control device added to the exhaust system to reduce pollutants from the exhaust gas stream. A single-bed converter design is used in combination with a three-way (reduction) catalyst. The catalytic coating on the three-way catalyst contains platinum and rhodium, which lowers the levels of oxides of nitrogen (NOx) as well as hydrocarbons (HC) and carbon monoxide (CO).

Check

2   The test equipment for a catalytic converter is expensive and highly sophisticated. If you suspect that the converter on your vehicle is malfunctioning, take it to a dealer or authorized emissions inspection facility for diagnosis and repair.
3   Whenever the vehicle is raised for servicing of underbody components, check the converter for leaks, corrosion and other damage. If damage is discovered, the converter should be replaced.

Replacement

4   Because the converter part of the exhaust system, converter replacement requires removal of the exhaust pipe assembly (see Chapter 4). Take the vehicle, or the exhaust system, to a dealer or a muffler shop.

---

9   Air Pump (AIR) system

The air pump system is used on some 1999 models to reduce emissions during start-up by blowing fresh air into the exhaust manifold for a maximum of 65 seconds. The system is made up of the air pump, a combination valve, a solenoid, a check valve, hoses and tubes.

Unlike conventional systems, the pump is not run from a drive belt, instead it is operated by its own electric motor as directed by the PCM.

Chapter 7 Part A
Manual transaxle

Contents

Specifications

Torque specifications

	Ft-lbs
Transaxle bracket-to-transaxle bolts (1992 and later)	40
Transaxle strut-to-cradle bracket fasteners (1992 and later)	52
Transaxle strut-to-transaxle bracket fasteners (1992 and later)	52
Cradle bracket-to-cradle fasteners (1992 and later)	52
Transaxle-to-engine bolts	
Upper	74
Lower	96
Transaxle lower mount-to-cradle fastener (1991)	35
Transaxle mount-to-cradle fastener (1992 and later)	52
Transaxle rear mount bracket fasteners (1991)	40
Rear transaxle mount pivot bolt (1991)	52
Front transaxle-to-cradle stud nuts (1991)	35
Right side mount-to-cradle stud nuts (1991)	40
Front transaxle mount-to-transaxle bolts (1991)	35

7A

1 General information

The vehicles covered by this manual are equipped with either a 5-speed manual or a 4-speed automatic transaxle. Information on the manual transaxle is included in this Part of Chapter 7. Service procedures for the automatic transaxle are contained in Chapter 7, Part B.

The manual transaxle is a compact, two-piece, lightweight aluminum alloy housing containing both the transmission and differential assemblies.

Because of the complexity, unavailability of replacement parts and special tools necessary, internal repair procedures for the manual transaxle are not recommended for the home mechanic.

2 Shift lever - removal and installation

1 Remove the console (see Chapter 11).
2 Remove the cable guide from the shift lever assembly.
3 Remove the single bolt and bracket that secures the two shift cables, just in front of the shift lever assembly.

4   Remove the four shift lever assembly retaining bolts.

5   Lay the shift lever assembly on its side and detach the selector cable from the shifter bellcrank: Use two flat-bladed screwdrivers and a twisting motion to pop the cable end off the bellcrank DO NOT try to pry off the cable with a single screwdriver - you could damage the bellcrank or the cable end.

6   To detach the other cable (shift cable) from the shifter assembly, grasp the cable end with a pair of channel-lock pliers and rotate the end until it pops loose from the shifter arm.

7   Installation is the reverse of removal. Use care when attaching the selector and shift cables - if the pushrods are bent, it will affect shifting. Make sure the cable retainer plate arrow faces toward the front of the vehicle.

3   Selector and shift cables - replacement

1   Remove the air intake duct (see Chapter 4).

2   Remove the retainer clips from the transaxle ends of the cables with a small screwdriver.

3   Pry off the cable retainer C-clips from the cable bracket.

4   Lift the cables out of the cable bracket. Don't pull them out of the bracket - dragging the boots through the bracket may damage them.

5   Working inside the vehicle, remove the console (see Chapter 11).

6   Remove the shift lever assembly and disconnect the cables from the shift lever (see Section 2).

7   Separate the carpeting at the front of the dash under the shift cables. The carpet is held together by Velcro.

8   Holding the sound-deadening material, remove the retainers on both sides of the cable opening.

9   Peel back the sound-deadening material at the front of the dash to gain access to the firewall cable grommet.

10   There are two kinds of grommets used on the vehicles covered by this manual. All 1991, 1992 and early 1993 models use a grommet and retaining plate attached to the floor pan by two retaining nuts located on the engine compartment side of the pan. Later models use a snap-in grommet. In either case, it's helpful to have an assistant to guide the cables and prevent them from becoming tangled:

a) *Raise the vehicle and place it securely on jackstands.*

b) *To remove the earlier-style grommet, remove the two retaining nuts, push the grommet and retaining plate into the vehicle and remove the cable assembly.*

c) *To remove the newer-style grommet, simply pry it out with a large screwdriver. You'll have to pry out the grommet from*

the driver's side - there's a locking tab on the passenger's side of the grommet. **Caution:** *Prying on the passenger's side of the grommet could damage the floor pan. Push the grommet through from the engine side to disengage it from the pan. Now shove the grommet back through the hole in the pan and remove the cable assembly.*

11   Installation is the reverse of removal. **Note:** *All replacement cable assemblies use the bolt-in type grommet.* Install it from the passenger's side of the pan.

12   Remove the jackstands and lower the vehicle.

4   Back-up light switch - check and replacement

Check

1   The back-up light switch is located on top of the transaxle.

2   Turn the ignition key to the On position and move the shift lever to the Reverse position. The back-up light switch should turn on the back-up lights.

3   If it doesn't, check the back-up light fuse (see Chapter 12).

4   If the fuse is okay, verify that there's voltage available on the battery side of the switch (with the ignition turned to On).

5   If there's no voltage on the battery side of the switch, check the wire between the fuse and the switch; if there is voltage, put the shift lever in Reverse and see if there's voltage on the ground side of the switch.

6   If there's no voltage on the ground side of the switch, replace the switch (see below); if there is voltage, note whether only one or both back-up light bulbs are out.

7   If only one bulb is out, replace it; if they're both out, it could be the bulbs but it's more likely that the wire between the switch and the bulbs has an open somewhere.

Replacement

8   Remove the air intake duct (see Chapter 4).

9   Unplug the electrical connector from the backup light switch.

10   Unscrew the switch.

11   Screw in the new switch and tighten it securely.

12   Plug in the connector.

13   Check the switch operation to ensure it's working properly.

5   Manual transaxle - removal and installation

Warning 1: *Gasoline is extremely flammable, so take extra precautions when disconnecting any part of the fuel system. Don't smoke or allow open flames or bare light bulbs in or near the work area and don't work in a garage where a natural gas appliance (such as a*

clothes dryer or water heater) is installed. If you spill gasoline on your skin, rinse it off immediately; latex gloves will protect your skin from exposure to gasoline. Have a fire extinguisher rated for gasoline fires handy and know how to use it!
Warning 2: *The air conditioning system is under high pressure - have a dealer service department or service station discharge the system before disconnecting any of the fittings or hoses.*

Removal

Caution: *On models equipped with a theft-deterrent radio, make sure you have the correct activation code, or the theft deterrent system is turned off, before disconnecting the battery.*

1   Disconnect the negative battery cable.

2   Loosen the front wheel lug nuts, raise the vehicle and place it securely on jackstands. Remove the front wheels and block the rear wheels.

3   Disconnect the selector and shift cables at the transaxle (see Section 3).

4   Unplug the electrical connectors for the back-up light switch (see Section 4) and the vehicle speed sensor.

5   Remove the left and right driveaxle assemblies (see Chapter 8) and the intermediate shaft.

6   Remove the transaxle mount(s) and (on 1992 and later models) the strut (see Chapter 7, Part B).

7   Disconnect all engine-related electrical connectors, wires, hoses, etc., install an engine hoist, then disconnect the mounts (refer to Chapter 2, Part B, Section 5).

8   Once everything is disconnected, carefully raise up the engine/transaxle assembly slightly, angle the transaxle end down and lift the engine/transaxle assembly out of the engine compartment.

9   With the engine/transaxle assembly out of the vehicle and on the ground, remove the transaxle-to-engine bolts and separate the transaxle from the engine.

10   While the engine and transaxle are removed from the vehicle, inspect all engine and transaxle mounts. If any of them are worn or damaged, replace them.

11   Now is also a good time to inspect and, if necessary, replace the clutch components (see Chapter 8).

Installation

12   Install any clutch components you removed or wish to replace (see Chapter 8).

13   Carefully slide the transaxle into position, engaging the input shaft with the clutch splines. Don't use excessive force to install the transaxle - if the input shaft doesn't slide into place, readjust the angle of the transaxle so it's level and/or turn the input shaft so the splines engage properly with the clutch.

14   Install the transaxle-to-engine bolts and tighten them to the torque listed in this Chapter's Specifications.

15   Raise the engine/transaxle assembly into position and reattach it to the cradle (see Chapter 2, Part B).

16   Install the transaxle mount(s) and (on 1992 and later models) the strut (see Chapter 7, Part B).

17   Install the left driveaxle assembly (see Chapter 8), the intermediate shaft and the right driveaxle assembly.

18   Plug in the electrical connectors for the back-up light switch and the vehicle speed sensor.

19   Connect the select and shift cables (see Section 3).

20   Install the clutch access plate and tighten the bolts securely.

21   Install the wheels, remove the jackstands and lower the vehicle. Tighten the wheel lug nuts to the torque listed in the Chapter 1 Specifications.

22   Check the transaxle lubricant level and add fluid to the specified level (see Chapter 1).

23   Drive the vehicle to an alignment shop and have the wheels aligned.

6   Manual transaxle overhaul - general information

1   Overhauling a manual transaxle is a difficult job for the do-it-yourselfer. It involves the disassembly and reassembly of many small parts. Numerous clearances must be precisely measured and, if necessary, changed with select fit spacers and snap-rings. As a result, if transaxle problems arise, it can be removed and installed by a competent do-it-yourselfer, but overhaul should be left to a transmission repair shop. Rebuilt transaxles may be available - check with your dealer parts department and auto parts stores. At any rate, the time and money involved in an overhaul is almost sure to exceed the cost of a rebuilt unit.

2   Nevertheless, it's not impossible for an inexperienced mechanic to rebuild a transaxle if the special tools are available and the job is done in a deliberate step-by-step manner so nothing is overlooked.

3   The tools necessary for an overhaul include internal and external snap-ring pliers, a bearing puller, a slide hammer, a set of pin punches, a dial indicator and possibly a hydraulic press. In addition, a large, sturdy workbench and a vise or transaxle stand will be required.

4   During disassembly of the transaxle, make careful notes of how each piece comes off, where it fits in relation to other pieces and what holds it in place. Noting how the parts are installed when you remove them will make it much easier to get the transaxle back together.

5   Before taking the transaxle apart for repair, it will help if you have some idea what area of the transaxle is malfunctioning. Certain problems can be closely tied to specific areas in the transaxle, which can make component examination and replacement easier. Refer to the *Troubleshooting* section at the front of this manual for information regarding possible sources of trouble.

7A

Notes

Chapter 7 Part B
Automatic transaxle

Contents

Specifications

Torque specifications

	Ft-lbs
Torque converter-to-driveplate bolts	52
Transaxle-to-engine mounting bolts	
Lower bolts	96
Upper bolts	74
Transaxle strut-to-cradle bracket fasteners (1992 and later)	52
Front transaxle mount-to-transaxle bolts (1991)	35
Transaxle mount-to-cradle fasteners (1992 and later)	52
Transaxle strut cradle bracket-to-cradle nut (1992 and later)	52
Front transaxle mount-to-transaxle bolts (1991)	35
Front transaxle-to-cradle stud nuts	35
Right side mount-to-cradle stud nuts (1991)	40
Rear pitch restrictor-to-transaxle bolts (1991)	40
Powertrain stiffening bracket bolts	35

1 General information

All vehicles covered in this manual come equipped with either a 5-speed manual or a 4-speed automatic transaxle. All information on the automatic transaxle is included in this Part of Chapter 7. Information for the manual transaxle can be found in Part A of this Chapter.

Due to the complexity of the automatic transaxles covered in this manual and to the specialized equipment necessary to perform most service operations, this Chapter contains only those procedures related to general diagnosis, routine maintenance, adjustment and removal and installation.

If the transaxle requires major repair work, it should be left to a dealer service department or an automotive or transmission repair shop. You can, however, remove and install the transaxle yourself and save the expense, even if the repair work is done by a transmission shop.

2 Diagnosis - general

Note: *Automatic transaxle malfunctions may be caused by five general conditions: poor engine performance, improper adjustments, hydraulic malfunctions, mechanical malfunctions or malfunctions in the computer or its signal network. Diagnosis of these problems should always begin with a check of the easily repaired items: fluid level and condition (see Chapter 1) and proper installation of the shift lever and shift cable. Next, perform a road test to determine if the problem has been corrected or if more diagnosis is necessary. If the problem persists after the preliminary tests and corrections are completed, additional diagnosis should be done by a dealer service department or transmission repair shop. Refer to the Troubleshooting section at the front of this manual for information on symptoms of transaxle problems.*

Preliminary checks

1 Drive the vehicle to warm the transaxle to normal operating temperature.
2 Check the fluid level as described in Chapter 1:

a) *If the fluid level is unusually low, add enough fluid to bring the level within the designated area of the dipstick, then check for external leaks (see below).*

b) *If the fluid level is abnormally high, drain off the excess, then check the drained fluid for contamination by coolant. The presence of engine coolant in the automatic transmission fluid indicates that a failure has occurred in the internal radiator walls that separate the coolant from the transmission fluid (see Chapter 3).*

c) *If the fluid is foaming, drain it and refill the transaxle, then check for coolant in the fluid, or a high fluid level.*

3 Check the engine idle speed. **Note:** *If the engine is malfunctioning, do not proceed*

with the preliminary checks until it has been repaired and runs normally.

4   Check the shift cable for freedom of movement and make sure it's properly installed (see Section 4).

5   Inspect the shift lever linkage (see Section 7) and the manual lever linkage (at the neutral-start switch) on the transaxle (see Chapter 5). Make sure both are operating properly and smoothly.

Fluid leak diagnosis

6   Most fluid leaks are easy to locate visually. Repair usually consists of replacing a seal or gasket. If a leak is difficult to find, the following procedure may help.

7   Identify the fluid. Make sure it's transmission fluid and not engine oil or brake fluid (automatic transmission fluid is a deep red color).

8   Try to pinpoint the source of the leak. Drive the vehicle several miles, then park it over a large sheet of cardboard. After a minute or two, you should be able to locate the leak by determining the source of the fluid dripping onto the cardboard.

9   Make a careful visual inspection of the suspected component and the area immediately around it. Pay particular attention to gasket mating surfaces. A mirror is often helpful for finding leaks in areas that are hard to see.

10   If the leak still cannot be found, clean the suspected area thoroughly with a degreaser or solvent, then dry it.

11   Drive the vehicle for several miles at normal operating temperature and varying speeds. After driving the vehicle, visually inspect the suspected component again.

12   Once the leak has been located, the cause must be determined before it can be properly repaired. If a gasket is replaced but the sealing flange is bent, the new gasket will not stop the leak. The bent flange must be straightened.

13   Before attempting to repair a leak, check to make sure that the following conditions are corrected or they may cause another leak. **Note:** *Some of the following conditions cannot be fixed without highly specialized tools and expertise. Such problems must be referred to a transmission shop or a dealer service department.*

Gasket leaks

14   Check all gasket mating surfaces of the top and side covers periodically. Make sure no bolts are missing, the gasket is in good condition and the cover is flat (dents in the top cover may indicate damage to the valve body inside).

15   If a cover gasket is leaking, the fluid level or the fluid pressure may be too high, the vent may be plugged, the cover bolts may be too tight, the cover sealing flange may be warped, the sealing surface of the transaxle housing may be damaged, the gasket may be damaged or the transaxle casting may be cracked or porous. If sealant instead of gasket material has been used to form a seal

between the cover and the transaxle housing, it may be the wrong sealant.

Seal leaks

16   If a transaxle seal is leaking, the fluid level or pressure may be too high, the vent may be plugged, the seal bore may be damaged, the seal itself may be damaged or improperly installed, the surface of the shaft protruding through the seal may be damaged or a loose bearing may be causing excessive shaft movement.

17   Make sure the dipstick tube seal is in good condition and the tube is properly seated. Periodically check the area around the speedometer gear or sensor for leakage. If transmission fluid is evident, check the O-ring for damage.

Case leaks

18   If the case itself appears to be leaking, the casting is porous and will have to be repaired or replaced.

19   Make sure the oil cooler hose fittings are tight and in good condition.

Fluid comes out vent pipe or fill tube

20   If this condition occurs, the transaxle is overfilled, there is coolant in the fluid, the case is porous, the dipstick is incorrect, the vent is plugged or the drain-back holes are plugged.

---

3   Brake Transaxle Shift Interlock (BTSI) system - description, check and component replacement

---

Warning: *On models equipped with a Supplemental Inflatable Restraint (SIR) system, always disable the airbag(s) before working in the vicinity of the steering wheel, instrument panel or SIR system components to avoid the possibility of accidental deployment of the airbag(s), which could cause personal injury (see Chapter 12).*

Description

1   The Brake Transaxle Shift Interlock (BTSI) system prevents the driver from taking the shift lever out of Park with the engine running unless the brake pedal is depressed. The 1991 BTSI system includes a solenoid, park sensor switch and brake light switch; the 1992 and later system includes a brake light switch (same as the 1991 switch) and a BTSI solenoid incorporated into the park lock cable.

BTSI system bypass

2   If, for some reason, application of the brakes should fail to release the shifter electronically, or if service or towing necessitates that the vehicle be shifted out of Park without operating the engine, you can bypass the BTSI system:

a) *Turn the ignition key to the Accessory position.*
b) *Apply the brakes*
c) *Shift to Neutral. The vehicle can now be operated normally.*

BTSI system shift lock override (1991 vehicles)

3   If the bypass procedure doesn't release the shift lever so it can be moved out of Park, there's also a manual override on these vehicles:

a) *Remove the ashtray.*
b) *Turn the ignition key to the Accessory position.*
c) *Apply the brakes.*
d) *Push down on the plastic release lever located under the ashtray on the left side.*
e) *While holding down on the release lever, move the shift lever to Neutral. The vehicle can now be operated normally.*

Component replacement

Solenoid (1991 models)

4   Remove the console, if you haven't already done so (see Chapter 11).

5   Unplug the electrical connector from the BTSI solenoid.

6   Remove the retaining nut from the BTSI solenoid and detach the solenoid from the shift lever assembly.

7   Installation is the reverse of removal. Before installing the console, however, test the system as follows: With the ignition on and the brake off, try to move the shift lever out of Park - it should be impossible to do so. Now, with the brake on, try to move the shift lever out of Park - you should be able to.

Solenoid (1992 through 1994 models)

8   The BTSI solenoid is part of the park lock cable assembly on these vehicles (see Section 3).

Solenoid (1994 and later models)

9   Remove the console, if you haven't already done so (see Chapter 11).

10   Unplug the electrical connector from the BTSI solenoid.

11   Using a small screwdriver, pry the retainers from the shift lever base and the park lock lever.

12   Installation is the reverse of removal. Check the system operation as in Step 15 above.

Park sensor switch (1991 models)

13   Remove the console, if you haven't already done so (see Chapter 11).

14   Unplug the electrical connector for the park sensor switch.

15   Remove the park sensor switch fastener and remove the switch.

16   Installation is the reverse of removal. Verify the operation of the BTSI system as described in Step 15.

4.2 Put the shift lever in 2nd gear and disconnect the cable end from the lever by popping it loose with a screwdriver

4.3 Depress the two tabs on the cable housing and remove the cable from the shift lever assembly

Brake switch

17   Before you replace the brake switch, make sure it's not simply out of adjustment:

a)   *With the brake pedal depressed, the switch contacts that input the signal to the PCM should be open (there should be no continuity); with the brake pedal released, these switch contacts should be closed (there should be continuity).*

b)   *To adjust the brake switch, loosen the bolt and slide it towards, or away from, the brake pedal. The plunger should be fully depressed with the brake pedal released.*

18   Before removing the brake switch, disconnect the cable from the negative battery terminal.

19   Unplug the brake switch electrical connector.

20   Remove the bolt from the brake switch and remove the brake switch.

21   Installation is the reverse of removal. There's a tang on the brake switch that fits into a slotted hole. This tab keeps the brake switch parallel to the brake pedal. Before you tighten the bolt, make sure the plunger is completely depressed when the brake pedal is released.

Brake light switch check and adjustment

22   The brake light switch sets pedal height. To check and adjust the brake light switch, refer to Chapter 9.

---

4   Shift cable - removal and installation

Warning: *On models equipped with a Supplemental Inflatable Restraint (SIR) system, always disable the airbag(s) before working in the vicinity of the steering wheel, instrument panel or SIR system components to avoid the possibility of accidental deployment of the airbag(s), which could cause personal injury (see Chapter 12).*

4.8 The forward end of the shift cable is attached to the neutral start switch on the back side of the transaxle; to disconnect the cable end from the shift lever on the transaxle, remove this nut (arrow)

Removal

Refer to illustrations 4.2, 4.3, 4.8 and 4.10

1   Remove the console (see Chapter 11).

2   Put the shift lever in 2nd gear and disconnect the cable end from the lever **(see illustration).**

3   Depress the two tabs on the cable housing **(see illustration)** and remove the cable from the shifter assembly.

4   Separate the carpeting at the front of the dash under the shift cables. The carpet is held together by Velcro.

5   Holding the sound deadening material, remove the retainers on both sides of the cable opening.

6   Peel back the sound deadening material at the front of the dash to gain access to the firewall cable grommet.

7   Remove the air intake duct assembly (see Chapter 4).

8   Disconnect the cable end from the shift lever on the transaxle **(see illustration).**

9   Depress the two shift cable housing retaining tabs and detach the cable housing

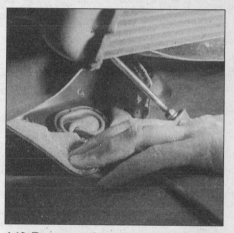

4.10 To remove the newer-style grommet, simply pry it out with a large screwdriver (pry the grommet from the driver's side - prying on the locking tab located on the passenger side of the grommet could damage the floor pan), push the grommet through from the engine side to disengage it from the pan, shove the grommet back through the hole in the pan and remove the cable assembly

from its bracket on the transaxle.

10   There are two kinds of grommets used on the vehicles covered by this manual. All 1991, 1992 and early 1993 models use a grommet and retaining plate attached to the floor pan by two retaining nuts located on the engine compartment side of the pan. Starting in late 1993, a snap-in grommet is used **(see illustration).** In either case, it's helpful to have an assistant to guide the cables and prevent them from becoming tangled:

a)   *Raise the vehicle and place it securely on jackstands.*

b)   *To remove the earlier-style grommet, remove the two retaining nuts, push the grommet and retaining plate into the vehicle and remove the cable assembly.*

c)   *To remove the newer-style grommet, simply pry it out with a large screwdriver.*

7B

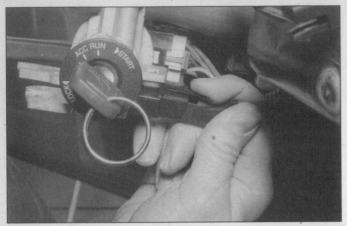

5.3 Turn the ignition to the On position, depress the retaining tab and disconnect the cable from the ignition-switch module

5.4 On 1992 through 1994 models, unplug the BTSI electrical connector

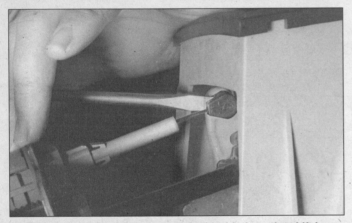

5.5 To unsnap the end of the park lock cable from the shift lever assembly, use a screwdriver to press the plastic clip rearward - DO NOT try to pry the clip off or you'll break it!

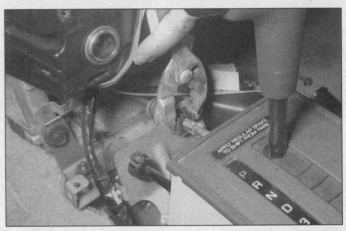

5.6 Compress the two tabs on the end of the cable housing and push the housing through the bracket

You'll have pry out the grommet from the driver's side - there's a locking tab on the passenger's side of the grommet. *Caution:* Prying on the passenger's side of the grommet could damage the floor pan. *Push the grommet through from the engine side to disengage it from the pan. Now shove the grommet back through the hole in the pan and remove the cable assembly.*

Installation

11   Installation is basically the reverse of removal, with the following addition.
12   Before reconnecting the cable to the transaxle shift lever, make sure the transaxle shift lever and the shift lever inside the vehicle are both in Park, then release the cable adjustment lock tab by prying up on the lock tab with a screwdriver until you can lift it with your hand .
13   After connecting the cable to the shift lever on the transaxle, move the cable housing back and forth in the adjuster and note the endplay: Center the cable housing in the middle of the endplay, press in the lock tab and verify that the cable operates correctly.
14   Installation is otherwise the reverse of removal. **Note:** *All replacement cable assem-*

blies use the bolt-in type grommet. Install it from the passenger's side of the pan.

---

5   Park lock cable - removal, installation and adjustment

Warning: *On models equipped with a Supplemental Inflatable Restraint (SIR) system, always disable the airbag(s) before working in the vicinity of the steering wheel, instrument panel or SIR system components to avoid the possibility of accidental deployment of the airbag(s), which could cause personal injury* (see Chapter 12).

Removal and installation

Refer to illustrations 5.3, 5.4, 5.5 and 5.6
Note: *If you're replacing the park lock cable on a 1992 or later vehicle because of an inoperative Brake Transaxle Shift Interlock (BTSI) system, make sure you adjust the park lock cable before you conclude that the BTSI solenoid is defective.*

1   Remove the console (see Chapter 11).
2   Remove the lower steering column fasteners and remove the steering column cover (see Chapter 11).

3   Turn the ignition to the On position, depress the retaining tab and disconnect the cable from the ignition switch module **(see illustration).**
4   On 1992 through 1994 models, unplug the BTSI electrical connector **(see illustration).**
5   Unsnap the end of the park lock cable from the plastic lock-out lever on the shifter **(see illustration).**
6   Depress the two tabs on the end of the cable housing and remove the cable from the shift lever assembly **(see illustration).**
7   Remove the cable assembly from the vehicle. Note the routing of the cable to ensure proper installation.
8   Installation is the basically the reverse of removal, with the following exceptions:
  a) *To reattach the cable to the ignition switch module, turn the ignition to the On position, insert the cassette end of the cable into the ignition switch module and turn the ignition to the Off position.*
  b) *Be sure to lift the lock tab on the cable end fitting to allow the housing to move freely in the end fitting, then depress the lock tab* **(see illustration 4.12).**
9   After the park lock cable is installed, be sure to adjust it.

Adjustment

1994 and earlier models

10 Remove the ashtray.

11 On 1991 vehicles, turn the ignition key to the Off position; on 1992 and later vehicles, turn it to the On position. Place the shift lever in Park, take your foot off the brake, depress the lock on the cable end fitting and remove the adjustment clip from the end terminal on the new cable.

12 To reconnect a cable without an adjustment clip, adjust it to provide a 0.05-inch gap between the end of the cable and the cable-to-park lock connector.

13 Verify the operation of the park lock cable as follows:

a) *Turn the ignition key to Off and put the shift lever in Park. Now try to shift out of Park. The shift lever should be impossible to shift out of Park.*

b) *Turn the ignition key to the On position. You should now be able to move the shift lever out of Park.*

c) *With the shift lever out of Park, turn the ignition key to the Off position. You should not be able to remove the key from the ignition.*

d) *Move the shift lever back to the Park position. You should now be able to remove the ignition key.*

14 If the park lock system doesn't operate as described, readjust it until it does.

1995 and later models

15 Place the shift lever in the park position, turn the ignition key to Lock and remove the key.

16 Remove the cupholder/ashtray assembly.

17 Using a small screwdriver, pry the cable locking tab up. Quickly depress and release the shift lever lock button three or four times to set the cable in the initial adjusting position.

18 Using a sharp awl, scribe a mark on the slider, at the lock body junction. Push the cable toward the shift lever until the mark is 0.050-inch from the lock body.

19 Hold the cable in this position and depress the lock tab.

20 Check the operation of the park lock system as described in Step 13 above. If the park lock system doesn't operate as described, readjust it until it does.

6 Neutral start switch - replacement and adjustment

Replacement

1 Remove the intake air duct (see Chapter 4). On MPFI models, remove the air cleaner housing as well. **Note:** *Some models require removal of the left front wheel and the left lower plastic splash shield to gain access to the switch from beneath. The air intake assembly can be left intact on these vehicles.*

2 Disconnect the shift cable from the control lever (see Section 4).

3 Unplug the electrical connectors from the switch

4 Remove the shift lever retaining nut and remove the shift lever.

5 Remove the switch retaining bolts and pull the switch off the shift shaft.

6 Installation is the reverse of removal. Don't tighten the switch retaining bolts or install the air cleaner and/or intake duct until the switch is adjusted.

Adjustment

Refer to illustration 6.8

7 Put the transaxle in Neutral.

8 Using an ohmmeter or continuity checker, check for continuity across the terminals of the switch electrical connector.

9 If there's no continuity, loosen the switch retaining bolts and rotate the switch until there is. The switch is now in the Neutral position.

10 Tighten the bolts securely.

11 The remainder of installation is the reverse of removal.

7 Shift lever - removal and installation

Warning: *On models equipped with a Supplemental Inflatable Restraint (SIR) system, always disable the airbag(s) before working in the vicinity of the steering wheel, instrument panel or SIR system components to avoid the possibility of accidental deployment of the airbag(s), which could cause personal injury (see Chapter 12).*

Caution: *During the following procedure, make sure the ignition remains in the Off position while the shift cable is disconnected. If the ignition key is turned on with the cable disconnected, the park lock cable could become disengaged from the ignition switch.*

1 Remove the console (see Chapter 11).

2 Disconnect the shift cable from the shift lever (see Section 4).

3 Disconnect the park lock cable from the shift lever (see Section 5).

4 Disconnect the brake transaxle shift

8.4b Using a small screwdriver, carefully pry out the seal retaining ring

interlock wiring and the illumination bulb.

5 Remove the mounting nuts or bolts

6 Installation is the reverse of removal.

8 Oil seal replacement

Side gear seals

Refer to illustrations 8.4a, 8.4b, 8.4c, 8.5, 8.6a and 8.6b

1 Transaxle oil leaks frequently occur due to wear of the differential side gear seals and/or the vehicle speed sensor O-ring. Replacement of these seals is relatively easy, since the repairs can be performed without removing the transaxle from the vehicle.

2 The differential side gear seals are located in the sides of the transaxle, where the splined inner ends of the driveaxles are inserted into the differential side gears. If leakage at either of these seals is suspected, raise the vehicle and support it securely on jackstands. If the seal is leaking, lubricant will be found on the side of the transaxle.

3 Remove the driveaxle (see Chapter 8).

4 Carefully pry off the oil seal cover, pry out the seal retaining ring and pry the seal out of its bore in the transaxle **(see illustrations)**.

8.4a Using a screwdriver or prybar, carefully pry the oil seal cover out of the transaxle bore

8.4c Using a screwdriver or prybar, pry the seal out of the bore - make sure you don't gouge or scratch the seal bore

8.5 If you're unable to remove the seal with a screwdriver or prybar, use a seal removal tool

8.6a Make sure the diameter of the seal installation tool (such as this aftermarket seal driver) matches the diameter of the new seal, so that the circumference of the seal installer seats against the small ridge around the circumference of the seal

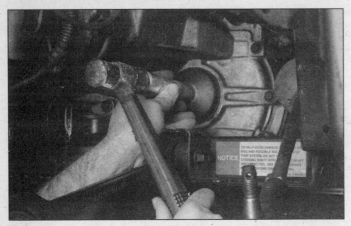

8.6b Drive the new seal into the bore squarely and make sure it's completely seated

8.8 Location of the vehicle speed sensor

5   If the oil seal cannot be removed with a screwdriver or a prybar, a special oil seal removal tool (available at auto parts stores) will be required **(see illustration)**.

6   Using a seal installer **(see illustration)**, a large section of pipe or a large deep socket as a drift, install the new oil seal **(see illustration)**. Drive it into the bore squarely and make sure that it is completely seated **(see illustration)**. Lubricate the lip of the new seal with multi-purpose grease.

7   Install the driveaxle (see Chapter 8). Be careful not to damage the lip of the new seal.

Vehicle speed sensor

Refer to illustration 8.8

8   The speed sensor **(see illustration)** is located on the transaxle housing. Look for lubricant around the cable housing to determine if the O-ring is leaking.

9   To replace the speed sensor O-ring, simply unplug the connector, unscrew the sensor, discard the old O-ring, install the new O-ring, screw in the sensor and plug in the connector.

9   Transaxle mounts - check and replacement

1   Transaxles on 1991 models use three mounts - front, lower left and rear. Transaxles on 1992 and later models use a lower left mount and a strut (instead of the front mount used on earlier units). There is no rear mount on 1992 and later transaxles.

Check

Refer to illustration 9.2

2   Insert a large screwdriver or prybar between each mount and/or strut, and the cradle and transaxle brackets between which it's installed, and pry up **(see illustration)**.

3   The transaxle should not move excessively away from the mount or strut. If it does, replace the mount or strut.

9.2 To check a transaxle mount or strut for wear, insert a large screwdriver or prybar between the mount/strut and the bracket and try to pry the transmission up and down or side-to-side; the transaxle shouldn't move very much; if it does, replace the mount or strut

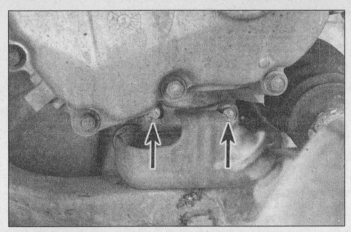

9.4a Left transaxle mount-to-transaxle bolts (arrows)

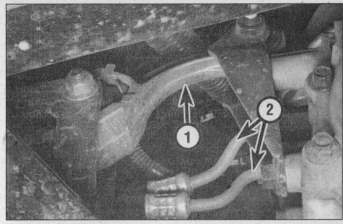

9.4b Left transaxle strut (1) and transaxle cooler lines (2)

10.8 After disconnecting the section of exhaust pipe that's bolted to the engine-to-transaxle stiffener plate, remove the stiffener bolts (arrows) and remove the stiffener, then remove the lower engine-to-transaxle bolt (upper right arrow, next to the axleshaft)

10.11 There are two upper and two lower engine-to-transaxle bolts

Replacement

Refer to illustrations 9.4a and 9.4b

4   To replace a mount or strut, support the transaxle with a jack, remove the nuts and bolts and remove the mount **(see illustrations)**. It may be necessary to raise the transaxle slightly to provide enough clearance to remove the mount.

5   Installation is the reverse of removal.

10 Automatic transaxle - removal and installation

Warning 1: *Gasoline is extremely flammable, so take extra precautions when disconnecting any part of the fuel system. Don't smoke or allow open flames or bare light bulbs in or near the work area and don't work in a garage where a natural gas appliance (such as a clothes dryer or water heater) is installed. If you spill gasoline on your skin, rinse it off immediately. Have a fire extinguisher rated for gasoline fires handy and know hot to use it!*

Warning 2: *The air conditioning system is under high pressure - have a dealer service department or service station discharge the system before disconnecting any of the fittings or hoses.*

Caution: *On models equipped with a theft-deterrent radio, make sure you have the correct activation code, or the theft deterrent system is turned off, before disconnecting the battery.*

Removal

Refer to illustrations 10.8 and 10.11

1   Disconnect the negative battery cable.

2   Loosen the front wheel lug nuts, raise the vehicle and place it securely on jackstands. Remove the front wheels and block the rear wheels.

3   Disconnect the shift cable (see Section 4).

4   Remove the clutch release cylinder and damper from the transaxle as a complete unit. Rotate the actuator 1/4 turn counterclockwise while pushing toward the housing to disengage the bayonet connector from the housing. Remove the two mounting nuts and slide the release cylinder and damper from the transaxle. Do not disconnect any hydraulic lines from the release cylinder system, because it is a sealed system (see Chapter 8).

5   Disconnect all engine-related electrical connectors, wires, hoses, etc., install an engine hoist, then disconnect the mounts and struts (refer to Chapter 2, Part B, Section 5). Also remove the torque converter access plate and the driveplate-to-torque converter bolts.

6   Disconnect the transaxle cooler hoses (the hoses are visible in illustration 9.4b). Plug the hoses to prevent contamination.

7   Remove the left and right driveaxle assemblies and the intermediate shaft (see Chapter 8).

8   Disconnect the exhaust pipe from the transaxle-to-engine stiffener (see Chapter 4). Remove the transaxle-to-engine stiffener plate **(see illustration)**. Then remove the lower engine-to-transaxle bolt (next to the axleshaft).

9   Remove the transaxle mount(s) and (on 1992 and later models) the strut (see Section 9).

10   Once everything is disconnected, carefully raise up the engine/transaxle assembly slightly, angle the transaxle end down and lift the engine/transaxle assembly out of the engine compartment.

11   With the engine/transaxle assembly out of the vehicle and on the ground, remove the transaxle-to-engine bolts **(see illustration)** and separate the transaxle from the engine. Keep the transaxle level, so the torque con-

7B

verter doesn't slide forward and fall out of the bellhousing.

12 While the engine and transaxle are removed from the vehicle, inspect all engine and transaxle mounts. If any of them are worn or damaged, replace them.

13 Remove the torque converter and discard the converter O-ring.

Installation

14 Install a new torque converter O-ring, apply a dab of grease to the nose of the torque converter and to the O-ring and slide the torque converter into position. Keeping the torque converter level, rotate it to align the input shaft spline and the pump drive slots. Then rotate the converter so that the paint dot faces down.

15 Make sure that the torque converter hub is securely engaged in the pump, then carefully guide the transaxle into place until the dowel pins and the torque converter are fully engaged. **Caution**: *Do not use the transaxle-to-engine bolts to force the engine and transaxle into alignment. You could crack or damage major components. If the transaxle does not easily mate up with the engine, the torque converter is probably not fully seated. Remove the transaxle and rotate the converter while pressing in - the converter should seat properly.*

16 Install the transaxle-to-engine bolts and tighten them to the torque listed in this Chapter's Specifications.

17 Raise the engine/transaxle assembly into position, angle the transaxle end down, then lower it into the engine compartment, align it properly and reattach it to the cradle (see Chapter 2, Part B).

18 Install the transaxle mount(s) and (on 1992 and later vehicles) the strut (see Section 9).

19 Install both left and right driveaxle assemblies (see Chapter 8) and the intermediate shaft.

20 Connect the transaxle cooler hoses.

21 Plug in the electrical connector for the vehicle speed sensor.

22 Connect the shift cable (see Section 4).

23 Install the torque converter-to-driveplate bolts and tighten them to the torque listed in that Chapter's Specifications. **Note:** *Install all of the bolts before tightening any of them. Install the torque converter access plate and tighten the bolts securely.*

24 Slide the damper onto the studs of the transaxle and the release cylinder, simultaneously. Install the actuator and rotate it 1/4 turn clockwise while pushing toward the housing.

25 Install the wheels, remove the jack stands and lower the vehicle. Tighten the wheel lug nuts to the torque listed in the Chapter 1 Specifications.

26 Check the transaxle fluid level and add fluid to the specified level (see Chapter 1).

Chapter 8
Clutch and driveaxles

Contents

Specifications

Clutch
Fluid type	See Chapter 1

Driveaxles
Inner CV joint length
Right	8-5/8 inches
Left	11-inches

Torque specifications
	Ft-lbs
Clutch pressure plate-to-flywheel bolts	18
Clutch hydraulic damper mounting nuts	18
Driveaxle/hub nut	145
Intermediate shaft support bracket bolts	40
Wheel lug nuts	See Chapter 1

8

1 General information

The information in this Chapter deals with the components from the rear of the engine to the front wheels, except for the transaxle, which is dealt with in Chapter 7A and 7B. For the purposes of this Chapter, these components are grouped into two categories: Clutch and driveaxles. Separate Sections within this Chapter offer general descriptions and checking procedures for both groups.

Since nearly all the procedures covered in this Chapter involve working under the vehicle, make sure it's securely supported on sturdy jackstands or a hoist where the vehicle can be easily raised and lowered.

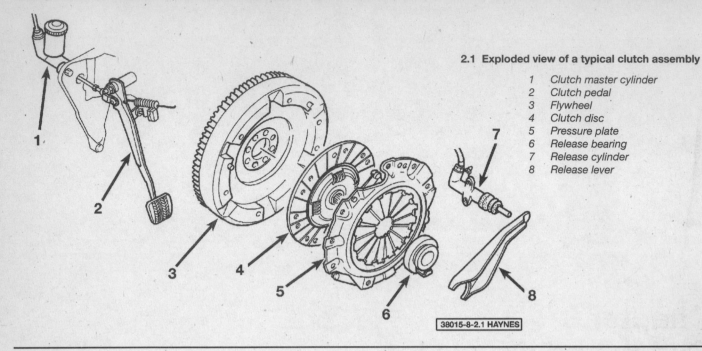

2.1 Exploded view of a typical clutch assembly

1 Clutch master cylinder
2 Clutch pedal
3 Flywheel
4 Clutch disc
5 Pressure plate
6 Release bearing
7 Release cylinder
8 Release lever

38015-8-2.1 HAYNES

2 Clutch - description and check

Refer to illustration 2.1

1 All vehicles with a manual transaxle use a single dry plate, diaphragm spring type clutch **(see illustration)**. The clutch disc has a splined hub which allows it to slide along the splines of the transaxle input shaft. The clutch and pressure plate are held in contact by spring pressure exerted by the diaphragm in the pressure plate.

2 The clutch release system is operated by hydraulic pressure. The hydraulic release system consists of the clutch pedal, a sealed hydraulic release system (which consists of a master cylinder and fluid reservoir, the hydraulic line and a slave cylinder) which actuates the clutch release lever and the clutch release (or throw-out) bearing.

3 When pressure is applied to the clutch pedal to release the clutch, hydraulic pressure is exerted against the outer end of the clutch release lever. As the lever pivots, it pushes the release bearing against the fingers of the diaphragm spring of the pressure plate assembly, which in turn releases the clutch plate.

4 Terminology can be a problem regarding the clutch components because common names have in some cases changed from that used by the manufacturer. For example, the driven plate is also called the clutch plate or disc, the pressure plate assembly is sometimes referred to as the clutch cover, the clutch release bearing is sometimes called a throw-out bearing, and the release cylinder is sometimes called the operating or slave cylinder.

5 Other than replacing components that have obvious damage, some preliminary checks should be performed to diagnose a clutch system failure.

a) *The first check should be of the fluid level in the clutch master cylinder (see Chapter 1). If the fluid level is low, add fluid as necessary and inspect the hydraulic system for leaks. If the master cylinder reservoir has run dry, bleed the system (see Section 6) and retest the clutch operation.*

b) *To check "clutch spin down time," run the engine at normal idle speed with the transaxle in Neutral (clutch pedal up - engaged). Disengage the clutch (pedal down), wait several seconds and shift the transaxle into Reverse. No grinding noise should be heard. A grinding noise would most likely indicate a problem in the pressure plate or the clutch disc.*

c) *To check for complete clutch release, run the engine (with the parking brake applied to prevent movement) and hold the clutch pedal approximately 1/2-inch from the floor. Shift the transaxle between 1st gear and Reverse several times. If the shift is not smooth, component failure is indicated. Check the release cylinder pushrod travel. With the clutch pedal depressed completely the release cylinder pushrod should extend substantially. If it doesn't, check the fluid level in the clutch master cylinder.*

d) *Visually inspect the clutch pedal bushing at the top of the clutch pedal to make sure there is no sticking or excessive wear.*

3 Clutch components - removal, inspection and installation

Warning: *Dust produced by clutch wear and deposited on clutch components may con-* tain asbestos, which is hazardous to your health. DO NOT blow it out with compressed air and DO NOT inhale it. DO NOT use gasoline or petroleum based solvents to remove the dust. Brake system cleaner should be used to flush the dust into a drain pan. After the clutch components are wiped clean with a rag, dispose of the contaminated rags and cleaner in a labeled, covered container.

Removal

Refer to illustration 3.6

1 Access to the clutch components is accomplished by removing the engine, rather than the usual method of removing the transaxle. If, of course, the engine is being removed for major overhaul, the opportunity should always be taken to check the clutch for wear and replace worn components as necessary. However, the relatively low cost of the clutch components compared to the time and labor involved in gaining access to them warrants their replacement any time the engine is removed, unless they are new or in near-perfect condition.

2 Remove the engine from the vehicle (see Chapter 2, Part B).

3 The release fork and release bearing can remain attached to the transaxle for the time being.

4 To support the clutch disc during removal, install a clutch alignment tool through the clutch disc hub.

5 Carefully inspect the flywheel and pressure plate for indexing marks. The marks are usually an X, an O or a white letter.

6 If they cannot be found, scribe marks yourself so the pressure plate and the flywheel will be in the same alignment during installation **(see illustration)**.

7 Slowly loosen the pressure plate-to-flywheel bolts. Work in a criss-cross pattern

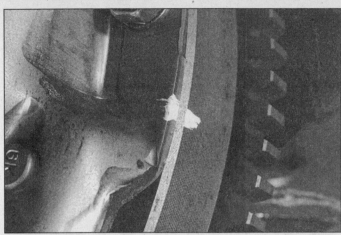

3.6 Mark the relationship of the pressure plate to the flywheel (in case you are going to reuse the same pressure plate)

3.10 Examine the clutch disc for evidence of excessive wear, such as smeared friction material, chewed-up rivets, worn hub splines and distorted damper cushions or springs

EXCESSIVE WEAR

NORMAL FINGER WEAR   **EXCESSIVE FINGER WEAR**   **BROKEN OR BENT FINGERS**

3.12a Replace the pressure plate if any of these conditions are noted

and loosen each bolt a little at a time until all spring pressure is relieved. Then hold the pressure plate securely and completely remove the bolts, followed by the pressure plate and clutch disc.

Inspection

Refer to illustrations 3.10, 3.12a and 3.12b

8   Ordinarily, when a problem occurs in the clutch, it can be attributed to wear of the clutch driven plate assembly (clutch disc). However, all components should be inspected at this time.

9   Inspect the flywheel for cracks, heat checking, score marks and other damage. If the imperfections are slight, a machine shop can resurface it to make it flat and smooth. Refer to Chapter 2, Part B for the flywheel removal procedure.

10   Inspect the lining on the clutch disc. There should be at least 1/16-inch of lining above the rivet heads. Check for loose rivets, distortion, cracks, broken springs and other obvious damage **(see illustration)**. As mentioned above, ordinarily the clutch disc is replaced as a matter of course, so if in doubt about the condition, replace it with a new one.

11   The release bearing should be replaced along with the clutch disc (see Section 4).

12   Check the machined surface and the diaphragm spring fingers of the pressure plate **(see illustrations)**. If the surface is grooved or otherwise damaged, replace the pressure plate assembly. Also check for obvious damage, distortion, cracking, etc. Light glazing can be removed with emery cloth or sandpaper. If a new pressure plate is indicated, new or factory rebuilt units are available.

Installation

Refer to illustration 3.14

13   Before installation, carefully wipe the flywheel and pressure plate machined surfaces clean. It's important that no oil or grease is on these surfaces or the lining of the clutch disc. Handle these parts only with clean hands.

14   Position the clutch disc and pressure plate against the flywheel, with the clutch

3.12b Examine the pressure plate friction surface for score marks, cracks and evidence of overheating (blue spots)

8

3.14 Center the clutch disc in the pressure plate with a clutch alignment tool

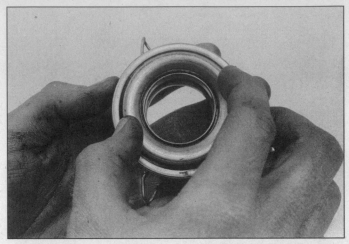

4.3 To check the operation of the bearing, hold it by the outer race and rotate the inner race while applying pressure - the bearing should turn smoothly - if it doesn't, replace it

held in place with an alignment tool **(see illustration)**. Make sure it's installed properly (most replacement clutch plates will be marked "flywheel side" or something similar - if not marked, install the clutch disc with the damper springs or cushions toward the transaxle).

15   Tighten the pressure plate-to-flywheel bolts only finger tight, working around the pressure plate.

16   Center the clutch disc by ensuring the alignment tool is through the splined hub and into the recess in the crankshaft. Wiggle the tool up, down or side-to-side as needed to bottom the tool. Tighten the pressure plate-to-flywheel bolts a little at a time, working in a criss-cross pattern to prevent distortion of the cover. After all of the bolts are snug, tighten them to the torque listed in this Chapter's Specifications. Remove the alignment tool.

17   Using high-temperature grease, *lightly* lubricate the splines of the transaxle input shaft. The manufacturer does not recommend lubricating the inner circumference of the release bearing.

18   Install the clutch release bearing, if removed (see Section 4).

19   Install the engine (see Chapter 2, Part B).

---

4   Clutch release bearing and lever - removal, inspection and installation

---

Warning: *Dust produced by clutch wear and deposited on clutch components may contain asbestos, which is hazardous to your health. DO NOT blow it out with compressed air and DO NOT inhale it. DO NOT use gasoline or petroleum-based solvents to remove the dust. Brake system cleaner should be used to flush it into a drain pan. After the clutch components are wiped clean with a rag, dispose of the contaminated rags and cleaner in a labeled, covered container.*

Removal

1   Remove the engine from the vehicle (see Chapter 2, Part B).

2   Remove the clutch release lever from the transaxle by unsnapping the lever from the ball stud, then remove the bearing from the lever.

Inspection

Refer to illustration 4.3

3   Hold the bearing by the outer race and rotate the inner race while applying pressure **(see illustration)**.

4   If the bearing doesn't turn smoothly or if it's noisy, replace it with a new one. Wipe the bearing with a clean rag and inspect it for damage, wear and cracks. Don't immerse the bearing in solvent - it's sealed for life and to do so would ruin it. Also check the release lever for cracks and bends.

Installation

5   Apply a light coat of high-temperature grease to the transaxle input shaft splines. **Note:** *The manufacturer recommends against lubricating the release bearing.*

6   Lubricate the release lever ball socket and release cylinder pushrod socket with high-temperature grease.

7   Attach the release bearing to the release lever.

8   Slide the release bearing onto the transaxle input shaft front bearing retainer. Push the clutch release lever onto the ball-stud until it's firmly seated.

9   The remainder of installation is the reverse of the removal procedure.

---

5   Clutch release system - removal and installation

Note: *The clutch hydraulic release system is a sealed unit and is not serviceable. The system must be replaced as a single unit, available at*

a dealer parts department. Aftermarket components, including their own bleeding instructions, are available at automotive parts stores.

Removal

1   It's a good idea to put a block of wood underneath the clutch pedal so it can't be depressed while the clutch hydraulic release system is removed. If you don't block up the pedal, make sure you do NOT depress it!

2   Remove the air intake duct (see Chapter 4).

3   Remove the two hydraulic damper-to-clutch bellhousing nuts and slide the damper and bracket off the studs.

4   While pushing toward the clutch bellhousing, rotate the release cylinder 1/4-turn counterclockwise to disengage the bayonet connector.

5   Remove the retainer clip that attaches the clutch master cylinder pushrod to the clutch pedal pin and disconnect the pushrod from the pedal.

6   Rotate the master cylinder about 1/8-turn clockwise and remove it from the firewall.

7   Remove the master cylinder, hydraulic line, damper and release cylinder from the vehicle as a single assembly.

Installation

8   Install the new master cylinder onto the firewall with the reservoir leaning out (away from the center of the vehicle), then rotate the master cylinder about 1/8-turn counterclockwise to lock it into position.

9   Insert the release cylinder into the clutch bellhousing with the hydraulic line facing down and rotate the release cylinder 1/4-turn clockwise while pushing toward the clutch bellhousing. **Note:** *Do not remove the plastic strap that holds the pushrod in position. It is designed to break off the first time the clutch pedal is depressed.*

10   Slide the damper and bracket onto the mounting studs on the clutch bellhousing and

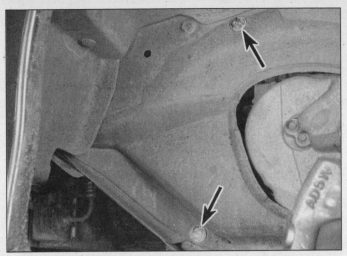

9.3a To remove the front inner splash shield, remove these two plastic rivets (arrows) by pulling out the center part . . .

9.3b . . . remove these three fasteners the same way (arrows) and pull the clip (lower arrow) off the cradle

install the nuts. Tighten the nuts evenly, a little at a time to the torque listed in this Chapter's Specifications.

11   Coat the clutch pedal pin with multi-purpose grease, then reattach the pushrod to the pedal and install the retainer.

12   If the vehicle is equipped with cruise control, check the adjustment of the cruise control clutch switch.

6   Clutch release system - bleeding

1   The clutch hydraulic release system is a sealed unit and is not serviceable. If leaks develop and clutch pedal pressure decreases, the system must be replaced as a single unit, available at a dealer parts department. Aftermarket components, including their own bleeding instructions, are available at automotive parts stores.

7   Clutch switch - check and adjustment

1   The clutch switch is located at the top of the clutch pedal.

2   To check the switch:

a) Unplug the switch electrical connector.

b) Depress the clutch pedal and check continuity between the switch connector terminals with an ohmmeter or continuity checker. There should be no continuity between the terminals when the pedal is depressed.

b) Release the clutch pedal. There should now be continuity between the switch connector terminals.

3   If the switch doesn't operate as described, adjust it by loosening the switch retaining bolt and sliding it towards or away from the clutch pedal. With the clutch pedal released, the plunger should protrude approximately 1/32-inch.

8   Driveaxles - general information and inspection

1   Power is transmitted from the transaxle to the wheels through a pair of driveaxles. The inner end of the left driveaxle is directly splined to the differential side gear. The inner end of the right driveaxle is splined to an intermediate shaft. The outer ends of the driveaxles are splined to the axle hubs and locked in place by a large nut.

2   The inner ends of the driveaxles are equipped with sliding constant velocity joints, which are capable of both angular and axial motion. Each inner joint assembly consists of a tri-pot bearing and a housing in which the joint is free to slide in-and-out as the driveaxle moves up-and-down with the wheel. These joints can be disassembled and cleaned in the event of a boot failure, but if any parts are damaged, the joints must be replaced as a unit (see Section 11).

3   Each outer joint, which consists of ball bearings running between an inner race and an outer cage, is capable of angular but not axial movement. These joints cannot be rebuilt, but they can be disassembled and cleaned and their boots can be replaced, in the event of a boot failure (see Section 11).

4   The boots should be inspected periodically for damage and leaking lubricant. Torn CV joint boots must be replaced immediately or the joints can be damaged. Boot replacement involves removal of the driveaxle (see Section 9). **Note:** Some auto parts stores carry "split" type replacement boots, which can be installed without removing the driveaxle from the vehicle. This is a convenient alternative; however, the driveaxle should be removed and the CV joint disassembled and cleaned to ensure the joint is free from contaminants such as moisture and dirt which will accelerate CV joint wear. The most common symptom of worn or damaged CV joints, besides lubricant leaks, is a clicking noise in turns, a clunk when accelerating after coasting and vibration at highway speeds. To check for wear in the CV joints and driveaxle shafts, grasp each axle (one at a time) and rotate it in both directions while holding the CV joint housings, feeling for play indicating worn splines or sloppy CV joints. Also check the driveaxle shafts for cracks, dents and distortion.

9   Driveaxle - removal and installation

Caution: *On models equipped with a theft-deterrent radio, make sure you have the correct activation code, or the theft deterrent system is turned off, before disconnecting the battery.*

Removal

Refer to illustrations 9.3a, 9.3b, 9.5, 9.7, 9.8a, 9.8b, 9.9 and 9.10

Note: *Not all of the steps in this procedure apply to all models. Read through the procedure carefully and determine which steps apply to the vehicle being worked on before actually beginning any work.*

1   Disconnect the cable from the negative terminal of the battery. Set the parking brake, loosen the front wheel lug nuts, raise the vehicle and support it securely on jackstands. Remove the wheel.

2   Carefully inspect both differential side gear seals for any signs of leakage. If either seal is leaking, replace it while the driveaxle is out (see Chapter 7B).

3   Remove the front inner splash shield **(see illustrations)**.

4   If you're removing the left driveaxle or the intermediate shaft, drain the transaxle lubricant (see Chapter 1). If you're removing the right driveaxle only, it's unnecessary to drain the transaxle.

5   Remove the driveaxle hub nut and washer. To prevent the hub from turning, wedge a prybar between two of the wheel

8

9.5 To prevent the hub from turning, wedge a prybar between two of the wheel studs and allow the prybar to rest against the ground or the floorpan of the vehicle

9.7 Place a prybar across the top of the tension strut and insert the tip under the cradle, then push down on the prybar to force the control arm down far enough to free the balljoint stud from the steering knuckle

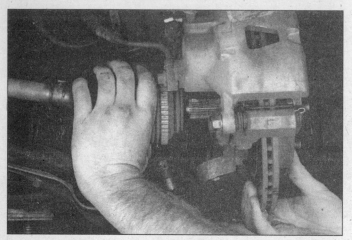

9.8a Swing the knuckle/strut assembly away from the vehicle and separate it from the outer end of the driveaxle

9.8b If the splined end of the driveaxle sticks in the hub, tap the end of the driveaxle with a soft-faced hammer or with a hammer and a brass punch or a block of wood

studs and allow the prybar to rest against the ground or the floorpan of the vehicle **(see illustration)**.

6   Break loose the control arm balljoint nut from the steering knuckle (see Chapter 10).

7   Place a prybar across the top of the tension strut and insert the tip under the cradle as shown, then push down on the prybar to move the control arm down far enough to free the balljoint stud from the steering knuckle **(see illustration)**. Make sure the knuckle doesn't damage the balljoint boot.

8   Swing the knuckle/strut assembly away from the vehicle and separate it from the outer end of the driveaxle **(see illustration)**. If the splined end of the driveaxle is difficult to pull out of the hub, tap the end of the driveaxle with a soft-faced hammer or a hammer and a brass punch **(see illustration)**. If the driveaxle is stuck in the hub splines and won't move, it may be necessary to remove the brake disc (see Chapter 9) and push it from the hub with a two-jaw puller.

9   If you're removing the left driveaxle, insert a large screwdriver or prybar between the transaxle and the inner CV joint housing and pry the splined inner end of the driveaxle out of the differential side gear **(see illustration)**.

10   If you're removing the right driveaxle, separate it from the intermediate shaft by tapping the axle with a hammer and a block of wood **(see illustration)**.

11   Should it become necessary to move the vehicle while the driveaxle is out, place a large bolt with two large washers (one on each side of the hub) through the hub and tighten the nut securely. If this isn't done, the hub bearings will be damaged.

12   Refer to Chapter 7 for the differential seal replacement procedure.

9.9 If you're removing the left driveaxle, insert a large screwdriver or prybar between the transaxle and the inner CV joint housing and pry the splined inner end of the driveaxle out of the differential side gear

9.10 If you're removing the right driveaxle, separate it from the intermediate shaft by tapping the axle with a hammer and a block of wood

10.6 Remove the intermediate shaft support bracket-to-engine block fasteners (arrows)

Installation

13   Installation is the reverse of the removal procedure, but with the following additional points:

a) *When installing the left driveaxle, make sure the retaining ring and the splines on the inner end of the CV joint don't damage the lips of the side gear seal.*

b) *Once the splines on the left driveaxle are properly aligned with the splines in the left side gear bore, tap the driveaxle sharply to seat the retaining ring on the end of the CV joint into its corresponding groove in the differential side gear bore.*

c) *When installing the right driveaxle, insert the inner end of the driveaxle onto the outer end of the intermediate shaft, then push firmly on the driveaxle to seat the retaining ring on the end of the intermediate shaft into its corresponding groove inside the end of the CV joint housing.*

d) *Tighten the driveaxle hub nut to the torque listed in this Chapter's Specifications.*

e) *Install the wheel and lug nuts, lower the vehicle and tighten the lug nuts to the torque listed in the Chapter 1 Specifications.*

f) *Refill the transaxle to the proper lubricant level (see Chapter 1).*

10   Intermediate shaft - removal and installation

Refer to illustration 10.6

1   Remove the right driveaxle (see Section 9).

2   Before removing the intermediate shaft, check the right differential seal for signs of leakage. Refer to Chapter 7B for the side gear seal replacement procedure.

3   Position a drain pan underneath the transaxle and drain the transaxle lubricant (see Chapter 1).

4   On vehicles with a DOHC engine, remove the intermediate shaft support bracket-

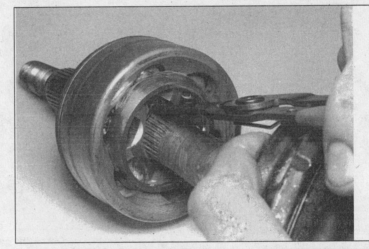

11.5 Use snap-ring pliers to expand the inner retaining ring

to-intake manifold bracket fasteners.

5   Remove the starter motor bracket or engine block to-intermediate shaft bracket fastener.

6   Remove the intermediate shaft support bracket-to-engine block fasteners **(see illustration)**.

7   Remove the intermediate shaft assembly.

8   Check the support bearing for smooth operation. If it feels rough or sticky it should be replaced. Take it to a dealer service department or other repair shop, as special tools are needed to perform this job.

9   Installation is the reverse of removal. Be sure to tighten the intermediate shaft support bracket-to-engine block fasteners to the torque listed in this Chapter's Specifications.

11   Driveaxle boot replacement and CV joint overhaul

Note: *If the CV joints must be overhauled (usually due to torn boots), explore all options before beginning the job. Complete rebuilt driveaxles are available on an exchange basis, which eliminates much time and work. Whichever route you choose to take, check*

on the cost and availability of parts before disassembling the vehicle.

Outer CV joint

Disassembly

Refer to illustrations 11.5, 11.7, 11.8, 11.10 and 11.11

1   Remove the driveaxle (see Section 9).

2   Mount the driveaxle in a vise with wood lined jaws (to prevent damage to the axle-shaft). Check the CV joint for excessive play in the radial direction, which indicates worn parts. Check for smooth operation throughout the full range of motion for each CV joint. If a boot is torn, the recommended procedure is to disassemble the joint, clean the components and inspect for damage due to loss of lubrication and possible contamination by foreign matter.

3   Remove the boot clamps using a hammer and chisel. Position the tip of the chisel under the end of the clamp, then strike the chisel with the hammer, which will pop the clamp open.

4   Using a screwdriver, pry up on the edge of the boot and push it away from the CV joint. Old and worn boots can be cut off.

5   Spread the inner retaining ring using snap-ring pliers **(see illustration)**, then slide

8

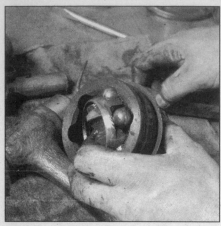

11.7 Tilt the inner race and cage assembly

11.8 Remove the ball bearings one at a time - you may have to pry them out with a dull screwdriver (don't scratch any of the components)

11.10 Align a large window of the cage with a land of the outer race, then lift the assembly from the outer race

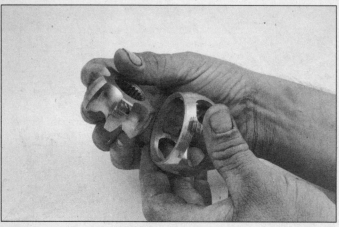

11.11 To remove the inner race from the cage, turn the inner race 90-degrees and place one of the inner race lands into one of the large windows of the cage, then swing the inner race out

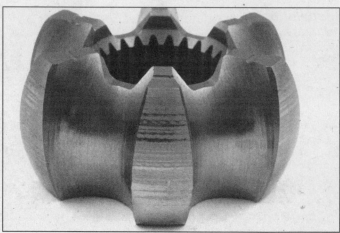

11.12a Check the inner race lands and grooves for score marks and pitting (also check for damaged splines)

the outer joint off the shaft.

6   Remove the boot. Mark the relationship of the inner race and cage. These marks must face up when the joint is reassembled.

7   Tilt the inner race and cage assembly **(see illustration)**.

8   Remove the ball bearings one at a time. Pry them out with a dull screwdriver, if necessary **(see illustration)**.

9   Tilt the inner race and cage assembly 90-degrees.

10   Align one of the large windows of the cage with one of the lands of the outer race. Lift the inner race and cage assembly out of the outer race **(see illustration)**.

11   Remove the inner race from the cage by turning the inner race 90-degrees and swinging it out **(see illustration)**.

Check

Refer to illustrations 11.12a and 11.12b

12   Clean all components with solvent to remove grease. Check for cracks, pitting, scoring and other signs of wear **(see illustrations)**.

11.12b Check the cage for cracks, pitting and score marks (shiny spots are normal and don't affect operation)

Reassembly

Refer to illustrations 11.13, 11.16, 11.17a, 11.17b, 11.18 and 11.19

13   Install the inner race into the cage **(see illustration)**.

11.13 Place one of the inner race lands into one of the large windows of the cage, then swing the inner race into the cage

14   Insert the inner race and cage assembly into the outer race **(see illustration 11.10)**.

15   Rotate the inner race and cage until it's flat within the outer race. The marks that were applied in Step 6 must both be facing up.

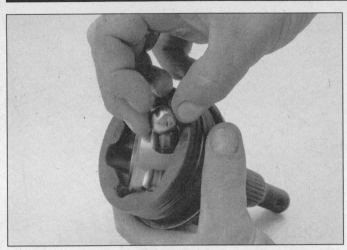

11.16 Tilt the inner race cage assembly to expose the openings in the cage and insert the ball bearings, one at a time, by hand

11.17a Insert grease through the splined hole . . .

11.17b . . . then push the grease down into the joint with your finger or a dowel (this may have to be done a few times until the joint is completely packed)

11.18 Wrap the axleshaft splines with tape to prevent damage to the boot when installing it

16   To install the ball bearings, tilt the inner race and cage assembly to expose the openings (see illustration).

17   Press the ball bearings in one at a time, turning the inner assembly so that each one can be installed. With all the ball bearings installed, the CV joint can now be packed with CV joint grease (see illustrations).

18   Wrap the splines of the axleshaft with tape to prevent damage to the boot (see illustration). Slide the new boot and small clamp onto the axleshaft. Install the CV joint assembly onto the axleshaft and, using a brass or plastic hammer, drive the joint onto the shaft until the retaining ring engages with the groove on the shaft. Pull on the joint to confirm this. Proceed to Step 26.

19   Slide the boot into position, making sure each end of the boot is properly seated in its groove. Slide a small, dull screwdriver under the edge of the boot to equalize the pressure inside, then install and tighten the boot clamps (see illustration).

11.19 tighten the boot clamps with a pair of special boot clamp pliers (available at most auto parts stores)

8

Inner CV joint
Disassembly
Refer to illustrations 11.21 and 11.22

20   Slide the outer race (joint housing) from the tripod.

21   Use a center-punch to mark the tripod and driveaxle to ensure that they are reassembled properly (see illustration). Also mark the relationship of the outer race (hous-

11.21 Use a center-punch to place marks (arrows) on the tripod and the driveaxle to ensure that they are reassembled properly

11.22 Expand the inner retaining ring and slide it toward the center of the shaft

11.25a Install the tripod assembly with the recessed portion (counterbore) of the splines facing the end of the axleshaft

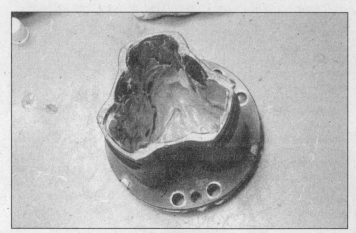

11.25b Place grease at the bottom of the housing

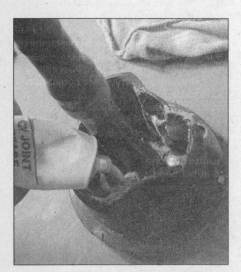

11.25c Insert the tripod into the housing, followed by the rest of the grease

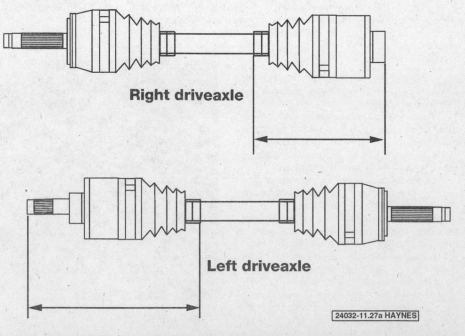

Right driveaxle

Left driveaxle

24032-11.27a HAYNES

11.27a Adjust the inner joint to the proper length (see the Specifications at the beginning of this Chapter for the correct dimension)

ing) to the axleshaft.

22   Expand the inner retaining ring with a pair of snap-ring pliers and slide it toward the center of the axleshaft approximately 1/2-inch **(see illustration)**. Push the tripod joint towards the center of the axleshaft to expose the outer retaining ring, then remove the outer retaining ring with a pair of snap-ring pliers.

23   Slide the tripod joint off the shaft. If it sticks, Use a hammer and a brass punch to drive it off. Remove the inner retaining ring and the old boot.

Check

24   Clean all components with solvent to remove the grease, then check for cracks, pitting, scoring and other signs of wear.

Reassembly

Refer to illustrations 11.25a, 11.25b, 11.25c, 11.27a and 11.27b

25   Tape the axleshaft splines to prevent damage to the new boot **(see illustration 11.18)**. Slide the clamps and boot onto the axleshaft, then remove the tape. Install the inner retaining ring and position it about 1/2-inch past its groove. Place the tripod assem-

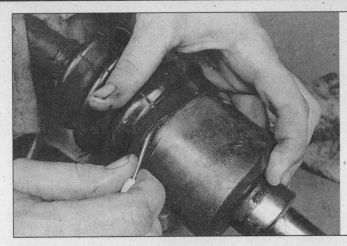

11.27b Equalize the pressure inside the boot by inserting a small, dull screwdriver between the boot and the outer race

bly on the shaft with the recessed portion, or counterbore, facing the end of the axleshaft **(see illustration)**. Also make sure the marks applied in Step 21 are aligned. Install the outer retaining ring, slide the tripod assembly against it, then seat the inner stop ring in its groove. Apply CV joint grease to the tripod assembly and inside the outer joint (housing) **(see illustrations)**. Insert the tripod

joint/axleshaft into the housing, making sure the match marks line up.

26   Slide the boot into place, making sure both ends seat in their grooves.

27   Adjust the joint to the length listed in this Chapter's Specifications **(see illustration)**. Equalize the pressure in the boot **(see illustration)**, then tighten the boot clamps **(see illustration 11.19)**.

8

Notes

Chapter 9 Brakes

Contents

Specifications

General
Brake fluid type ... See Chapter 1

Disc brakes
Minimum brake pad thickness ... See Chapter 1
Front disc thickness*
- Standard ... 0.710 inch
- Minimum after turning ... 0.633 inch
- Discard thickness ... 0.625 inch

Rear disc thickness (models with rear disc brakes)*
- Standard ... 0.430 inch
- Minimum after turning ... 0.370 inch
- Discard thickness ... 0.350 inch

Maximum allowable scoring depth (front and rear) ... 0.060 inch
Maximum lateral runout ... 0.0024 inch
Disc thickness variation limit (front and rear) ... 0.0005 inch

If different specifications are cast into the disc, they supersede information printed here.

Drum brakes
Drum inside diameter*
- Standard ... 7.870 inches
- Maximum ... 7.900 inches
- Discard ... 7.930 inches

Maximum allowable scoring depth ... 0.040 inch

If different specifications are cast into the drum, they supersede information printed here.

Torque specifications

	Ft lbs (unless otherwise indicated)
Caliper lock/guide pins (front and rear calipers)	27
Caliper torque plate bolts	
Front	81
Rear	63
Brake hose-to-caliper banjo bolt	36
Wheel cylinder mounting bolts	89 in-lbs
Brake hose-to-wheel cylinder bolt	36
Master cylinder-to-brake booster nuts	20
Power brake booster mounting nuts	20
Wheel lug nuts	See Chapter 1

1   General information

The vehicles covered by this manual are equipped with hydraulically operated front and rear brake systems. The front brakes are disc type and the rear brakes are either drum or disc type. Both the front and rear brakes are self adjusting. The disc brakes automatically compensate for pad wear, while the drum brakes incorporate an adjustment mechanism which is activated as the parking brake is applied.

Hydraulic system

The hydraulic system consists of two separate circuits. The master cylinder has separate reservoirs for the two circuits, and, in the event of a leak or failure in one hydraulic circuit, the other circuit will remain operative. A dual proportioning valve on the firewall provides brake balance between the front and rear brakes.

Power brake booster

The power brake booster, utilizing engine manifold vacuum and atmospheric pressure to provide assistance to the hydraulically operated brakes, is mounted on the firewall in the engine compartment.

Parking brake

The parking brake operates the rear brakes only, through cable actuation. It's activated by a lever mounted in the center console.

Service

After completing any operation involving disassembly of any part of the brake system, always test drive the vehicle to check for proper braking performance before resuming normal driving. When testing the brakes, perform the tests on a clean, dry, flat surface. Conditions other than these can lead to inaccurate test results.

Test the brakes at various speeds with both light and heavy pedal pressure. The vehicle should stop evenly without pulling to one side or the other. Avoid locking the brakes, because this slides the tires and diminishes braking efficiency and control of the vehicle.

Tires, vehicle load and wheel alignment are factors which also affect braking performance.

2   Anti-lock Brake System (ABS) - general information

1   The anti-lock brake system is designed to maintain vehicle steerability, directional stability and optimum deceleration under severe braking conditions and on most road surfaces. It does so by monitoring the rotational speed of each wheel and controlling the brake line pressure to each wheel during braking. This prevents the wheels from locking up.

Components

Brake control assembly

Refer to illustration 2.2

2   The brake control assembly consists of the master cylinder, the motor pack assembly, the modulator assembly and a pair of solenoid valves **(see illustration)**. The purpose of these components is to modulate brake hydraulic pressure when the antilock brake control module senses a locked-up wheel or wheels.

a) *The electric pump provides hydraulic pressure to charge the reservoirs in the actuator, which supplies pressure to the*

2.2 The brake control assembly consists of the master cylinder, the motor pack assembly, the modulator assembly and a pair of solenoid valves

2.3a Each front wheel has a speed sensor (upper arrow) installed in the steering knuckle and a sensor ring (lower arrow) pressed onto the end of the outer CV joint

2.3b The rear wheel speed sensor (arrow) is an integral part of the rear wheel hub and bearing assembly (the sensor ring, not visible in this photo, is also an integral part of the hub/bearing assembly)

3.5 Before removing a front brake caliper, use a C-clamp to squeeze the piston back into the caliper (don't try this on a rear caliper, however, or you'll damage the actuating mechanism for the parking brake; to retract a rear caliper piston, see illustration 3.6l)

braking system. The pump and reservoirs are housed in the actuator assembly
b) The solenoid valves modulate brake line pressure during ABS operation. The body contains four valves - one for each wheel.

Wheel speed sensors

Refer to illustrations 2.3a and 2.3b

3   The wheel speed sensors are located at each wheel. The front sensors **(see illustration)** are installed in the steering knuckles; the rear sensors **(see illustration)** are an integral component of the rear wheel hub/bearing assemblies. The wheel speed sensors generate a frequency and voltage signal that's proportional to actual wheel speed. Here's how they work: Each sensor consists of wire windings around a magnet; when voltage is available and the wheel is turning, the sensor generates a magnetic field. A 47-tooth sensor ring - pressed onto the outer CV joint housing up front **(see illustration 2.3a)** and integral with the hub/bearing assembly at the rear - interrupts this magnetic field, alternately increasing and decreasing the field strength. This fluctuating field strength produces a variable voltage and frequency signal that varies in accordance with wheel speed.

Anti-lock brake control module

4   The anti-lock brake control module (computer) is mounted under the left end of the dashboard. The module is the 'brain' of the ABS system. Its function is to process information from the wheel speed sensors, control the hydraulic line pressure and avoid wheel lock up. Besides the wheel speed sensors, other inputs to the module include the brake switch, the ignition switch and battery voltage through the ABS enable relay.

5   The anti-lock brake control module also constantly monitors itself and the rest of the ABS system, even under normal driving conditions, to find faults within the system. If a problem develops within the system, an amber 'ANTILOCK' light on the instrument cluster will flash or light continuously.

a) If the problem has no immediate effect on ABS operation, the amber light will flash. However, the ABS system should still be repaired immediately.
b) If the problem will have a serious effect on ABS operation, the amber light will light continuously. If this condition occurs, the ABS system will not function properly. The brake system will function like a conventional (non-ABS) brake system. But the ABS system should be repaired as soon as possible.

6   A red brake warning light on the instrument cluster may also come on if the brake fluid level in the master cylinder is low or if the parking brake switch is closed. The brake warning light can also be activated, in conjunction with the amber warning light, by the anti-lock brake control module if serious malfunctions - the type that can seriously affect ABS operation - should occur.

Diagnosis and repair

7   When an ABS malfunction occurs, the anti-lock brake control module stores a diagnostic code which, when accessed with the proper scan tool, provides information regarding the location and/or cause of the problem, and will also indicate whether the problem is continuous or intermittent. There are 54 of these diagnostic codes, but they can only be read and cleared with a scan tool. We therefore strongly recommend taking your vehicle to a dealer immediately should a problem occur in the ABS system.

8   If either dashboard warning light comes on and stays on while the vehicle is in operation, the ABS system requires immediate attention. Although a scan tool is necessary to properly diagnose the system, the home mechanic can perform a few preliminary checks before taking the vehicle to a dealer service department or other repair shop which is equipped with a tester.

a) Check the brake fluid level in the reservoir.

b) Check that all electrical connectors are securely connected.
c) Check the fuses.

9   If the above preliminary checks do not rectify the problem, the vehicle should be diagnosed and repaired by a dealer service department or other repair shop.

---

3   Disc brake pads - replacement

Refer to illustrations 3.5 and 3.6a through 3.6o

Warning: *Disc brake pads must be replaced on both front or rear wheels at the same time - never replace the pads on only one wheel. Also, the dust created by the brake system may contain asbestos, which is harmful to your health. Never blow it out with compressed air and don't inhale any of it. An approved filtering mask should be worn when working on the brakes. Do not, under any circumstances, use petroleum-based solvents to clean brake parts. Use brake system cleaner only! When servicing the disc brakes, use only high-quality, nationally recognized brand-name pads.*

Note: *This procedure applies to both the front and rear disc brakes.*

1   Remove the cap from the brake fluid reservoir.
2   Loosen the wheel lug nuts, raise the front or rear of the vehicle and support it securely on jackstands. Block the wheels at the opposite end.
3   Remove the wheels. Work on one brake assembly at a time, using the assembled brake for reference if necessary.
4   Inspect the brake disc carefully as outlined in Section 5. If machining is necessary, follow the information in that Section to remove the disc, at which time the pads can be removed as well.
5   Push the piston back into its bore to provide room for the new brake pads. A C-clamp can be used to accomplish this on the

9

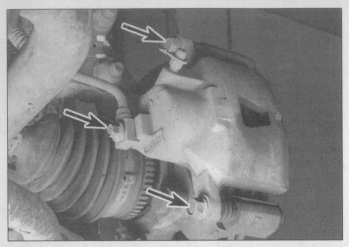

3.6a If you're changing the brake pads on a front caliper, remove the caliper lock pin (lower arrow) and swing the caliper up and out of the way; if you're removing the front caliper, you'll have to remove the guide pin (upper arrow); and if you're replacing the caliper, you'll need to remove the brake hose banjo bolt (left arrow) as well

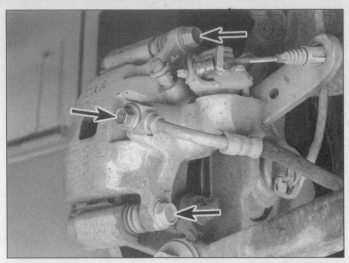

3.6b On a rear caliper, you must remove the guide pin (upper arrow) and the lock pin (lower arrow) - if you're replacing the caliper, you'll also have to remove the brake hose banjo bolt (middle arrow)

3.6c If you do remove both caliper bolts, don't allow the caliper to hang by the brake hose - hang it from the strut coil spring with a piece of wire

3.6d Remove the inner brake pad

3.6e Remove the outer brake pad

3.6f Remove the lower anti-rattle clip

3.6g Remove the upper anti-rattle clip

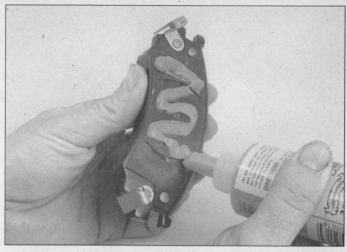

3.6h Apply anti-squeal compound to the back sides of both
new brake pads

3.6i Install the new upper anti-rattle clip . . .

3.6j . . . and the new lower anti-rattle clip onto the torque plate -
make sure they're fully seated against the torque plate

3.6k Install the new brake pads - make sure their upper and
lower notches are properly engaged with the projections on the
torque plate

3.6l Don't try to push down the piston in a rear caliper with a C-
clamp. Instead - using a pin spanner (shown), needle-nose pliers
or a special tool designed for the job - rotate the piston clockwise
until it's bottomed in the bore . . .

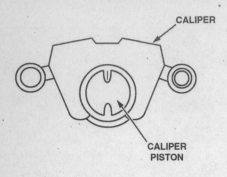

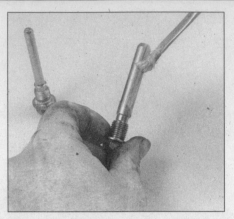

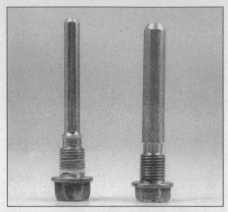

3.6m ... then turn the piston so the cutouts are aligned like this

3.6n Wipe off the guide and lock pins and apply a thin coat of multi-purpose grease to their sliding surfaces

3.6o Note that the lock pin (left) and the guide pin (right) are not the same size - be sure to install them in their proper positions

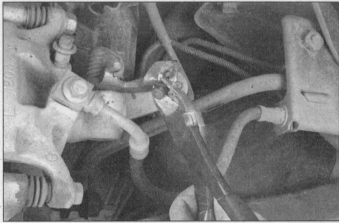

4.3a To disconnect the parking brake cable from the rear caliper, squeeze the locking tangs together to release the cable housing from the cable bracket . . .

4.3b . . . then remove the end of the cable out of the parking brake lever on the caliper (make sure the parking brake lever inside the vehicle is fully released and, if necessary, loosen the cable adjusting nut at the forward end of the cable as shown in illustration 12.3)

front caliper only **(see illustration)**. As the piston is depressed to the bottom of the caliper bore, the fluid in the master cylinder will rise. Make sure that it doesn't overflow. If necessary, siphon off some of the fluid.

6   Follow the accompanying photos, beginning with **illustration 3.6a,** for the actual pad replacement procedure. Be sure to stay in order and read the caption under each illustration.

7   When reinstalling the caliper, be sure to tighten the mounting bolts to the torque listed in this Chapter's Specifications. After the job has been completed, firmly depress the brake pedal a few times to bring the pads into contact with the disc. Check the level of the brake fluid, adding some if necessary. Check the operation of the brakes carefully before placing the vehicle into normal service.

4   Disc brake caliper - removal, overhaul and installation

Warning 1: *Dust created by the brake system*

may contain asbestos, which is harmful to your health. Never blow it out with compressed air and don't inhale any of it. An approved filtering mask should be worn when working on the brakes. Do not, under any circumstances, use petroleum-based solvents to clean brake parts. Use brake system cleaner only!

Warning 2: *This procedure should not be undertaken on the rear caliper of a model equipped with an Anti-lock Brake System (ABS), since a special "scan" tool is needed to properly bleed the rear brakes. Take the vehicle to a dealer service department or other repair shop that has the proper tools.*

Note: *If an overhaul is indicated (usually because of fluid leakage), explore all options before beginning the job. New and factory rebuilt calipers are available on an exchange basis, which makes this job quite easy. If it's decided to rebuild the calipers, make sure a rebuild kit is available before proceeding. Always rebuild the calipers in pairs - never rebuild just one of them.*

Removal

Refer to illustrations 4.3a and 4.3b

1   Loosen the wheel lug nuts, raise the vehicle, place it securely on jackstands and remove the wheel.

2   Remove the brake hose banjo bolt and disconnect the brake hose from the caliper **(see illustration 3.6a or 3.6b)**. Plug the brake hose to keep contaminants out of the brake system and to prevent losing any more brake fluid than is necessary.

3   Refer to Section 3 for both front and rear caliper removal procedures (removing the caliper is part of the brake pad replacement procedure). If you're removing a rear caliper, you'll also need to disconnect the parking brake cable from the cable bracket and parking brake lever on the caliper **(see illustrations)**.

Overhaul

Refer to illustrations 4.4, 4.5 and 4.7

4   To overhaul the caliper, remove the boot set ring and the boot **(see illustration)**.

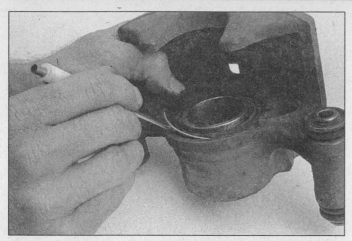

4.4 Using a screwdriver, remove the boot set ring

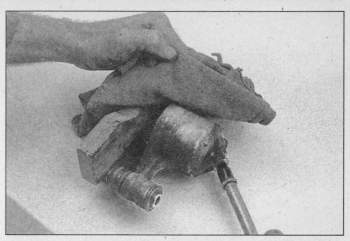

4.5 With the caliper padded to catch the piston, use compressed air to force the piston out of its bore - make sure your hands or fingers are not between the piston and caliper

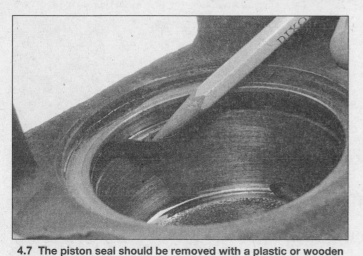

4.7 The piston seal should be removed with a plastic or wooden tool to avoid damage to the bore and seal groove - a pencil will do the job

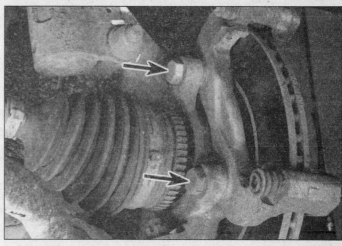

5.2a To remove the torque plate for the front caliper, remove these two bolts (arrows)

Before you remove the piston, place a wood block between the piston and caliper to prevent damage as it is removed.

5   To remove the piston from the caliper, apply compressed air to the brake fluid hose connection on the caliper body **(see illustration)**. Use only enough pressure to ease the piston out of its bore. **Warning**: *Be careful not to place your fingers between the piston and the caliper as the piston may come out with some force.*

6   Inspect the mating surfaces of the piston and caliper bore wall. If there is any scoring, rust, pitting or bright areas, replace the complete caliper unit with a new one.

7   If these components are in good condition, remove the piston seal from the caliper bore using a wooden or plastic tool **(see illustration)**. Metal tools may damage the cylinder bore.

8   Wash all the components in clean brake fluid or brake cleaner.

9   Pull the guide and lock pin dust boots out of the torque plate and inspect them for cracks, tears and other deterioration. If they're worn or damaged, replace them.

10   To reassemble the caliper, you should already have the correct rebuild kit for your vehicle. Submerge the new piston seal in clean brake fluid and install it into the groove in the caliper bore, making sure it isn't twisted.

11   Lubricate the sides of the piston with clean brake fluid and install it in the bore. Do not force the piston into the bore, but make sure that it is squarely in place, then apply firm (but not excessive) pressure to install it.

12   Install the new piston dust boot and set ring.

13   Lubricate the straight portions of the lock pin and guide pin with silicone-based grease (supplied in the kit).

Installation

14   Install the caliper by reversing the removal procedure. Remember to replace the copper sealing washers (gaskets) at the brake hose-to-caliper connection (new washers normally come with the rebuild kit). Tighten the banjo fitting bolt to the torque listed in this Chapter's Specifications.

15   Bleed the brake circuit according to the procedure in Section 10. Make sure there are

no leaks from the hose connections. Test the brakes carefully before returning the vehicle to normal service.

5   Brake disc - inspection, removal and installation

Note: *This procedure applies to both front and rear brake discs.*

Inspection

Refer to illustrations 5.2a, 5.2b, 5.3, 5.4a, 5.4b, 5.5a and 5.5b

1   Loosen the wheel lug nuts, raise the vehicle and support it securely on jackstands. Remove the wheel and install two lug nuts to hold the disc in place. If the rear brake disc is being worked on, release the parking brake.

2   Remove the brake caliper as outlined in Section 4. It isn't necessary to disconnect the brake hose. After removing the caliper bolts, suspend the caliper out of the way with a piece of wire **(see illustration 3.6c)**. Remove the two torque plate-to-steering knuckle or

9

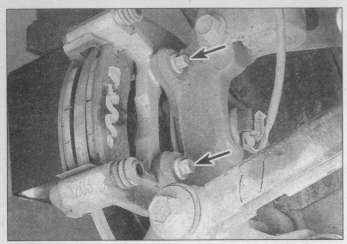

5.2b To remove the torque plate for the rear caliper, remove these two bolts (arrows)

5.3 The brake pads on this vehicle were obviously neglected, as they wore down to the rivets and cut deep grooves into the disc - wear this severe means the disc must be replaced

5.4a To check disc runout, mount a dial indicator as shown and rotate the disc

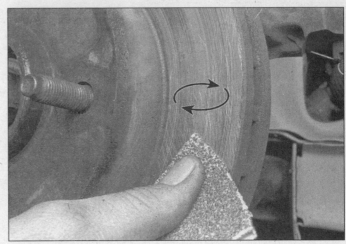

5.4b Using a swirling motion, remove the glaze from the disc with sandpaper or emery cloth

5.5a The minimum wear dimension is cast into the disc (typical disc shown)

cator reading should not exceed the specified allowable runout limit. If it does, the disc should be refinished by an automotive machine shop. **Note:** *The discs should be resurfaced regardless of the dial indicator reading, as this will impart a smooth finish and ensure a perfectly flat surface, eliminating any brake pedal pulsation or other undesirable symptoms related to questionable discs. At the very least, if you elect not to have the discs resurfaced, remove the glaze from the surface with emery cloth using a swirling motion* **(see illustration)**.

5   It's absolutely critical that the disc not be machined to a thickness under the specified minimum allowable disc refinish thickness. The minimum wear (or discard) thickness is cast into the inside of the disc **(see illustration)**. The disc thickness can be checked with a micrometer **(see illustration)**.

Removal

Refer to illustration 5.6

6   Remove the lug nuts which were put on to hold the disc in place and remove the disc from the hub. If the disc is stuck to the hub

rear suspension knuckle bolts **(see illustrations)** and detach the torque plate.

3   Inspect the disc surface for score marks and other damage. Light scratches and shallow grooves are normal after use and may not always be detrimental to brake operation, but deep scoring - over 0.060-inch - requires disc removal and refinishing by an automo-

tive machine shop. Be sure to check both sides of the disc **(see illustration)**. If pulsating has been noticed during application of the brakes, suspect disc runout.

4   To check disc runout, place a dial indicator at a point about 1/2-inch from the outer edge of the disc **(see illustration)**. Set the indicator to zero and turn the disc. The indi-

5.5b Use a micrometer to measure disc thickness

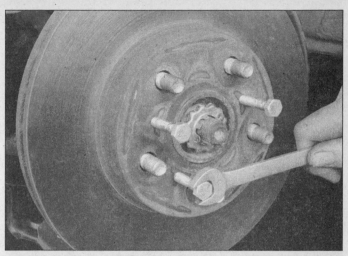

5.6 To free the disc, thread the appropriate size self-tapping bolts into the holes provided in the disc. Alternate tightening the bolts until the disc is free

6.4a Remove the lower return spring (don't stretch any of the springs any more than necessary to disconnect them, or you may weaken them)

6.4b Remove the adjuster spring

and won't come off, thread bolts into the holes provided **(see illustration)** and tighten them. Alternate between the bolts, turning them 1/4-turn at a time, until the disc is free.

Installation

7   Place the disc in position over the threaded studs.

8   Install the torque plate and caliper assembly over the disc and position it on the steering knuckle. Tighten the torque plate bolts to the torque listed in this Chapter's Specifications.

9   Install the wheel, then lower the vehicle to the ground. Tighten the lug nuts to the torque listed in the Chapter 1 Specifications. Depress the brake pedal a few times to bring the brake pads into contact with the disc. Bleeding won't be necessary unless the brake hose was disconnected from the caliper. Check the operation of the brakes carefully before driving the vehicle.

6   Drum brake shoes - replacement

Refer to illustrations 6.4a through 6.4u and 6.5

Warning: *Drum brake shoes must be replaced on both wheels at the same time - never replace the shoes on only one wheel. Also, the dust created by the brake system may contain asbestos, which is harmful to your health. Never blow it out with compressed air and don't inhale any of it. An approved filtering mask should be worn when working on the brakes. Do not, under any circumstances, use petroleum-based solvents to clean brake parts. Use brake system cleaner only!*

Caution: *Whenever the brake shoes are replaced, the return and hold-down springs should also be replaced. Due to the continuous heating/cooling cycle the springs are subjected to, they lose tension over a period of time and may allow the shoes to drag on*

the drum and wear at a much faster rate than normal. When replacing the rear brake shoes, use only high-quality, nationally recognized brand-name parts.

1   Loosen the wheel lug nuts, raise the rear of the vehicle and support it securely on jackstands. Block the front wheels to keep the vehicle from rolling.

2   Release the parking brake.

3   Remove the wheel. **Note:** *All four rear brake shoes must be replaced at the same time, but to avoid mixing up parts, work on only one brake assembly at a time.*

4   Follow the accompanying illustrations for the brake shoe replacement procedure **(see illustrations 6.4a through 6.4u)**. Be sure to stay in order and read the caption under each illustration. **Note:** *If the brake drum cannot be easily pulled off the axle and shoe assembly, make sure the parking brake is completely released. If the drum still cannot be pulled off, the brake shoes will have to be retracted. This is done by first removing the*

9

6.4c Remove the leading (front) brake shoe hold-down spring, cup (the washer on top of the spring) and the pin (which can be pulled out from the backside of the backing plate), then disengage the shoe from the upper return spring and the adjuster assembly (arrow) with a twisting motion and remove the shoe

6.4d Remove the adjuster lever from the trailing (rear) shoe, disengage the adjuster assembly (arrow) from its notch in the trailing shoe, unhook the upper return spring (right above the adjuster notch) and remove the adjuster and the spring

6.4e Remove the trailing shoe hold-down cup, spring and pin, pull the trailing shoe off the backing plate and flip it upside down so you can get at the parking brake cable

6.4f To disconnect the parking brake cable from the parking brake lever, push the end of the lever against the spring to remove tension while simultaneously disengaging the plug on the end of the cable from the end of the lever

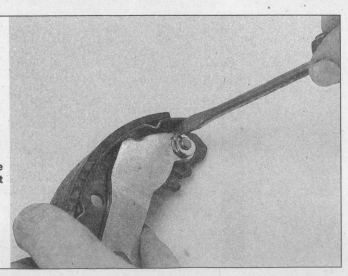

6.4g To separate the parking brake lever from the trailing shoe, remove this C-clip and the wave washer underneath (installing the lever on the new shoe is the reverse - be sure to lubricate the pivot post and use a new C-clip)

plug from the backing plate. With the plug removed, turn the adjuster wheel with a brake adjustment tool, moving the shoes away from the drum. The drum should now come off. **Warning:** *After removing the drum and before removing any brake components, clean the brake assembly with brake system cleaner and allow it to dry. Position a drain pan under the brake to catch the fluid and residue. DO NOT USE COMPRESSED AIR TO BLOW THE DUST FROM THE PARTS!*
5   Before reinstalling the drum, it should be checked for cracks, score marks, deep scratches and hard spots, which will appear as small discolored areas. If the hard spots cannot be removed with fine emery cloth or if any of the other conditions listed above exist, the drum must be taken to an automotive

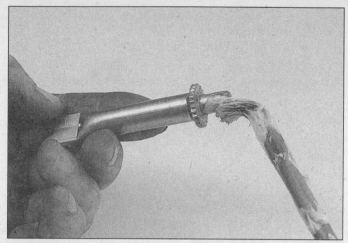

6.4h Lubricate the adjuster assembly with high-temperature grease

6.4i Also lightly lubricate the six raised shoe-contact pads on the backing plate with the same grease

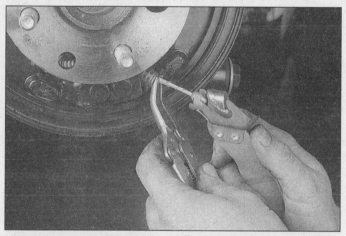

6.4j To reattach the parking brake cable to the parking brake lever, compress the spring far enough to expose at least an inch of cable, hold the spring with a pair of small vise-grip pliers as shown, engage the end of the cable with the lever and release the spring

6.4k Install the trailing shoe, pin, hold-down spring and cup

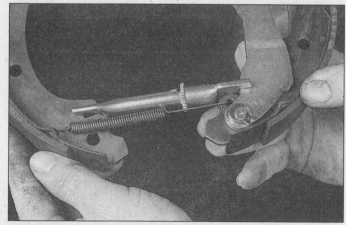

6.4l This is how the adjuster assembly should look from the back side when it's properly assembled, so refer to this photo as you perform the next steps

6.4m Connect the adjuster assembly to the trailing shoe and engage the inner notch with the parking brake lever

9

machine shop to have it turned. **Note:** *Professionals recommend resurfacing the drums each time a brake job is done. Resurfacing will eliminate the possibility of out-of-round* drums. If the drums are worn so much that they can't be resurfaced without exceeding the maximum allowable diameter (stamped into the drum), then new ones will be required (see illustration). *At the very least, if you elect not to have the drums resurfaced, remove the glaze from the surface with emery cloth using a swirling motion.*

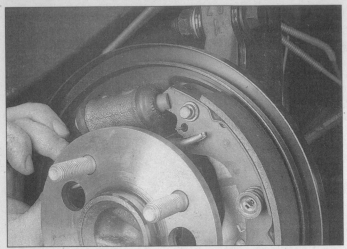

6.4n Install the upper return spring with its long straight end toward the trailing shoe (note how the spring is hooked to the trailing shoe through the back side - don't install it through the front or you'll have problems)

6.4o Hook the upper return spring to the leading shoe

6.4p Engage the leading shoe with the wheel cylinder as shown

6.4q Push the lower end of the leading shoe forward and engage the adjuster assembly with the shoe

6.4r Align the leading shoe and install the pin, hold-down spring and cup

6.4s Install the lower return spring as shown

6.4t Install the adjuster lever and spring as shown (note that the lever is engaged with the outer notch in the rear of the adjuster assembly and the lower leg engages the teeth of the star wheel)

6.4u This is what it should look like when you're done

6.5 The maximum drum diameter is cast into the inside of the rear drums

7.4 To remove the wheel cylinder, remove the brake hose banjo bolt (lower arrow) and remove the two wheel cylinder mounting bolts (upper arrows)

6   Install the brake drum on the axle flange. Turn the drum by hand - it should turn freely, without any binding, but you should hear a slight dragging of the brake shoes. If you don't hear any dragging, turn the star wheel on the adjuster until you do, then back-off the adjuster until the drag is barely perceptible.

7   Mount the wheel, install the lug nuts, then lower the vehicle. Tighten the lug nuts to the torque listed in this Chapter's Specifications.

8   Make a number of forward and reverse stops and operate the parking brake to adjust the brakes until satisfactory pedal action is obtained.

9   Check the operation of the brakes carefully before driving the vehicle.

---

7   Wheel cylinder - removal, and installation

Note: *If the wheel cylinders leak, they must*

replaced with new ones - the manufacturer does not recommend rebuilding them.

Removal

Refer to illustration 7.4

1   Raise the rear of the vehicle and support it securely on jackstands. Block the front wheels to keep the vehicle from rolling.

2   Remove the brake shoe assembly (see Section 6).

3   Remove all dirt and foreign material from around the wheel cylinder.

4   Disconnect the brake hose banjo bolt **(see illustration)**.

5   Remove the wheel cylinder mounting bolts.

6   Detach the wheel cylinder from the brake backing plate. Immediately plug the brake hose to prevent fluid loss and contamination.

Installation

7   Place the wheel cylinder in position and install the bolts, tightening them to the torque

listed in this Chapter's Specifications.

8   Install the brake hose banjo bolt, using new sealing washers. Tighten the bolt to the torque listed in this Chapter's Specifications.

9   Install the brake shoe assembly. Bleed the brakes (see Section 10).

10   Check the operation of the brakes carefully before driving the vehicle.

---

8   Master cylinder - removal, overhaul and installation

Warning: *This procedure does not apply to vehicles with anti-lock brakes. When the brake lines are disconnected from the ABS master cylinder, air immediately gets into the hydraulic control unit. Because the hydraulic unit cannot be bled without special tools, this task is beyond the scope of the home mechanic. However, if you need to replace the power brake booster on an ABS-equipped vehicle, you can remove the master*

9

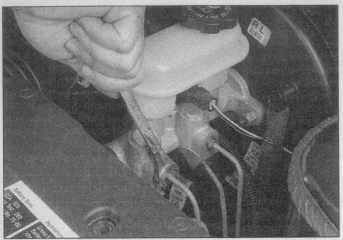

8.3 Disconnect the brake line fittings with a flare nut wrench, unplug the electrical connector for the brake fluid level sensor and remove the two mounting nuts

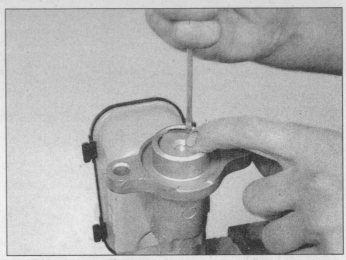

8.9 Press down on the piston and remove the retaining ring

8.10 Remove the primary piston assembly

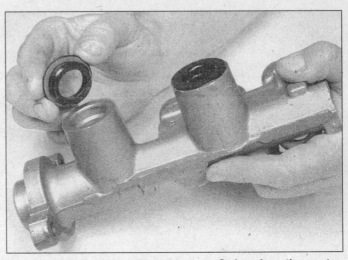

8.13 Remove the reservoir grommets or O-rings from the master cylinder body

cylinder mounting nuts and pull the master cylinder/hydraulic unit assembly far enough forward to remove the booster.

Note: *Before deciding to overhaul the master cylinder, check on the availability and cost of a new or factory rebuilt unit and the availability of a rebuild kit.*

1   A master cylinder overhaul kit should be purchased before beginning this procedure. The kit will include all the replacement parts necessary for the overhaul procedure. The rubber replacement parts, particularly the seals, are the key to fluid control within the master cylinder. As such, it is very important that they be installed securely and facing in the proper direction. Be careful during the rebuild procedure that no grease or petroleum-based solvents come in contact with the rubber parts.

Removal

Refer to illustration 8.3

2   Completely cover the front fender and

cowling area of the vehicle, as brake fluid can ruin painted surfaces.

3   Disconnect the brake line connections at the master cylinder, using a flare nut wrench, if available **(see illustration)**. Rags or newspapers should be placed under the master cylinder to soak up the fluid that will drain out. Unplug the electrical connector for the brake fluid level sensor.

4   Remove the two master cylinder mounting nuts. Remove the master cylinder from the vehicle. Immediately cap the brake lines to prevent fluid loss or contamination.

Overhaul

Refer to illustrations 8.9, 8.10, 8.13, 8.17, 8.19 and 8.20

5   Remove the reservoir cover or cap and reservoir diaphragm, then discard any remaining fluid in the reservoir.

6   Clamp the master cylinder flange in a vise. Don't apply pressure to the master cylinder body.

7   If equipped, remove the brake fluid level sensor. Use needle-nose pliers to compress the locking tabs at the inner side of the master cylinder.

8   Drive out the spring pins with a 1/8-inch punch. Be careful not to damage the reservoir or master cylinder body when driving out the pins.

9   Remove the primary piston retaining ring by depressing the piston and prying the ring out with a screwdriver **(see illustration)**.

10   Remove the primary piston assembly **(see illustration)**.

11   Remove the secondary piston assembly. It may be necessary to remove the cylinder from the vise, invert it and tap it against a wood block.

12   Remove the reservoir by pulling it straight up.

13   Remove the reservoir grommets or O-rings from the master cylinder body **(see illustration)**.

14   Inspect the reservoir for cracks and distortion. Replace it if damage is noted.

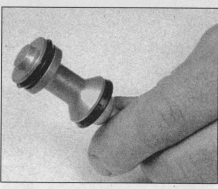

8.17 The secondary piston seals must be installed with the lips facing outwards

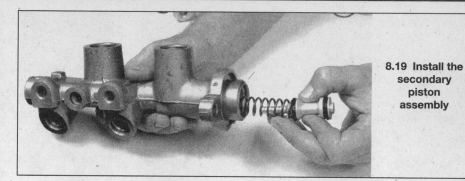

8.19 Install the secondary piston assembly

8.20 Install the primary piston assembly in the body

15 Be sure to note the installed direction of the old seal lips so the new seals can be installed the same way. The primary piston assembly is serviced as an assembly, while the secondary piston seals can be serviced separately. Clean all parts with brake system cleaner and dry them with unlubricated compressed air. Lubricate all rubber parts with clean brake fluid to ease reassembly.

16 Inspect the cylinder bore for corrosion and damage. If any corrosion or damage is found, replace the master cylinder with a new one, as abrasives cannot be used on the bore.

17 Remove the old seals from the secondary piston assembly and install the new seals so the lips face out **(see illustration)**.

18 Attach the spring retainer to the secondary piston assembly.

19 Lubricate the cylinder bore and secondary piston assembly with clean brake fluid and install the spring and secondary piston assembly into the cylinder **(see illustration)**.

20 Lubricate the primary piston assembly and install it in the cylinder bore. Depress it and install the lock ring **(see illustration)**.

21 Lubricate the new reservoir grommets (or O-rings) with clean brake fluid and press the grommets into the master cylinder body, making sure they are properly seated.

22 Lay the reservoir on a hard surface and press the master cylinder body onto the reservoir, either by using a rocking motion or by pressing it straight down onto the master cylinder body.

23 Drive in the spring pins to retain the reservoir, using care not to damage the reservoir or master cylinder body.

24 **Note**: *Whenever the master cylinder is removed, the complete hydraulic system must be bled.* The time required to bleed the system can be reduced if the master cylinder is filled with fluid and bench bled (refer to Steps 25 through 28) before the master cylinder is installed on the vehicle.

25 Insert threaded plugs of the correct size into the cylinder outlet holes and fill the reservoir with brake fluid. The master cylinder should be supported in such a manner that brake fluid will not spill during the bench bleeding procedure.

26 Loosen one plug at a time and push the

piston assembly into the bore to force air from the master cylinder. To prevent air from being drawn back into the cylinder, the appropriate plug must be replaced before allowing the piston to return to its original position.

27 Stroke the piston three or four times for each outlet to ensure that all air has been expelled.

28 Refill the master cylinder reservoirs and install the cap.

Installation

29 Install the master cylinder by reversing the removal steps, then bleed the brake system (see Section 10).

30 Test the brakes carefully before driving the vehicle in traffic.

9 Brake hoses and lines - inspection and replacement

Warning: *This procedure should not be undertaken on the rear brake circuit of a model equipped with an Anti-lock Brake System (ABS), since a special "scan" tool is needed to properly bleed the rear brakes. Take the vehicle to a dealer service department or other repair shop that has the proper tools.*

Inspection

1 About every six months, with the vehicle raised and supported securely on jackstands, the rubber hoses which connect the steel brake lines with the front and rear brake assemblies should be inspected for cracks, chafing of the outer cover, leaks, blisters and other damage. These are important and vulnerable parts of the brake system and

inspection should be complete. A light and mirror will be helpful for a thorough check. If a hose exhibits any of the above conditions, replace it with a new one.

Replacement

Front brake hose

Refer to illustrations 9.3 and 9.4

2 Loosen the wheel lug nuts, raise the vehicle and support it securely on jackstands. Remove the wheel.

3 At the frame bracket, unscrew the brake line fitting from the hose **(see illustration)**. Use a flare-nut wrench to prevent rounding off the corners.

9.3 If you're replacing a front brake hose, unscrew the brake line fitting from the hose at the frame bracket (use a flare-nut wrench to prevent rounding off the corners) - if the bracket start to bend as the fitting is loosened, place a backup wrench on the hose end

9

9.4 Pry the U-clip off and detach the hose from the bracket

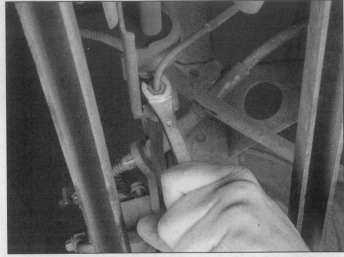

9.12a If you're replacing a rear brake hose, unscrew the brake line fitting from the hose at the crossmember bracket . . .

4   Pry the U-clip off the fitting with a screwdriver **(see illustration)** and detach the hose from the metal line.

5   At the caliper end of the hose, remove the banjo fitting bolt, then separate the hose from the caliper **(see illustrations 3.6a and 3.6b)**. Note that there are two copper sealing washers on either side of the fitting - they should be replaced with new ones during installation.

6   Pry the U-clip from the strut bracket, then feed the hose through the bracket.

7   To install the hose, pass the caliper fitting end through the strut bracket, then connect the fitting to the caliper with the banjo bolt and copper washers. Make sure the locating lug on the fitting is engaged with the hole in the caliper, then tighten the bolt to the torque listed in this Chapter's Specifications.

8   Push the metal support into the strut bracket and install the U-clip. Make sure the hose isn't twisted between the caliper and the strut bracket.

9   Route the hose into the frame bracket, again making sure it isn't twisted, then connect the brake line fitting, starting the threads by hand. Install the clip and E-ring, if equipped, then tighten the fitting securely.

10   Bleed the caliper (see Section 10).

11   Install the wheel and lug nuts, lower the vehicle and tighten the lug nuts to the torque specified in Chapter 1.

Rear brake hose

Refer to illustrations 9.12a and 9.12b

12   Perform Steps 2, 3, 4 and 5 **(see illustrations)**. Be sure to bleed the wheel cylinder (or caliper) (see Section 10).

Metal brake lines

13   When replacing brake lines, be sure to use the correct parts. Don't use copper tubing for any brake system components. Purchase steel brake lines from a dealer or auto parts store.

14   Prefabricated brake line, with the tube ends already flared and fittings installed, is

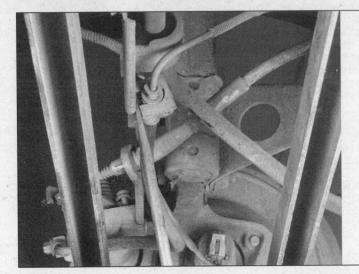

9.12b . . . then pry off the U-clip with a screwdriver

available at auto parts stores and dealer parts departments. These lines are also bent to the proper shapes.

15   When installing the new line, make sure it's securely supported in the brackets and has plenty of clearance between moving or hot components.

16   After installation, check the master cylinder fluid level and add fluid as necessary. Bleed the brake system (see Section 10) and test the brakes carefully before driving the vehicle in traffic.

10   Brake hydraulic system - bleeding

Refer to illustrations 10.8

Warning 1: *Wear eye protection when bleeding the brake system. If the fluid comes in contact with your eyes, immediately rinse them with water and seek medical attention.*

Warning 2: *This procedure should not be undertaken on the rear calipers of a model equipped with an Anti-lock Brake System (ABS), since a special "scan" tool is needed*

to properly bleed the rear brakes. Take the vehicle to a dealer service department or other repair shop that has the proper tools.

Note: *Bleeding the hydraulic system is necessary to remove any air that manages to find its way into the system when it's been opened during removal and installation of a hose, line, caliper or master cylinder.*

1   You'll probably have to bleed the system at all four brakes if air has entered it due to low fluid level, or if the brake lines have been disconnected at the master cylinder.

2   If a brake line was disconnected only at a wheel, then only that caliper or wheel cylinder must be bled.

3   If a brake line is disconnected at a fitting located between the master cylinder and any of the brakes, that part of the system served by the disconnected line must be bled.

4   Remove any residual vacuum from the brake power booster by applying the brake several times with the engine off.

5   Remove the master cylinder reservoir cover and fill the reservoir with brake fluid. Reinstall the cover. **Note:** *Check the fluid level often during the bleeding operation and*

10.8 When bleeding the brakes, a hose is connected to the bleed screw at the caliper or wheel cylinder and then submerged in brake fluid - air will be seen as bubbles in the tube and container (all air must be expelled before moving to the next wheel)

12.3 To tighten the parking brake cable, turn the adjusting nut (arrow) in (clockwise); to loosen the cable, turn the nut out (counterclockwise) - early model shown

add fluid as necessary to prevent the fluid level from falling low enough to allow air bubbles into the master cylinder.

6 Have an assistant on hand, as well as a supply of new brake fluid, a clear plastic container partially filled with clean brake fluid, a length of 3/16-inch plastic, rubber or vinyl tubing to fit over the bleeder valve and a wrench to open and close the bleeder valve.

7 Beginning at the right rear wheel, loosen the bleeder valve slightly, then tighten it to a point where it's snug but can still be loosened quickly and easily.

8 Place one end of the tubing over the bleeder valve and submerge the other end in brake fluid in the container **(see illustration)**.

9 Have the assistant pump the brakes slowly a few times to get pressure in the system, then hold the pedal down firmly.

10 While the pedal is held down, open the bleeder valve just enough to allow a flow of fluid to leave the valve. Watch for air bubbles to exit the submerged end of the tube. When the fluid flow slows after a couple of seconds, close the valve and have your assistant release the pedal slowly.

11 Repeat Steps 9 and 10 until no more air is seen leaving the tube, then tighten the bleeder valve and proceed to the left front wheel, the left rear wheel and the right front wheel, in that order, and perform the same procedure. Be sure to check the fluid in the master cylinder reservoir frequently.

12 Never use old brake fluid. It contains moisture which will deteriorate the brake system components.

13 Refill the master cylinder with fluid at the end of the operation.

14 Check the operation of the brakes. The pedal should feel solid when depressed, with no sponginess. If necessary, repeat the entire process. **Warning:** *Do not operate the vehicle if you're in doubt about the effectiveness of the brake system.*

11 Power brake booster - check, removal and installation

Operating check

1 Depress the brake pedal several times with the engine running and make sure there's no change in the pedal reserve distance.

2 Depress the pedal and start the engine. If the pedal goes down slightly, operation is normal.

Airtightness check

3 Start the engine and turn it off after one or two minutes. Depress the brake pedal slowly several times. If the pedal depresses less each time, the booster is airtight.

4 Depress the brake pedal while the engine is running, then stop the engine with the pedal depressed. If there's no change in the pedal reserve travel after holding the pedal for 30 seconds, the booster is airtight.

Removal

Note: *Power brake booster units shouldn't be disassembled. They require special tools not normally found in most automotive repair stations or shops. They're fairly complex and, because of their critical relationship to brake performance, should be replaced with a new or rebuilt one. Also, the vacuum booster and pushrod are a matched set, i.e. they're not interchangeable - they must always be installed as a set.*

5 Detach the cable from the negative battery terminal.

6 Remove the air cleaner assembly (see Chapter 4).

7 On ABS-equipped vehicles, remove the battery, battery box and battery tray (see Chapter 5).

8 Detach the brake master cylinder and

move it far enough forward to clear the booster (see Section 8).

9 Disconnect the vacuum line from the vacuum check valve.

10 Remove the left side knee bolster (see Chapter 11). Locate the booster pushrod retainer and washer and remove the pushrod from the brake pedal pin.

11 Remove the four nuts and washers holding the brake booster to the firewall (you may need a light to see them) **(see illustration)**.

12 Slide the booster straight out from the firewall until the studs clear the holes.

Installation

13 Installation procedures are basically the reverse of removal. Tighten the clevis locknut securely and the booster mounting nuts to the torque listed in this Chapter's Specifications.

14 Check and adjust the brake light switch (see Section 14).

12 Parking brake - adjustment

Refer to illustration 12.3

1 The parking brake lever, when properly adjusted, should travel four to seven clicks when a moderate pulling force is applied. If it travels less than four clicks, there's a chance the parking brake might not be releasing completely and might be dragging on the drum or disc. If the lever can be pulled up more than eight clicks, the parking brake may not hold adequately on an incline, allowing the car to roll.

2 To gain access to the parking brake cable adjuster on a 1991 model, remove the center console (see Chapter 11). On 1992 through 1994 models, the adjuster can be found by removing the rear ashtray. And on 1995 and later models, the adjuster can be

9

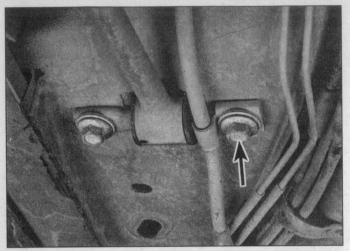

13.4 To detach the parking brake cable assembly from the underside of the vehicle, remove this trailing arm mounting bolt (arrow)

14.1 The brake light switch (shown here as seen from above, with the dashboard removed for clarity) is located at the upper end of the brake pedal; in this photo, the plunger (arrow) is extended (i.e. the brake pedal is depressed)

found by removing the screw on the right-hand side of the parking brake cover and sliding the cover foward.

3   To provide less parking brake cable travel, tighten the cable by turning the adjusting nut **(see illustration)** clockwise with a wrench; to provide more cable travel, back off the nut (turn it counterclockwise).

4   To check parking brake operation, raise the rear of the vehicle and support it securely on jackstands. With the parking brake lever at the first click, there should be moderate drag when the rear wheels are rotated by hand. There must be zero drag when the lever is released.

5   Lower the vehicle and replace the console when the adjustment is correct.

13   Parking brake cables - replacement

Refer to illustration 13.4

1   Loosen the rear wheel lug nuts, raise the rear of the vehicle and support it securely on jackstands. Block the front wheels. Remove the wheel. Make sure the parking brake is completely released.

2   On vehicles with rear drum brakes, remove the brake drum, remove the brake shoes and disconnect the cable from the parking brake lever (see Section 6).

3   On vehicles with rear disc brakes, detach the parking brake cable from the bracket on the caliper, then disconnect the cable from the lever on the caliper **(see illustrations 4.3a and 4.3b)**.

4   Unbolt the cable bracket from the rear suspension trailing arm **(see illustration)**. If the cable is also retained by a nylon cable tie, cut and remove the cable tie.

5   Remove the adjusting nut from the forward end of the cable at the equalizer bar **(see illustration 12.3)**.

6   Pry the cable grommet from the pan and remove the cable assembly.

7   Installation is the reverse of removal. Be sure to tighten the trailing arm-to-body bolt to the torque listed in the Chapter 10 Specifications.

8   When you're done, be sure to adjust the cable freeplay (see Section 12).

14   Brake light switch - check and adjustment

Check

Refer to illustration 14.1

1   The brake light switch **(see illustration)** is located on a bracket at the top of the brake pedal. When the brake pedal is depressed, the switch activates the brake lights.

2   Watch the brake light switch while pulling up on the brake pedal assembly with moderate force.

3   If you note any movement, adjust the switch (see Step 6).

4   If there's no movement, release the pedal and inspect the switch plunger (it's green, and located in the center of the switch).

a) *If 0.040-inch or less of the plunger is visible between the switch and the switch actuator pad on the brake pedal arm (see illustration 14.1), the switch is correctly adjusted. (The height of the rounded crown of the plunger itself is about 0.040-inch.)*

b) *If more than 0.040-inch of the plunger is visible, adjust the switch.*

Adjustment

5   Loosen the brake light switch mounting nut enough to allow the switch to move back and forth in its adjustment slot.

6   Insert a brake light switch adjustment gage or a 0.040-inch feeler gauge, between the switch and the switch actuator pad on the brake pedal arm. Make sure the plunger protrudes through the slot in the gage.

7   Pull the brake pedal with moderate force, pushing the switch forward against the gauge. Make sure you hold the switch perpendicular to the actuator pad on the brake pedal arm to prevent the switch from rotating out of alignment when tightening the mounting nut.

8   Tighten the switch mounting nut.

9   Release the brake pedal assembly.

10   With the pedal released and the adjustment gage still in position, swing the gauge back-and-forth.

a) *If the gauge swings freely from side-to-side, go to the next Step.*

b) *If the gauge sticks or doesn't swing freely, repeat Steps 5 through 9.*

11   With the adjustment gauge in position, pull up on the brake pedal with very light force and tap the gauge from side-to-side.

a) *If the gauge swings freely, repeat Steps 5 through 9.*

b) *If the gauge sticks and/or doesn't swing freely, go to the next Step.*

12   Check the switch plunger:

a) *If there's 0.040-inch or less of the plunger is visible between the switch and the switch actuator pad on the brake pedal arm, the switch is correctly adjusted. (The height of the rounded crown of the plunger is about 0.040-inch).*

b) *If there's more than 0.040-inch of plunger visible, repeat the adjustment procedure.*

Chapter 10
Suspension and steering systems

Contents

Specifications

Torque specifications — Ft-lbs

Front suspension

Strut
Strut upper mounting nuts ... 21
Strut-to-suspension support (damper shaft) nut ... 37
Strut-to-steering knuckle bolts/nuts
1995 and earlier ... 148
1996 ... 126
Front tension strut
Tension strut-to-control arm nut ... 106
Tension strut-to-cradle bracket bolts ... 103
Control arm
Control arm balljoint-to-steering knuckle nut ... 55
Control arm-to-cradle fasteners
Bolt ... 92
Nut ... 74

10

Rear suspension	Ft-lbs
Strut assembly	
Strut upper mounting nuts	21
Strut-to-suspension knuckle bolts/nuts	
1995 and earlier	148
1996	126
Trailing arm	
Trailing arm-to-knuckle nut	74
Trailing arm-to-body bolts..	89
Lateral links	
Lateral link-to-crossmember bolts/nuts	
Front..	126
Rear...	89
Lateral link-to-suspension knuckle	122

Hub/brake backing plate-to-knuckle bolts

(all models, drum or disc)..	63
Stabilizer bar	
Stabilizer bar-to-crossmember bracket bolts	41
Stabilizer bar-to-strut bracket nut	30

Steering system

Steering wheel nut...	30
Intermediate shaft U-joint-to-steering gear pinch bolt	35
Steering gear-to-cradle mounting bolts	37
Tie-rod end-to-steering knuckle nut	33
Power steering pump	
Pump-to-engine block mounting bolts	28
DOHC engines only	
Pump bracket-to-engine block bracket bolts..............................	22
Pump-to-intake manifold bracket bolts......................................	22
Wheel lug nuts	See Chapter 1

1   General information

Refer to illustrations 1.1 and 1.2

The front suspension is a MacPherson strut design. The upper end of each strut is attached to the vehicle. The lower end of the strut is connected to the upper end of the steering knuckle. The steering knuckle is attached to a balljoint in the outer end of the lower control arm. A front tension strut - attached to the crossmember and to both control arms - performs the job of a stabilizer bar and a strut rod **(see illustration)**.

The rear suspension also utilizes strut/coil spring assemblies. The upper end of each strut is attached to the vehicle body; the lower end of each strut is attached to a suspension knuckle. The knuckle is located by a pair of lateral links and a longitudinally mounted trailing arm between the body and each knuckle. A stabilizer bar is attached to the crossmember and to a bracket on each strut **(see illustration)**.

The rack-and-pinion steering gear is located behind the engine/transaxle assembly on the crossmember and actuates the tie-rods, which are attached to the steering knuckles. The steering column is designed to collapse in the event of an accident.

Frequently, when working on the sus-pension or steering system components, you may come across fasteners which seem impossible to loosen. These fasteners on the underside of the vehicle are continually subjected to water, road grime, mud, etc., and can become rusted or "frozen," making them extremely difficult to remove. In order to unscrew these stubborn fasteners without damaging them (or other components), be sure to use lots of penetrating oil and allow it to soak in for a while. Using a wire brush to clean exposed threads will also ease removal of the nut or bolt and prevent damage to the threads. Sometimes a sharp blow with a hammer and punch will break the bond between a nut and bolt threads, but care must be taken to prevent the punch from slipping off the fastener and ruining the threads. Heating the stuck fastener and surrounding area with a torch sometimes helps too, but isn't recommended because of the obvious dangers associated with fire. Long breaker bars and extension, or "cheater," pipes will increase leverage, but never use an extension pipe on a ratchet - the ratcheting mechanism could be damaged. Sometimes tightening the nut or bolt first will help to break it loose. Fasteners that require drastic measures to remove should always be replaced with new ones.

Since most of the procedures dealt with in this Chapter involve jacking up the vehicle and working underneath it, a good pair of jackstands will be needed. A hydraulic floor jack is the preferred type of jack to lift the vehicle, and it can also be used to support certain components during various operations. **Warning:** *Never, under any circumstances, rely on a jack to support the vehicle while working on it. Whenever any of the suspension or steering fasteners are loosened or removed they must be inspected and, if necessary, replaced with new ones of the same part number or of original equipment quality and design. Torque specifications must be followed for proper reassembly and component retention. Never attempt to heat or straighten any suspension or steering components. Instead, replace any bent or damaged part with a new one.*

2   Front strut assembly - removal, inspection and installation

Removal

Refer to illustrations 2.2, 2.4 and 2.6

1   Loosen the front wheel lug nuts. Raise the vehicle and support it securely on jackstands. Remove the front wheel(s).

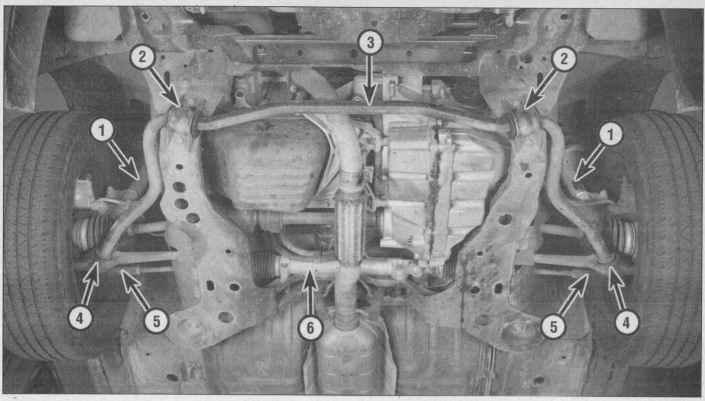

1.1 Underside view of the front suspension components

1	Strut assembly	3	Tension strut	5	Tie-rod end
2	Tension strut bracket	4	Lower control arm	6	Steering gear assembly

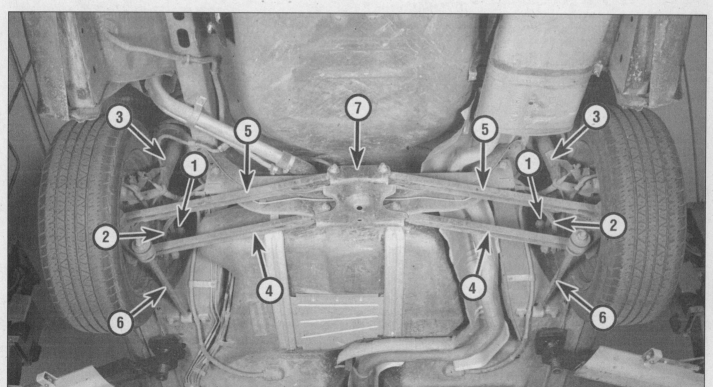

1.2 Underside view of the rear suspension components

1	Stabilizer bar link	3	Strut assembly	5	Rear lateral link	7	Crossmember
2	Stabilizer bar	4	Front lateral link	6	Trailing arm		

10

2.2 If you're replacing the strut on an ABS-equipped vehicle, drill out this rivet head (arrow) to detach the speed sensor wire harness bracket from the strut; when you install the new strut, reattach the bracket with a new rivet or self-tapping screw (if you're planning to reuse the same strut, don't drill out the rivet - just thread the harness out of the bracket)

2   If you're going to replace the strut on an ABS-equipped vehicle, drill off the rivet head which attaches the ABS wire harness to the strut **(see illustration)**.

3   If you're removing the strut, but you're going to reuse it, simply disconnect the ABS wire harness from the bracket and leave the bracket attached to the strut.

4   Apply reference marks on the area where the strut contacts the knuckle and remove the strut-to-knuckle bolts **(see illustration)**. **Note:** *It is necessary to apply reference marks only if the strut has been modified to allow for camber adjustment.*

5   Separate the strut from the steering knuckle.

6   Support the strut and spring assembly and remove the three strut upper mounting nuts **(see illustration)**. Remove the assembly out from the fenderwell.

2.4 Remove the strut-to-knuckle nuts and bolts (arrows)

Inspection

7   Check the strut body for leaking fluid, dents, cracks and other obvious damage which would warrant repair or replacement.

8   Check the coil spring for chips or cracks in the spring coating (this will cause premature spring failure due to corrosion). Inspect the spring seat for cuts, hardness and general deterioration.

9   If any undesirable conditions exist, proceed to the strut disassembly procedure (see Section 3).

Installation

10   Guide the strut assembly up into the fenderwell and insert the three upper mounting studs through the holes in the strut tower. Once the three studs protrude from the strut tower, install the nuts so the strut won't fall back through. This is most easily accomplished with the help of an assistant, as the strut is quite heavy and awkward.

11   Slide the steering knuckle into the strut flange and insert the two bolts. Install the nuts, align the previously applied marks and tighten the nuts to the torque listed in this

Chapter's Specifications.

12   If you installed a new strut, attach the ABS wire harness bracket with a new rivet or self-tapping screw; if you're installing the old strut, thread the ABS harness through the bracket.

13   Install the wheel(s) and lug nuts, then lower the vehicle and tighten the lug nuts to the torque listed in the Chapter 1 Specifications.

14   Tighten the three upper mounting nuts to the torque listed in this Chapter's Specifications.

15   Drive the vehicle to an alignment shop and have the wheel alignment checked, and if necessary, adjusted.

3   Strut/spring assembly - replacement

1   If the struts or coil springs exhibit the telltale signs of wear (leaking fluid, loss of damping capability, chipped, sagging or cracked coil springs) explore all options before beginning any work. The strut assemblies are not serviceable and must be replaced if a problem develops. However, strut assemblies complete with springs may be available on an exchange basis, which eliminates much time and work. Whichever route you choose to take, check on the cost and availability of parts before disassembling your vehicle. **Warning:** *Disassembling a strut is a potentially dangerous undertaking and utmost attention must be directed to the job, or serious injury may result. Use only a high quality spring compressor and carefully follow the manufacturer's instructions furnished with the tool. After removing the coil spring from the strut assembly, set it aside in a safe, isolated area.*

Disassembly

Refer to illustrations 3.3, 3.4, 3.5, 3.6 and 3.7

2   Remove the strut and spring assembly following the procedure described in the previous Section. Mount the strut assembly in a

2.6 Support the strut from below and remove these three strut retaining nuts (arrows)

3.3 Install the spring compressor according to the tool manufacturer's instructions and compress the spring until all pressure is relieved from the upper spring seat

3.4 Remove the damper shaft nut

3.5 Lift the suspension support off the damper shaft

3.6 Remove the spring seat from the damper shaft

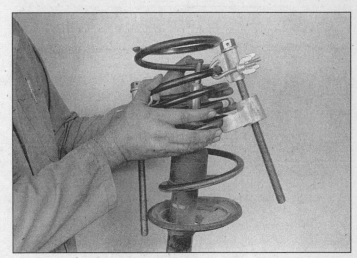

3.7 Remove the compressed spring assembly - keep the ends of the spring pointed away from your body

vise. Line the vise jaws with wood or rags to prevent damage to the unit and don't tighten the vise excessively.

3   Following the tool manufacturer's instructions, install the spring compressor (which can be obtained at most auto parts stores or equipment yards on a daily rental basis) on the spring and compress it sufficiently to relieve all pressure from the upper spring seat **(see illustration)**. This can be verified by wiggling the spring.

4   Loosen the damper shaft nut with a socket wrench **(see illustration)**.

5   Remove the nut, washer and suspension support **(see illustration)**. Inspect the bearing in the suspension support for smooth operation. If it doesn't turn smoothly, replace the suspension support. Check the rubber portion of the suspension support for cracking and general deterioration. If there is any separation of the rubber, replace it.

6   Lift the spring seat and upper insulator from the damper shaft **(see illustration)**. Check the rubber spring seat for cracking and hardness, replacing it if necessary.

7   Carefully lift the compressed spring

from the assembly **(see illustration)** and set it in a safe place. **Warning:** *Never place your head near the end of the spring!*

8   Slide the rubber bumper off the damper shaft.

9   Check the lower insulator for wear, cracking and hardness and replace it if necessary.

Reassembly

Refer to illustration 3.11

10   If the lower insulator is being replaced, set it into position with the dropped portion seated in the lowest part of the seat. Extend the damper rod to its full length and install the rubber bumper.

11   Carefully place the coil spring onto the lower insulator, with the end of the spring resting in the lowest part of the insulator **(see illustration)**.

12   Install the upper insulator and spring seat.

13   Install the suspension support to the damper shaft.

14   Install the washer and nut and tighten

3.11 When installing the spring, make sure the end fits into the recessed portion of the lower seat (arrow)

10

the nut to the torque listed in this Chapter's Specifications.

15   Install the strut/coil spring assembly following the procedure outlined previously (see Section 3).

4.2 To knock the balljoint stud loose from the steering knuckle, give the knuckle a few sharp raps with a hammer

4.4 Use a prybar to separate the control arm from the steering knuckle

4.5 Remove the nut and bolt (arrows) that attach the inner end of the control arm to the cradle

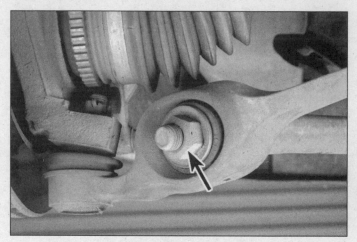

4.6 Remove the nut (arrow) and washer from the tension strut stud

4   Lower control arm - removal, inspection and installation

Removal

Refer to illustrations 4.2, 4.4, 4.5 and 4.6

1   Loosen the front wheel lug nuts. Raise the vehicle and support it securely on jackstands. Remove the front wheel(s).

2   Remove the cotter pin from the control arm-to-steering knuckle balljoint nut, loosen the nut and knock the stud loose by rapping sharply on the knuckle with a hammer **(see illustration)**. Remove the nut. **Caution:** *Be careful not to damage the speed sensor ring on ABS equipped vehicles. Minor damage could cause ABS malfunctions.*

3   Remove the front inner splash shield **(see illustrations 9.3a and 9.3b in Chapter 8)**.

4   Use a prybar to disconnect the control arm from the steering knuckle **(see illustration)**.

5   Remove the nut and bolt that attach the inner end of the lower control arm to the cradle **(see illustration)**.

6   Remove the nut and washer from the tension strut stud **(see illustration)** and separate the control arm from the tension strut.

7   Remove the control arm.

Inspection

8   Check the control arm for distortion and the bushings for wear. If the bushings are worn or the control arm is bent, replace the control arm. Don't try to straighten a bent control arm.

Installation

9   Installation is the reverse of removal. Tighten all of the nuts to the torque listed in this Chapter's Specifications after the control arm is reconnected to the steering knuckle and to the tension strut. And be sure to install the large washer with its concave (dished) side facing towards the nut. If it is necessary to turn the nut after tightening to line up the hole, always tighten the nut, never loosen it.

10   Install the wheel(s) and lug nuts, lower the vehicle and tighten the lug nuts to the torque listed in the Chapter 1 Specifications.

11   Drive the vehicle to an alignment shop and have the front wheel alignment checked, and if necessary, adjusted.

5   Tension strut - removal and installation

Removal

Refer to illustrations 5.5 and 5.6

1   Loosen the left (driver's side) front wheel lug nuts. Raise the vehicle and support it securely on jackstands. Remove the left front wheel.

2   Separate the left control arm from the steering knuckle as described in the previous Section.

3   Remove the left front inner splash shield **(see illustrations 9.3a and 9.3b in Chapter 8)**.

4   At the left side of vehicle, remove the nut and bolt that attach the lower control arm to the cradle **(see illustration 4.5)**.

5   At the right side of the vehicle, remove the tension strut-to-lower control arm nut and washer **(see illustration)**.

6   Remove the tension strut-to-cradle

5.5 At the right lower control arm, remove the tension strut-to-control arm nut (arrow) and washer

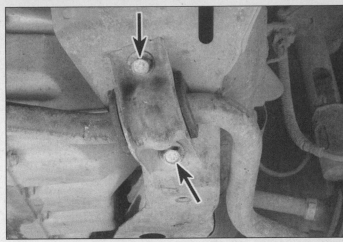

5.6 Remove the tension strut-to-cradle bracket bolts (arrows) (left bracket shown, right bracket identical)

bracket bolts **(see illustration).** Note: *These bolts aren't threaded into the cradle itself - they're screwed into nuts welded to the cradle. These may break loose from the cradle, making it difficult to remove the bolts. If any of the tension strut bracket bolts turn but don't back out, insert a wrench or socket into the access hole on the upper side of the cradle and hold the nut while breaking the bolt loose.*

7   Remove the tension strut and the left lower control arm as a single assembly.

8   Remove the left lower control arm from the tension strut (see Section 4).

9   Remove the bracket bushings and inspect them for cracks, tears and other deterioration. If they're damaged or worn, replace them.

Installation

10   Reattach the left lower control arm to the tension strut, but don't tighten the nut yet.

11   Install the tension strut-to-cradle bushings onto the tension strut. Position the tension strut bushings with the slits facing forward. To simplify reassembly, lubricate the inside and outside of the new bushings with silicone grease. **Caution:** *Don't use petroleum or mineral-based lubricants or brake fluid - they will lead to deterioration of the bushings.*

12   Install the tension strut by positioning the right end of the tension strut into the right control arm, then position the left control arm into the cradle. Do not install the fasteners at this time.

13   Apply threadlocking compound the threads of the tension strut bushing bracket bolts. Install the brackets and tighten the bolts to the torque listed in this Chapter's Specifications.

14   Install the left wheel and tighten, but do not torque, the lug nuts. With the help of an assistant, push the bottom of the wheel in (toward the center of the vehicle). This will move the left control arm into position, allow-

ing you to install the control arm-to-cradle bolt and nut. Install the bolt and nut and tighten to the torque listed in this Chapter's Specifications.

15   Install the right side tension strut-to-control arm nut and washer (concave side facing toward the nut) and tighten to the specified torque. Tighten the left side tension strut-to-control arm nut to the torque listed in this Chapter's Specifications.

16   Connect the left control arm balljoint stud to the steering knuckle. Install the castle nut and tighten it to the torque listed in this Chapter's Specifications. Install a new cotter pin. If it is necessary to turn the castle nut to line up the hole in the stud, always tighten the nut, never loosen it.

17   The remainder of installation is the reverse of removal. Re-tighten all nuts and bolts to the specified torque values, including the wheel lug nuts (see Chapter 1).

18   Drive the vehicle to an alignment shop and have the wheel alignment checked, and if necessary adjusted.

6   Balljoints - check and replacement

1   Loosen the front wheel lug nuts. Raise the vehicle and support it securely on jackstands. Remove the front wheel(s).

2   Remove the lower control arm (see Section 4).

3   Try to wiggle the balljoint stud. It should be firm and difficult to move. If it's loose, replace the lower control arm. The balljoints on these vehicles are neither removable nor rebuildable.

4   Install the lower control arm (see Section 4).

5   Install the wheels, tighten but don't torque the wheel lug nuts, remove the jackstands, lower the vehicle and tighten the wheel lug nuts to the torque listed · in the Chapter 1 Specifications.

7   Steering knuckle and hub - removal and installation

Warning: *Dust created by the brake system may contain asbestos, which is harmful to your health. Never blow it out with compressed air and don't inhale any of it. Do not, under any circumstances, use petroleum-based solvents to clean brake parts. Use brake system cleaner only.*

Removal

Refer to illustrations 7.2a and 7.2b

1   Loosen the front wheel lug nuts. Raise the vehicle and support it securely on jackstands. Remove the wheel(s).

2   Remove the brake caliper mounting bolts (see Chapter 9), remove the caliper assembly and support it with a piece of wire, remove the brake disc from the hub. Remove the driveaxle/hub nut (see Chapter 8). On ABS-equipped vehicles, unplug the electrical connector for the wheel speed sensor and

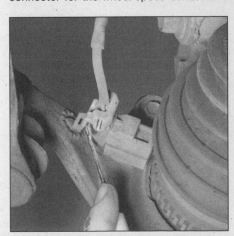

7.2a Pry open the small locking tang on the side of the electrical connector for the ABS wheel speed sensor and unplug the connector

10

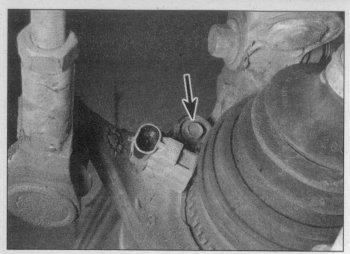

7.2b Remove this bolt (arrow) and remove the wheel speed sensor

9.8 Remove the nuts and bolts (arrows) that attach the lower end of the rear strut to the rear suspension knuckle

remove the sensor **(see illustrations)**.

3   Loosen, but do not remove, the strut-to-steering knuckle bolts **(see illustration 2.4)**.

4   Separate the tie-rod end from the steering knuckle (see Section 17).

5   Separate the lower control arm from the steering knuckle (see Section 4).

6   Push the driveaxle from the hub as described in Chapter 8. Support the end of the driveaxle with a piece of wire.

7   Carefully remove the strut-to-steering knuckle bolts and separate the steering knuckle from the strut and lower arm.

Installation

8   Guide the knuckle and hub assembly into position, inserting the driveaxle into the hub.

9   Push the knuckle into the strut flange and install the bolts and nuts, but don't tighten them yet.

10   Attach the control arm to the steering knuckle (see Section 4).

11   Attach the tie-rod end to the steering knuckle arm (see Section 17).

12   Match up the alignment marks and tighten the strut-to-knuckle bolts and nuts, the control arm balljoint stud-to-steering knuckle castle nut and the tie-rod end castle nut to the torque listed in this Chapter's Specifications. Install new cotter pins in the balljoint and tie-rod end castle nuts. If it is necessary to turn the castle nut to line up the hole in the stud, always tighten the nut, never loosen it.

13   Place the brake disc on the hub and install the caliper assembly, tighten the caliper mounting bolts to the torque listed in the Chapter 9 specifications.

14   Install the speed sensor and connect the electrical connector.

15   Install the hub nut and tighten it to the torque listed in the Chapter 8 Specifications. To hold the wheel with the vehicle off the ground, place a large screwdriver or prybar between the wheel studs **(see illustration 9.5 in Chapter 8)**.

16   Install the wheel(s) and lug nuts.

17   Lower the vehicle and tighten the lug nuts to the torque listed in the Chapter 1 Specifications.

8   Front hub and bearing assembly - removal and installation

Due to the special tools and expertise required to press the hub and bearing from the steering knuckle, this job should be left to a professional mechanic. However, the steering knuckle and hub may be removed and the assembly taken to a dealer service department or other repair shop. See Section 7 for the steering knuckle and hub removal procedure.

9   Rear strut assembly - removal, inspection and installation

Removal

Refer to illustrations 9.8 and 9.10

1   On coupe models, remove the rear seat

cushion bottom, the left or right interior rocker panel molding and the left or right interior sail panel, then remove the screws fastening the package tray to the side of the cargo area.

2   On sedan models, remove the left or right C-pillar interior molding, fold down the rear seat backs and remove the rear seat side bolsters from the vehicle (see Chapter 11).

3   On all models, remove the speaker grille fasteners and the speaker grilles (see Chapter 11).

4   Remove the seatbelt bezel and separate the seat belts from the package tray.

5   Remove the package tray carpeting.

6   Loosen the rear wheel lug nuts. Raise the vehicle and support it securely on jackstands. Remove the rear wheel(s).

7   If you're replacing the strut on an ABS-equipped vehicle, drill off the rivet head which attaches the ABS harness bracket to the strut, and remove the speed sensor harness bracket. If you're going to reuse the strut on an ABS-equipped vehicle, disconnect the speed sensor wire harness from the bracket; don't remove the bracket itself.

8   Mark the relationship of the strut to the knuckle, then loosen the two strut-to-knuckle

9.10 After disconnecting and raising or removing the rear package tray, you can remove the rear strut upper mounting nuts (arrows)

10.3 Remove the lateral link-to-rear knuckle fasteners (upper arrows); if you're removing the trailing arm, remove the nut (lower arrow) from the knuckle

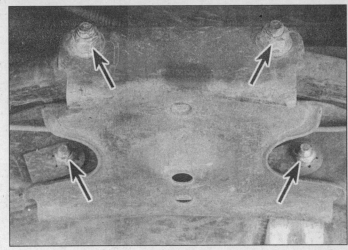

10.4 Remove the nut (arrows) and bolt from the inner end of the lateral link you're replacing

nuts and bolts **(see illustration)**. **Note:** *It is necessary to apply reference marks only if the strut has been modified to allow for camber adjustment.*

9   Place a floor jack underneath the knuckle. Raise the floor jack high enough to support the knuckle.

10   Remove the three upper strut-to-body retaining nuts **(see illustration)**.

11   Slowly and carefully lower the jack until the strut-to-knuckle bolts can be removed. Separate the strut from the knuckle and guide the strut out of the fenderwell.

Inspection

12   Follow the inspection procedures described in Section 2. If it is determined that the strut assembly must be disassembled for replacement of the strut or the coil spring, refer to Section 3.

Installation

13   Maneuver the assembly up into the fenderwell and insert the mounting studs through the holes in the body. Install the upper mounting nuts, but don't tighten them yet.

14   Push the lower end of the strut into the knuckle, install the strut-to-knuckle bolts and nuts, match up the alignment marks you made before disassembly and tighten the fasteners to the torque listed in this Chapter's Specifications.

15   If the strut was replaced on an ABS-equipped vehicle, attach the speed sensor wire harness bracket to the strut with a new rivet or a self-tapping screw; if the strut wasn't replaced, thread the wire harness through the bracket.

16   Install the wheel(s) and lug nuts, lower the vehicle and tighten the lug nuts to the torque listed in the Chapter 1 Specifications.

17   Tighten the three strut upper mounting nuts to the torque listed in this Chapter's Specifications.

18   Install any interior components that

were removed (see Chapter 11).

19   Drive the vehicle to an alignment shop and have the wheel alignment checked, and if necessary, adjusted.

10   Lateral links - removal and installation

Removal

Refer to illustrations 10.3 and 10.4

1   Loosen the rear wheel lug nuts. Raise the vehicle and support it securely on jackstands. Remove the rear wheel(s).

2   If you're removing a front lateral link on an early model, remove the fuel tank (refer to Chapter 4). On later models, support the rear crossmember with a floor jack, then disconnect the brake tubes from the floor pan. Remove the crossmember bolts and lower the crossmember to provide access to the front lateral link bolts.

3   Remove the lateral link-to-knuckle fasteners **(see illustration)**.

4   Remove the lateral link-to-crossmember fasteners **(see illustration)**.

5   Remove the lateral link from the vehicle.

Installation

6   Installation is the reverse of removal, with the following provisions:

a) *The inner (crossmember) ends of both front lateral links on 1991 and 1992 models have a welded-on washer. This special washer isn't used on 1993 and later models; they use a special truss-headed bolt instead. This truss headed bolt can be used on any year model, but the standard hex bolt used on 1991 and 1992 models cannot be used on 1993 and later models.*

b) *Don't tighten any fasteners to the torque listed in this Chapter's Specifications until all fasteners have been installed.*

7   If you replaced either front lateral link, install the fuel tank (see Chapter 4).

8   Install the wheel(s) and lug nuts, then lower the vehicle to the ground. Tighten the lug nuts to the torque listed in the Chapter 1 Specifications.

9   Drive the vehicle to an alignment shop and have the wheel alignment checked, and if necessary, adjusted.

11   Trailing arm - removal and installation

Refer to illustration 11.3

1   Loosen the rear wheel lug nuts. Raise the vehicle and support it securely on jackstands. Remove the rear wheel(s).

2   Remove the trailing arm-to-knuckle nut **(see illustration 10.3)**.

3   Remove the trailing arm-to-body bolts **(see illustration)** and detach the trailing arm from the vehicle.

4   Installation is the reverse of the removal

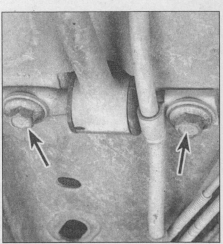

11.3 Remove the trailing arm-to-body bolts (arrows)

10

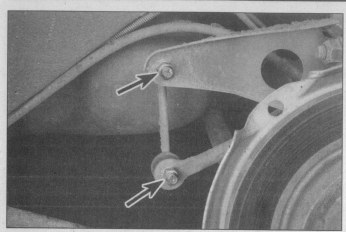

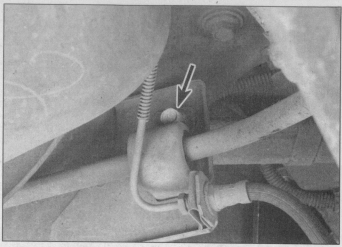

12.3 Remove the upper nut (upper arrow) that attaches the upper end of the stabilizer bar link to the bracket (if you want to inspect and/or replace the bushing at the lower end of the link, remove the lower nut as well)

12.4 Remove both stabilizer bar bracket bolts (arrow points to left bracket bolt) from the crossmember

procedure. Be sure to tighten the bolts to the torque listed in this Chapter's Specifications.

12 Rear stabilizer bar and bushings - removal and installation

Refer to illustrations 12.3 and 12.4

Warning: *This procedure should not be performed on a model equipped with an Antilock Brake System (ABS), since a special "scan" tool is needed to properly bleed the rear brakes. Take the vehicle to a dealer service department or other repair shop that has the proper tools.*

1   Loosen the right rear wheel lug nuts, raise the vehicle, place it securely on jackstands and remove the right rear wheel.

2   Place a container under the left rear brake line/brake hose junction and disconnect the hose from the line (see Chapter 9). Plug the brake line to prevent contamination and fluid loss.

3   Remove the left and right stabilizer bar link-to-knuckle bracket fasteners **(see illustration)**.

4   Unbolt the left and right stabilizer bar brackets from the crossmember **(see illustration)**.

5   Loosen - but don't remove at this time - the lateral link-to-left knuckle fastener.

6   Remove the left trailing arm-to-knuckle nut **(see illustration 10.3)**.

7   Remove the two left trailing arm-to-body fasteners **(see illustration 11.3)**.

8   Slide the left trailing arm out of the knuckle.

9   Remove the left lateral link-to-knuckle fastener **(see illustration 10.3)** and swing the left lateral links down, out of the way.

10   Disconnect the brake line fasteners and detach the brake line from the crossmember.

Warning: *Trying to remove the stabilizer bar without disconnecting the brake lines from the crossmember could result in bent or broken brake lines.*

11   The stabilizer bar can now be removed from the vehicle. Pull the brackets off the stabilizer bar (if they haven't already fallen off) using a rocking motion.

12   Check the bracket bushings for wear, hardness, distortion, cracking, and other signs of deterioration, replacing them if necessary. Also check the link bushings for these signs.

13   Using a wire brush, clean the areas of the bar where the bushings ride. Installation is the reverse of the removal procedure. If

necessary, use a light coat of silicone grease to ease bushing and U-bracket installation (don't use petroleum based lubricants or brake fluid, as these will damage the rubber).

14   Installation is the reverse of removal. Be sure to tighten the stabilizer bar fasteners and any other suspension fasteners you disconnected, to the torque listed in this Chapter's Specifications.

15   Install the right rear wheel and lug nuts. Lower the vehicle and tighten the lug nuts to the torque listed in Chapter 1 Specifications.

13 Rear hub and bearing assembly - removal and installation

Warning: *Dust created by the brake system may contain asbestos, which is harmful to your health. Never blow it out with compressed air and don't inhale any of it. Do not, under any circumstances, use petroleum-based solvents to clean brake parts. Use brake system cleaner only.*

Note: *The hub and bearing assembly cannot be serviced separately; if the bearing is defective or worn out, the complete hub and bearing assembly must be replaced.*

Removal

Refer to illustrations 13.2 and 13.4

1   Loosen the rear wheel lug nuts. Raise the vehicle and support it securely on jackstands. Remove the wheel(s).

2   On ABS-equipped vehicles, unplug the electrical connector from the wheel speed sensor **(see illustration)**.

3   Remove the brake drum, or if equipped with rear disc brakes, remove the caliper mounting bolts (see Chapter 9). Remove the caliper assembly and support it out of the way. Remove the brake disc.

4   Remove the four hub-to-knuckle bolts. You can reach these bolts through the large circular cutouts in the hub flange **(see illustration)**.

5   Remove the hub and bearing assembly.

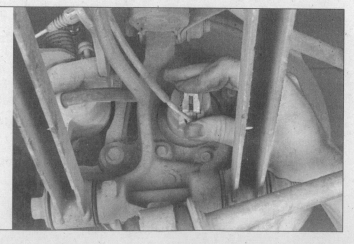

13.2 On ABS-equipped vehicles, unplug the electrical connector for the wheel speed sensor from the back side of the suspension knuckle

13.4 Access the bolts retaining the hub and bearing assembly through the holes in the hub flange

16.7 Once you have removed the steering wheel retaining nut and marked the relationship of the wheel to the steering shaft, install a steering wheel puller and remove the wheel

Installation

6   Position the hub and bearing assembly on the knuckle and align the holes in the backing plate. Install the bolts (a magnet is useful in guiding the bolts through the hub flange and into position). After all four bolts have been installed, tighten them to the torque listed in this Chapter's Specifications.

7   Install the brake drum, or brake disc and caliper assembly. Tighten the caliper mounting bolts to the torque listed in Chapter 9 Specifications. Install the wheel(s). Lower the vehicle and tighten the lug nuts to the torque listed in the Chapter 1 Specifications.

14   Rear suspension knuckle - removal and installation

Warning: *Dust created by the brake system may contain asbestos, which is harmful to your health. Never blow it out with compressed air and don't inhale any of it. Do not, under any circumstances, use petroleum-based solvents to clean brake parts. Use brake system cleaner only.*

Removal

1   Loosen the rear wheel lug nuts. Raise the vehicle and support it on jackstands. Remove the rear wheel(s).

2   Remove the rear hub and bearing assembly (see Section 13).

3   On models equipped with rear drum brakes, suspend (with wire) the backing plate and brake assembly out of the way.

4   On models equipped with rear disc brakes, remove the caliper and disc (see Chapter 9).

5   Loosen - but don't disconnect - the lateral link-to-knuckle fasteners (see Section 9).

6   Loosen - but don't disconnect - the fasteners that attach the lower end of the rear strut to the knuckle (see Section 9).

7   Remove the nut that attaches the rear end of the trailing arm to the knuckle and the two bolts that attach the forward end of the arm to the vehicle body and remove the trailing arm (see Section 11).

8   Remove the lateral link-to-knuckle fasteners and disconnect the outer ends of both lateral links from the knuckle (see Section 10).

9   Remove the strut-to-knuckle fasteners and disconnect the lower end of the strut fro the knuckle (see Section 9). The knuckle can now be removed from the vehicle.

Installation

10   Installation is the reverse of removal. Make sure you tighten all fasteners to the torque listed in this Chapter's Specifications.

11   Install the wheel(s) and lug nuts. Lower the vehicle and tighten the lug nuts to the torque listed in the Chapter 1 Specifications.

12   Drive the vehicle to an alignment shop and have the wheel alignment checked and, if necessary, adjusted.

15   Steering system - general information

All models are equipped with rack-and-pinion steering. The steering gear is bolted to the firewall and operates the steering arms via tie-rods. The inner ends of the tie-rods are protected by rubber boots which should be inspected periodically for secure attachment, tears and leaking lubricant.

The power assist system consists of a belt-driven pump and associated lines and hoses. The fluid level in the power steering pump reservoir should be checked periodically (see Chapter 1).

The steering wheel operates the steering shaft, which actuates the steering gear through universal joints. Looseness in the steering can be caused by wear in the steering shaft universal joints, the steering gear, the tie-rod ends and loose retaining bolts.

16   Steering wheel - removal and installation

Warning: *On models equipped with a Supplemental Inflatable Restraint (SIR) system, always disable the airbag(s) before working in the vicinity of the steering wheel, instrument panel or SIR system components to avoid the possibility of accidental deployment of the airbag(s), which could cause personal injury (see Chapter 12).*

Caution: *On models equipped with a theft-deterrent radio, make sure you have the correct activation code, or the theft deterrent system is turned off, before disconnecting the battery.*

Removal

Refer to illustration 16.7

1   If the vehicle is equipped with a Supplemental Inflatable Restraint (SIR) system, (i.e. an airbag), disable the system before proceeding (see Chapter 12).

2   Disconnect the cable from the negative battery terminal.

3   On vehicles without an airbag, pull firmly on the edge of the horn pad to detach it from the steering wheel. Disconnect the wires attached to the back of the horn pad and remove the pad.

4   On vehicles with an airbag, remove the four screws from the back side of the steering wheel (the side of the wheel facing the dash), pull out the inflator module, unplug the electrical connector and remove the module. Lay the module down carefully, with the driver's side facing up.

5   Disconnect the wiring from the cruise control switch and the horn switch.

6   Remove the steering wheel retaining nut (some models also have a clip that must be removed before the nut can be removed), then mark the relationship of the steering shaft to the hub (if marks don't already exist or don't line up) to simplify installation and ensure steering wheel alignment. Discard the nut as a new one must be installed.

7   Use a puller to disconnect the steering wheel from the shaft **(see illustration)**. Don't beat on the shaft to detach the wheel.

8   To avoid rotating the air bag coil while

10

17.3a Loosen the tie-rod end jam nut, back it off far enough that some threads are showing . . .

17.3b . . . and make an alignment mark on the threads to ensure that the new tie-rod end is installed with exactly the same number of threads showing (otherwise, you'll upset the toe-in adjustment)

the steering wheel is removed, use tape to firmly secure the coil.

Installation

9   To install the wheel, align the mark on the steering wheel hub with the mark on the shaft and slip the wheel onto the shaft. Install the new nut and tighten it to the torque listed in this Chapter's Specifications.

10   On vehicles without an airbag, connect the horn wire and install the horn pad.

11   On vehicles with an airbag, plug in the inflator module electrical connector, install the module and install the four module retaining screws from the back side of the steering wheel.

12   Connect the negative battery cable.

13   On vehicles with an airbag, enable the SIR system (see Chapter 12).

17   Tie-rod ends - removal and installation

Removal

Refer to illustrations 17.3a, 17.3b and 17.4

1   Loosen the front wheel lug nuts. Raise the vehicle and support it securely on jackstands. Remove the front wheel(s).

2   Remove and discard the cotter pin and loosen the nut on the tie-rod end stud.

3   Loosen the jam nut and back it off far enough to mark the position of the tie-rod end in relation to the threads **(see illustrations)**.

4   Disconnect the tie-rod from the steering knuckle arm with a puller or a balljoint separator tool **(see illustration)**. Remove the nut and separate the tie-rod.

5   Unscrew the tie-rod end from the tie-rod.

Installation

6   Thread the tie-rod end on to the marked

17.4 Install a small puller or balljoint separator tool and separate the tie-rod end from the steering knuckle

position and insert the tie-rod stud into the steering knuckle arm. Tighten the jam nut securely.

7   Install the castellated nut on the stud and tighten it to the torque listed in this Chapter's Specifications. Install a new cotter pin. If it is necessary to turn the castle nut to line up the hole in the stud, always tighten the nut, never loosen it.

8   Install the wheel(s) and lug nuts. Lower the vehicle and tighten the lug nuts to the torque listed in the Chapter 1 Specifications.

9   Drive the vehicle to an alignment shop and have the wheel alignment checked, and if necessary, adjusted.

18   Steering gear boots - replacement

1   Loosen the front wheel lug nuts. Raise the vehicle and support it securely on jackstands. Remove the wheel(s).

2   Remove the tie-rod end and jam nut (see Section 17).

3   Remove the steering gear boot clamps

and slide the boot off.

4   Before installing the new boot, wrap the threads and serrations on the end of the steering rod with a layer of tape so the small end of the new boot isn't damaged.

5   Slide the new boot into position on the steering gear until it seats in the groove in the steering rod and install new clamps.

6   Remove the tape and install the tie-rod end (see Section 17).

7   Install the wheel(s) and lug nuts. Lower the vehicle and tighten the lug nuts to the torque listed in the Chapter 1 Specifications.

8   Drive the vehicle to an alignment shop and have the wheel alignment checked, and if necessary, adjusted.

19   Steering gear - removal and installation

Warning 1: *On models equipped with a Supplemental Inflatable Restraint (SIR) system, always disable the airbag(s) before working in the vicinity of the steering wheel, instrument panel or SIR system components to avoid the*

19.4 Mark the relationship of the intermediate shaft universal joint to the steering gear input shaft and remove the pinch bolt (arrow)

19.7 Remove these two bolts (arrows) to detach the steering gear from the cradle

possibility of accidental deployment of the airbag(s), which could cause personal injury (see Chapter 12).

Warning 2: *On models equipped with an airbag, DO NOT allow the steering column shaft to rotate with the steering gear removed or damage to the airbag coil could occur. To prevent the shaft from turning, turn the ignition key to the Lock position before beginning work, or wrap the seat belt around the steering wheel rim and buckle the belt in place.*

Removal

Refer to illustrations 19.4 and 19.7

1   Park the vehicle with the wheels pointing straight ahead. Loosen the front wheel lug nuts. Raise the vehicle and support it securely on jackstands. Remove the front wheels. Remove the engine under covers on models so equipped.

2   Separate the tie-rod ends from the steering knuckle arms (see Section 17).

3   Remove the left inner fender splash shield **(see illustrations 9.3a and 9.3b in Chapter 8).**

4   Detach the intermediate shaft cover from the steering gear and pull it up far enough to expose the intermediate shaft universal joint and pinch bolt **(see illustration).** Mark the relationship of the lower universal joint to the steering gear input shaft. Remove the lower intermediate shaft pinch bolt.

5   On 1991 models equipped with power steering, unplug the power steering pressure switch from the steering gear.

6   Place a drain pan under the steering gear. Detach the power steering pressure and return lines and cap the ends to prevent excessive fluid loss and contamination.

7   Support the steering gear and remove the steering gear-to-cradle mounting bolts **(see illustration).** Separate the steering gear input shaft from the intermediate shaft U-joint and pull the steering gear out through the left fender well.

Installation

8   Raise the steering gear into position and connect the input shaft to the U-joint, aligning the marks. **Note:** *Be sure the steering gear is centered before connecting the input shaft.*

9   Install the mounting bolts and tighten them to the torque listed in this Chapter's Specifications.

10   Connect the tie-rod ends to the steering knuckles. Tighten the castle nut to the torque listed in this Chapter's Specifications. Install a new cotter pin. If it is necessary to turn the castle nut to align the holes in the stud, always tighten the nut, never loosen it.

11   Install the U-joint pinch bolt and tighten it to the torque listed in this Chapter's Specifications.

12   Connect the power steering pressure and return hoses to the steering gear and fill the power steering pump reservoir with the recommended fluid (see Chapter 1).

13   Lower the vehicle and bleed the steering system (see Section 21).

---

20   Power steering pump - removal and installation

Caution: *On models equipped with a theft-deterrent radio, make sure you have the correct activation code, or the theft deterrent system is turned off, before disconnecting the battery.*

Removal

1   Disconnect the cable from the negative battery terminal.

2   Using a large syringe or suction gun, suck as much fluid out of the power steering fluid reservoir as possible.

3   Raise the vehicle and support it securely on jackstands.

4   Place a drain pan under the vehicle to catch any fluid that spills out when the hoses

are disconnected. Disconnect the power steering hoses from the steering gear and allow the system to drain. **Note:** *Do NOT turn the steering wheel while the hoses are disconnected, or you'll pump fluid out of the steering gear.*

5   Remove the serpentine drive belt from the power steering pump pulley.

6   On vehicles with a DOHC engine, remove the power steering pump-to-intake manifold bolts and the bracket, then remove the power steering pump bracket-to-engine block bolts and bracket.

7   Remove the three pump bracket-to-block mounting bolts and raise the pump enough to unplug the electrical connector from the pump. Do not remove the brackets from the pump at this time.

8   Remove the pump and hoses from the vehicle.

9   Disconnect the pressure and return hoses from the pump.

Installation

10   To install the pump, reverse the removal procedure. Tighten the mounting bolts to the torque listed in this Chapter's Specifications. Adjust the drivebelt tension following the procedure described in Chapter 1.

11   Top up the fluid level in the reservoir (see Chapter 1) and bleed the system (see Section 21).

---

21   Power steering system - bleeding

10

1   Following any operation in which the power steering fluid lines have been disconnected, the power steering system must be bled to remove all air and obtain proper steering performance.

2   With the front wheels in the straight ahead position, check the power steering fluid level and, if low, add fluid until it reaches the FULL mark on the dipstick.

3   Set the parking brake and block the rear

wheels. Raise the front of the vehicle until the front wheels are just off the ground.

4　Bleed the system by turning the wheels from side to side, without hitting the stops. This will work the air out of the system (fluid with air in it has a light tan appearance). Keep the reservoir full of fluid as this is done. It may take turning the wheels from side to side several times to remove all the air.

5　When the air is worked out of the system, return the wheels to the straight ahead position, lower the vehicle, and start the engine. Leave the vehicle running for several minutes before shutting it off.

6　Road test the vehicle to be sure the steering system is functioning normally and noise free.

7　Recheck the fluid level to be sure it is up to the FULL mark on the dipstick while the engine is at normal operating temperature. Add fluid if necessary (see Chapter 1).

22　Wheels and tires - general information

Refer to illustration 22.1

1　All vehicles covered by this manual are equipped with metric-sized steel belted radial tires **(see illustration)**. Use of other size or type of tires may affect the ride and handling of the vehicle. Don't mix different types of tires, such as radials and bias belted, on the same vehicle as handling may be seriously affected. It's recommended that tires be replaced in pairs on the same axle, but if only one tire is being replaced, be sure it's the same size, structure and tread design as the other.

2　Because tire pressure has a substantial effect on handling and wear, the pressure on all tires should be checked at least once a month or before any extended trips (see Chapter 1).

3　Models are equipped with either steel or aluminum wheels. Wheels must be replaced if they are bent, dented, leak air, have elongated bolt holes, are heavily rusted, out of vertical symmetry or if the lug nuts won't stay tight. Wheel repairs that use welding or peening are not recommended. Replacement wheel must be of the same specifications as the original. Do not mix wheels of different size or construction on the vehicle.

4　Tire and wheel balance is important in the overall handling, braking and performance of the vehicle. Unbalanced wheels can adversely affect handling and ride characteristics as well as tire life. Whenever a tire is installed on a wheel, the tire and wheel should be balanced by a shop with the proper equipment.

5　All vehicles covered by this manual are equipped with a high pressure compact spare tire mounted on a narrow (4 inch) wheel. The spare tire should be checked periodically for the correct pressure (60 psi). Do not attempt to mount standard tires, wheel covers or accessories the compact spare wheel.

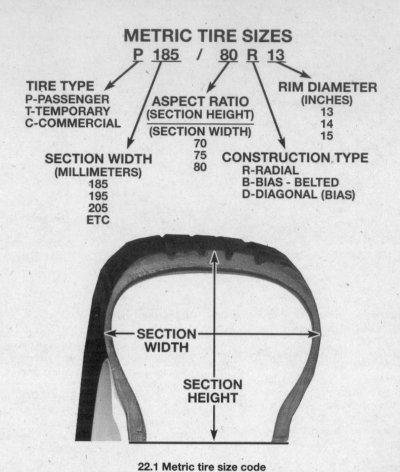

METRIC TIRE SIZES

22.1 Metric tire size code

23　Wheel alignment - general information

Refer to illustration 23.1

　A wheel alignment refers to the adjustments made to the wheels so they are in proper angular relationship to the suspension and the ground. Wheels that are out of proper alignment not only affect vehicle control, but also increase tire wear. The front end angles normally measured are camber, caster and toe-in **(see illustration)**. Camber and toe-in are adjustable at the rear as well.

　Getting the proper wheel alignment is a very exacting process, one in which complicated and expensive machines are necessary to perform the job properly. Because of this, you should have a technician with the proper equipment perform these tasks. We will, however, use this space to give you a basic idea of what is involved with a wheel alignment so you can better understand the process and deal intelligently with the shop that does the work.

　Toe-in is the distance a wheel is turned in or out from the straight ahead direction. The purpose of a toe specification is to ensure parallel rolling of the wheels. In a vehicle with zero toe-in, the distance between the front edges of the wheels will be the same as the distance between the rear edges of the wheels. The actual amount of toe-in is normally only a fraction of an inch. On the front end, toe-in is controlled by the tie-rod end position on the tie-rod. On the rear end, it's controlled by the position of the rear lateral link. Incorrect toe-in will cause the tires to wear improperly by making them scrub against the road surface.

　Camber is the tilting of the wheels from vertical when viewed from one end of the vehicle. When the wheels tilt out at the top, the camber is said to be positive (+). When the wheels tilt in at the top the camber is negative (-). The amount of tilt is measured in degrees from vertical and this measurement is called the camber angle. This angle affects the amount of tire tread which contacts the road and compensates for changes in the suspension geometry when the vehicle is cornering or traveling over an undulating surface. An incorrect camber angle will cause the tire(s) to wear on either the inside or outside edge of the tire.

　Caster is the tilting of the front steering axis from the vertical. A tilt toward the rear is positive caster and a tilt toward the front is negative caster. An incorrect caster angle will not cause tire wear, but will cause the vehicle to lead (pull) to one side or the other.

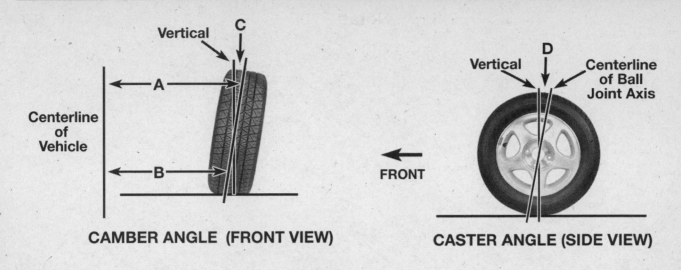

CAMBER ANGLE (FRONT VIEW)

CASTER ANGLE (SIDE VIEW)

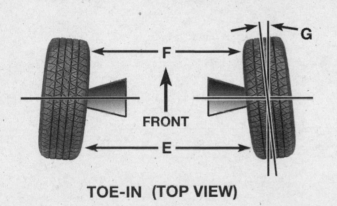

TOE-IN (TOP VIEW)

23.1 **Typical front end alignment details**
A minus B = C (degrees camber)
E minus F = toe-in (measured in inches)
G = toe-in (expressed in degrees)

10

Notes

Chapter 11   Body

Contents

1   General information

These models are of unitized construction. The frame consists of a floorpan with front and rear frame rails which support the body components, front and rear suspension systems and other mechanical components.

All vertical external body panels (i.e., fenders, doors, quarter and rocker panels) are rigid thermoplastic. Front and rear bumper covers are ThermoPlastic Olefin (TPO). The hood, roof and rear deck (trunk lid) are steel. Thermoplastic and TPO panels are rustproof and designed to withstand minor impacts without damage.

Only general body maintenance practices and body panel repair procedures within the scope of the average home mechanic are included in this Chapter.

2   Body – maintenance

1   The condition of your vehicle's body is very important, because the resale value depends a great deal on it. It's much more difficult to repair a neglected or damaged body than it is to repair mechanical components. The hidden areas of the body, such as the wheel wells and the engine compartment, are equally important, although they don't require as frequent attention as the rest of the body.

2   Once a year, or every 12,000 miles, it's a good idea to have the underside of the body steam cleaned. All traces of dirt and oil will be removed and the area can then be inspected carefully for rust, damaged brake lines, frayed electrical wires, damaged cables and other problems. The front suspension components should be greased after completion of this job.

3   At the same time, clean the engine and the engine compartment with a steam cleaner or water soluble degreaser.

4   The wheel wells should be given close attention, since undercoating can peel away and stones and dirt thrown up by the tires can cause the paint to chip and flake, allowing rust to set in. If rust is found, clean down to the bare metal and apply an anti-rust paint.

5   The body should be washed about once a week. Wet the vehicle thoroughly to soften the dirt, then wash it down with a soft sponge and plenty of clean soapy water. If the surplus dirt is not washed off very carefully, it can wear down the paint.

6   Spots of tar or asphalt thrown up from the road should be removed with a cloth soaked in solvent.

11

7   Once every six months, wax the body and chrome trim. If a chrome cleaner is used to remove rust from any of the vehicle's plated parts, remember that the cleaner also removes part of the chrome, so use it sparingly.

3   Upholstery and carpets – maintenance

1   Every three months remove the carpets or mats and clean the interior of the vehicle (more frequently if necessary). Vacuum the upholstery and carpets to remove loose dirt and dust.
2   In areas where the interior of the vehicle is subject to bright sunlight, cover leather seats with a sheet if the vehicle is to be left out for any length of time.

4   Vinyl trim – maintenance

Don't clean vinyl trim with detergents, caustic soap or petroleum-based cleaners. Plain soap and water works just fine, with a soft brush to clean dirt that may be ingrained. Wash the vinyl as frequently as the rest of the vehicle.

After cleaning, application of a high quality rubber and vinyl protectant will help prevent oxidation and cracks. The protectant can also be applied to weatherstripping, vacuum lines and rubber hoses, which often fail as a result of chemical degradation, and to the tires.

5   Hinges and locks – maintenance

Once every 3000 miles, or every three months, the hinges and latch assemblies on the doors, hood and trunk should be given a few drops of light oil or lock lubricant. The door latch strikers should also be lubricated with a thin coat of grease to reduce wear and ensure free movement. Lubricate the door and trunk locks with spray-on graphite lubricant.

6   Body repair – minor damage

TPO flexible panels (front and rear bumper covers)
Note 1: *The following repair procedure applies to the bumper covers, fenders, doors, quarter panels and rocker panels, all of which are made of this material.*
Note 2: *Below is a list of the equipment and materials necessary to perform the following repair procedures. Although a specific brand of material may be mentioned, it should be noted that equivalent products from other manufacturers may be used instead.*

Wax, grease and silicone removing solvent
Cloth-backed body tape
Sanding discs
Drill motor with three-inch disc holder
Hand sanding block
Rubber squeegees
Sandpaper
Non-porous mixing palette
Wood paddle or putty knife
Curved tooth body file
Flexible parts repair material

1   Remove the damaged panel, if necessary or desirable. In most cases, repairs can be carried out with the panel installed.
2   Clean the area(s) to be repaired with a wax, grease and silicone removing solvent applied with a water-dampened cloth.
3   If the damage is structural, that is, if it extends through the panel, clean the backside of the panel area to be repaired as well. Wipe dry.
4   Sand the rear surface about 1-1/2 inches beyond the break.
5   Cut two pieces of fiberglass cloth large enough to overlap the break by about 1-1/2 inches. Cut only to the required length.
6   Mix the adhesive from the 3M #5900 kit according to the instructions included with the kit, and apply a layer of the mixture approximately 1/8-inch thick on the backside of the panel. Overlap the break by at least 1-1/2 inches
7   Apply one piece of fiberglass cloth to the adhesive and cover the cloth with additional adhesive. Apply a second piece of fiberglass cloth to the adhesive and immediately cover the cloth with additional adhesive insufficient quantity to fill the weave.
8   Allow the repair to cure for 20 to 30 minutes at 60-degrees to 80-degrees F.
9   If necessary, trim the excess repair material at the edge.
10   Remove all of the paint film over and around the area(s) to be repaired. The repair material should not overlap the painted surface.
11   With a drill motor and a sanding disc (or a rotary file), cut a "V" along the break line approximately 1/2-inch wide. Remove all dust and loose particles from the repair area.
12   Mix and apply the repair material. Apply a light coat first over the damaged area; then continue applying material until it reaches a level slightly higher than the surrounding finish.
13   Cure the mixture for 20 to 30 minutes at 60-degrees to 80-degrees F.
14   Roughly establish the contour of the area being repaired with a body file. If low areas or pits remain, mix and apply additional adhesive.
15   Block sand the damaged area with sandpaper to establish the actual contour of the surrounding surface.
16   If desired, the repaired area can be temporarily protected with several light coats of primer. Because of the special paints and techniques required for flexible body panels, it is recommended that the vehicle be taken to a paint shop for completion of the body repair.

Rigid plastic panels (fenders, doors, quarter and rocker panels)

Note 1: *Repairs are most effective when the temperature is approximately 60 to 80-degrees F.*
Note 2: *The following repair procedure is for minor scratches and gouges. Repair of more serious damage should be left to a dealer service department or qualified auto body shop.*
Note 3: *Below is a list of the equipment and materials necessary to perform the following repair procedures. Although a specific brand of material may be mentioned, it should be noted that equivalent products from other manufacturers may be used instead.*

Sanding discs
Drill motor with three-inch disc holder
Hand sanding block
Rubber squeegees
Sandpaper
Non-porous mixing palette
Wood paddle or putty knife
Rigid plastic repair material

17   Clean the repair area. Start with soap and water, then finish with isopropyl alcohol.
18   Using 80-grit sandpaper, taper the scratch or gouge about 1-1/2 inches around the scratch or gouge. Be sure to penetrate well into the plastic material; a D-A-type power sander and/or coarser-grit sandpaper may help.
19   Using 220-grit sandpaper, feather-edge the paint around the repair area. Sand the repair area with 180-grit sandpaper.
20   Wipe the repair area clean with a clean tack rag or other clean, dry rag. **Note:** *The filler will adhere better to a smooth surface than a rough surface.*
21   According to label instructions, mix 3M Rigid Plastic Repair Material (RPRM), or equivalent, and apply it to the repair area with a squeegee or plastic spreader. Build the material slightly higher than the surrounding undamaged surface. Allow the material to cure approximately 20 to 30 minutes.
22   Using a sanding block, sand the repair area with 180-grit sandpaper, followed by 240-grit sandpaper.
23   Fill scratches and pinholes in the repair area with a light coat of 3M RPRM and sand lightly, as described in the previous Step.
24   Apply a double wet coat of Glasurit Glassohyd Sealer, or equivalent, over the repair area. Allow it to dry for 1 to 2 hours. Sand lightly with 320-grit sandpaper, then wipe with a dry cloth. Because of the special paints used on these panels, it's recommended that you allow a dealer service department or qualified auto body shop to apply the final paint.

Steel panels (hood, trunk lid and roof)

Repair of rust holes or gashes

25   Remove all paint from the affected area and from an inch or so of the surrounding

metal using a sanding disk or wire brush mounted in a drill motor. If these are not available, a few sheets of sandpaper will do the job just as effectively.

26   With the paint removed, you will be able to determine the severity of the corrosion and decide whether to replace the whole panel, if possible, or repair the affected area. New body panels are not as expensive as most people think and it is often quicker to install a new panel than to repair large areas of rust.

27   Remove all trim pieces from the affected area except those which will act as a guide to the original shape of the damaged body, such as headlight shells, etc. Using metal snips or a hacksaw blade, remove all loose metal and any other metal that is badly affected by rust. Hammer the edges of the hole in to create a slight depression for the filler material.

28   Wire brush the affected area to remove the powdery rust from the surface of the metal. If the back of the rusted area is accessible, treat it with rust inhibiting paint.

29   Before filling is done, block the hole in some way. This can be done with sheet metal riveted or screwed into place, or by stuffing the hole with wire mesh.

30   Once the hole is blocked off, the affected area can be filled and painted. See the following subsection on *filling and painting.*

Filling and painting

31   Many types of body fillers are available, but generally speaking, body repair kits which contain filler paste and a tube of resin hardener are best for this type of repair work. A wide, flexible plastic or nylon applicator will be necessary for imparting a smooth and contoured finish to the surface of the filler material. Mix up a small amount of filler on a clean piece of wood or cardboard (use the hardener sparingly). Follow the manufacturer's instructions on the package, otherwise the filler will set incorrectly.

32   Using the applicator, apply the filler paste to the prepared area. Draw the applicator across the surface of the filler to achieve the desired contour and to level the filler surface. As soon as a contour that approximates the original one is achieved, stop working the paste. If you continue, the paste will begin to stick to the applicator. Continue to add thin layers of paste at 20-minute intervals until the level of the filler is just above the surrounding metal.

33   Once the filler has hardened, the excess can be removed with a body file. From then on, progressively finer grades of sandpaper should be used, starting with a 180-grit paper and finishing with 600-grit wet-or-dry paper. Always wrap the sandpaper around a flat rubber or wooden block, otherwise the surface of the filler will not be completely flat. During the sanding of the filler surface, the wet-or-dry paper should be periodically rinsed in water. This will ensure that a very smooth finish is produced in the final stage.

34   At this point, the repair area should be

surrounded by a ring of bare metal, which in turn should be encircled by the finely feathered edge of good paint. Rinse the repair area with clean water until all of the dust produced by the sanding operation is gone.

35   Spray the entire area with a light coat of primer. This will reveal any imperfections in the surface of the filler. Repair the imperfections with fresh filler paste or glaze filler and once more smooth the surface with sandpaper. Repeat this spray-and-repair procedure until you are satisfied that the surface of the filler and the feathered edge of the paint are perfect. Rinse the area with clean water and allow it to dry completely.

36   The repair area is now ready for painting. Spray painting must be carried out in a warm, dry, windless and dust free atmosphere. These conditions can be created if you have access to a large indoor work area, but if you are forced to work in the open, you will have to pick the day very carefully. If you are working indoors, dousing the floor in the work area with water will help settle the dust which would otherwise be in the air. If the repair area is confined to one body panel, mask off the surrounding panels. This will help minimize the effects of a slight mismatch in paint color. Trim pieces such as chrome strips, door handles, etc., will also need to be masked off or removed. Use masking tape and several thicknesses of newspaper for the masking operations.

37   Before spraying, shake the paint can thoroughly, then spray a test area until the spray painting technique is mastered. Cover the repair area with a thick coat of primer. The thickness should be built up using several thin layers of primer rather than one thick one. Using 600-grit wet-or-dry sandpaper, rub down the surface of the primer until it is very smooth. While doing this, the work area should be thoroughly rinsed with water and the wet-or-dry sandpaper periodically rinsed as well. Allow the primer to dry before spraying additional coats.

38   Spray on the top coat, again building up the thickness by using several thin layers of paint. Begin spraying in the center of the repair area and then, using a circular motion, work out until the whole repair area and about two inches of the surrounding original paint is covered. Remove all masking material 10 to 15 minutes after spraying on the final coat of paint. Allow the new paint at least two weeks to harden, then use a very fine rubbing compound to blend the edges of the new paint into the existing paint. Finally, apply a coat of wax.

7   Body repair – major damage

1   Major damage must be repaired by an auto body/frame repair shop with the necessary welding and hydraulic straightening equipment.

2   If the damage has been serious, it is vital that the structure be checked for proper

alignment or the vehicle handling characteristics may be adversely affected. Other problems, such as excessive tire wear and wear in the driveline and steering may occur.

3   Due to the fact that all of the major body components (hood, fenders, etc.) are separate and replaceable units, any seriously damaged components should be replaced rather than repaired. Sometimes these components can be found in a wrecking yard that specializes in used vehicle components, often at considerable savings over the cost of new parts.

8   Windshield and fixed glass – replacement

Replacement of the windshield and fixed glass requires the use of special fast-setting adhesive/caulk materials and some specialized tools. It is recommended that these operations be left to a dealer or a shop specializing in glass work.

9   Hood – removal, installation and adjustment

Refer to illustration 9.1

Note: *The hood is heavy and somewhat awkward to remove and install at least two people should perform this procedure.*

Removal and installation

1   Make marks around the bolt heads to ensure proper alignment during installation **(see illustration).**

2   Use blankets or pads to cover the cowl area of the body and fenders. This will protect the body and paint as the hood is lifted off.

3   Disconnect any cables or wires that will interfere with removal.

4   Have an assistant support the hood. Remove the hinge-to-hood bolts.

5   Lift off the hood.

6   Installation is the reverse of removal.

9.1 Use a marking pen to mark around the hood bolts

9.10 **Adjust the hood closed height by loosening the bolts (arrows) and moving the latch up-or-down**

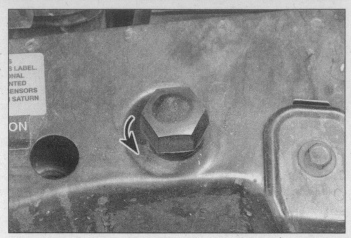

9.11 **Screw the rubber bumper in-or-out to make fine adjustment to the hood closed height**

Adjustment

Refer to illustrations 9.10 and 9.11

7   Fore-and-aft and side-to-side adjustment of the hood is done by moving the hinge plate slot after loosening the bolts.

8   Scribe or draw a line around the bolt heads and the entire hinge plate so you can judge the amount of movement **(see illustration 9.1)**.

9   Loosen the bolts and move the hood into correct alignment. Move it only a little at a time. Tighten the hinge bolts or nuts and carefully lower the hood to check the position.

10   If necessary after installation, the entire hood latch assembly can be adjusted up-and-down as well as from side-to-side on the radiator support so the hood closes securely, flush with the fenders. To make the adjustment, scribe a line around the hood latch mounting bolts to provide a reference point, then loosen them and reposition the latch assembly, as necessary **(see illustration)**. Following adjustment, retighten the mounting bolts.

11   Finally, adjust the hood bumpers on the radiator support so the hood, when closed, is flush with the fenders **(see illustration)**.

12   The hood latch assembly, as well as the hinges, should be periodically lubricated with lithium-base grease to prevent binding and wear.

10   Fender – removal and installation

Refer to illustration 10.1

Warning: *On models equipped with a Supplemental Inflatable Restraint (SIR) system, always disable the airbag(s) before working in the vicinity of the steering wheel, instrument panel or SIR system components to avoid the possibility of accidental deployment of the airbag(s), which could cause personal injury* (see Chapter 12).

1   Working inside the wheel well, remove the wheel well liner by pulling out the centers

of the plastic retainers and removing the retainers; then detach the liner **(see illustration)**.

2   Remove the fasteners, then lower the front of the lower rocker panel cover.

3   Remove the triangular windshield side frame cover by removing the bolt and then sliding the cover upward.

4   Slide the front bumper fascia forward to disconnect it from the fender bolts.

5   Remove the fender bolts from all points.

6   Installation is the reverse of removal. Tighten the fender bolts securely, a little at a time, working from bolt to bolt. After placing the fender liner in position, install the plastic retainers and lock them in place by pushing the centers in.

11   Liftgate – removal, installation and adjustment

Note: *The liftgate is heavy and somewhat awkward to remove and install – at least two people should perform this procedure.*

1   Open the liftgate and cover the edges of the rear compartment with pads or cloths to protect the painted surfaces when the liftgate is removed.

2   Disconnect any cables or wire harness connectors attached to the liftgate that would interfere with removal.

3   Make alignment marks around the hinge bolt mounting flanges.

4   Have an assistant support the liftgate and detach the support struts (see Section 12).

5   While an assistant supports the liftgate, remove the bolts on both sides, then lift it off.

6   Installation is the reverse of removal.

Note: *When reinstalling the liftgate, align the lid-to-hinge bolts with the marks made during removal.*

7   After installation, close the liftgate and make sure it's in proper alignment with the surrounding body.

8   Forward-or-backward and side-to-side

10.1 **Use wire cutters to gently pull out the center of each plastic retainer (don't cut it off!), then pry the retainer out**

adjustments are made by detaching the headliner for access, then loosening the hinge to liftgate nuts and gently moving the liftgate into correct alignment. Tighten the nuts after adjusting.

9   The lock striker can be adjusted by loosening the mounting bolts and gently tapping it with a plastic hammer.

12   Liftgate support strut – replacement

Refer to illustrations 12.1a and 12.1b

Warning: *The support strut is filled with pressurized gas – do not disassemble this component (if it is faulty replace it with a new one).*

Note: *The trunk lid/rear liftgate is heavy and somewhat awkward to hold securely while replacing the struts – at least two people should perform this procedure.*

1   With the liftgate supported in the open position for the removal procedure, use a

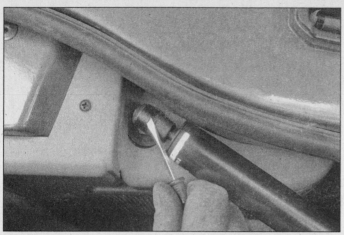

12.1a Use a screwdriver to pry out the support strut release clip . . .

12.1b . . . then, if necessary, pry the end of the strut off the ball stud

13.3 Mark around the trunk lid bolts (arrows) so the trunk lid will be correctly aligned on reinstallation

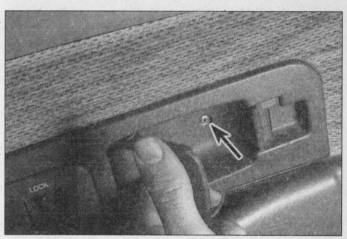

14.4 Pull the door handle out for access to this screw (arrow)

screwdriver to pry out the clips and detach the ends of the struts **(see illustrations)**.
2   Installation is the reverse of the removal procedure.

13   Trunk lid – removal, installation and adjustment

Refer to illustration 13.3
1   Open the trunk lid and cover the edges of the trunk compartment with pads or cloths to protect the painted surfaces when the lid is removed.
2   Disconnect any cables or electrical connectors attached to the trunk lid that would interfere with removal.
3   Use a marking pen to make alignment marks around the hinge bolts **(see illustration)**.
4   With the help of an assistant, remove the hinge bolts on both sides and lift the trunk lid off.
5   Installation is the reverse of the removal procedure. When reinstalling the lid, align the heads of the the bolts with the marks made

during removal.
6   After installation, close the lid and make sure it's in proper alignment with the surrounding body panels. Fore-and-aft and up-and-down adjustments of the lid are controlled by the position of the bolts in the hinge arms. To adjust it, loosen the hinge bolts, reposition the lid and retighten the bolts.
7   Side-to-side movement of the lid is adjusted by loosening the hinge-to-body nuts under the rear package shelf, moving the lid and then tightening the nuts.
8   The height of the lid in relation to the body can be adjusted by loosening the latch striker bolts, repositioning the striker and retightening the bolts.

14   Door trim panel – removal and installation

Refer to illustrations 14.4, 14.5 and 14.8
Caution: *On models equipped with a theft-deterrent radio, make sure you have the correct activation code, or the theft deterrent system is turned off, before disconnecting the battery.*

1   Disconnect the negative cable from the battery.
2   If equipped with manual windows, remove the window crank by working a cloth back-and-forth behind the crank handle to dislodge the retainer. A standard handle removal tool can also be used. With the retainer removed, pull off the handle. If equipped with power door locks, remove the door lock switch by prying at the top of the switch housing with a small screwdriver, then unplugging the electrical connector **(see illustrations 16.1a and 16.1b)**. **Note:** *Wrap the tip of the screwdriver with tape to prevent scratching the plastic.*
3   Remove the mirror inner trim plate by pulling it firmly to disengage the clips.
4   Remove the screw from behind the door handle **(see illustration)**. Pull the handle forward and then use pliers to disconnect the control rod from it.
5   Remove any door trim panel retaining screws and door pull/armrest assemblies. Plastic push pin rivets are used at some locations.
6   Insert a wide putty knife or a thin prybar between the trim panel and door to disen-

11

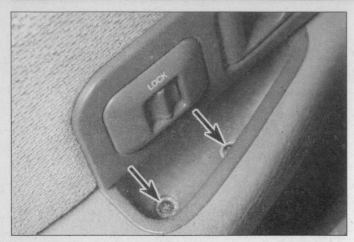

14.5 Two trim panel screws are located in the door pull recess (arrows)

14.8 Peel the water deflector carefully away from the door, taking care not to tear or distort it

15.5 Remove the bolt and detach the door check link

gage the retaining clips. Work around the outer edge until the panel is free.

7   Once all of the clips are disengaged, detach the trim panel, unplug any electrical connectors and remove the trim panel from the vehicle by gently pulling it up and out.

8   For access to the inner door, peel back the plastic water deflector, taking care not to tear it **(see illustration)**. To install the trim panel, first press the water deflector into place.

9   Prior to installation of the door panel, be sure to reinstall any clips in the panel which may have come out during the removal procedure and stayed in the door.

10 Plug in any electrical connectors and place the panel in position. Press it into place until the clips are seated and install any retaining screws and armrest/door pulls. Install the manual regulator window crank.

---

15 Door – removal, installation and adjustment

Removal and installation

Refer to illustrations 15.5, 15.7 and 15.8

1   Remove the fender (refer to Section 10).

2   Remove the outer door panel screws and lift off the panel (refer to Section 18).

3   Remove the inner door insulator.

4   Disconnect and label all wiring from the door assembly. Pull the harnesses out of the door.

5   Remove the bolt and detach the door check link **(see illustration)**.

6   Use a marking pen to mark around the door bolts. This allows you to easily adjust the door when reinstalling it.

7   Remove the hinge-to-door bolts and carefully detach the door **(see illustration)**. Installation is the reverse of removal.

Adjustment

8   Following installation, make sure the door is aligned properly. Adjust it if necessary as follows:

a) *Up-and-down and forward-and-backward adjustments are made by loosening the hinge-to-body bolts and moving the door, as necessary. A special offset tool may be required to reach some of the bolts.*

b) *In-and-out and up-and-down adjustments are made by loosening the door side hinge bolts and moving the door, as necessary.*

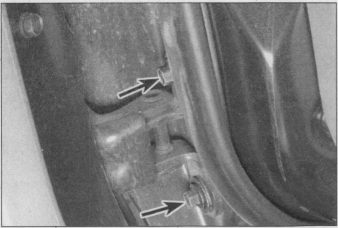

15.7 With the door supported, remove the bolts (arrows) from each hinge

15.8 Adjust the door closed position by loosening the Torx screws (arrows), then move the latch by tapping it with a mallet

16.1a Insert a small screwdriver into the notch at the top and pry the power door lock switch out

16.1b Unplug the door lock switch and remove it

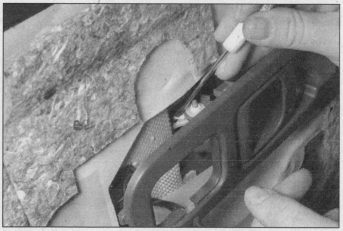

16.2 Push the retainer off with a small screwdriver, then detach the link and withdraw the door handle assembly (door panel removed for clarity)

16.4 Outside door handle installation details

c) *The door lock striker can also be adjusted both up-and-down and sideways to provide a positive engagement with the locking mechanism. This is done by loosening the screws and moving the striker, as necessary* **(see illustration).**

16 Door latch, lock cylinder and handle – removal and installation

Inside handle

Refer to illustrations 16.1a, 16.1b and 16.2

1 Remove the door lock switch (if equipped) **(see illustrations).**

2 Pull the handle out and use a small screwdriver to disconnect the control link from the lock cylinder and outside handle **(see illustration).**

3 Installation is the reverse of removal.

Outside handle

Refer to illustration 16.4

4 Pull the handle out and push the centers of the retaining pins through the pins, then remove them and detach the handle **(see illustration).**

5 Remove the door outer panel (see Section 18) for access to the handle pivot assembly.

6 Remove the handle pivot assembly by detaching the rod at the latch, then compressing the pin at the front of the pivot, lifting the assembly out for access and detaching the rod clip.

7 Installation is the reverse of removal.

Door latch and lock cylinder

8 Remove the door outer panel (see Section 18).

Door latch

Refer to illustration 16.10

9 Disconnect the link rods from the latch.

10 Remove the latch retaining screws from the end of the door **(see illustration).**

11 Remove the door latch.

12 Installation is the reverse of removal.

Lock cylinder

13 Detach the link rod and unplug any elec-

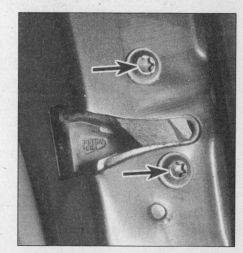

16.10 The door latch is held in place by three screws (arrows)

trical connectors.

14 Detach the clip and remove the lock cylinder from the door.

15 Installation is the reverse of removal.

11

18.5 Use a Torx head tool to remove the outer door panel screws

17 Door window glass – removal and installation

1   Remove the door outer panel (see Section 18).
2   Lower the window glass.
3   Remove the two nuts and one screw securing the glass.
4   Remove the glass by lifting it up carefully and rotating it inward.
5   Installation is the reverse of the removal procedure.

18 Door outer panel – removal and installation

Refer to illustrations 18.5, 18.7a and 18.7b

1   Remove the door trim panel (see Section 14).
2   Remove the outside mirror (Section 21). Remove the foam block from the mirror opening.
3   Remove the outside door handle (Section 16).

4   Open the door and pry the outer sealing strip out of the window glass opening.
5   Remove the door panel screws **(see illustration)**.
6   Detach the front and rear of the outer panel, then lift it toward the rear to remove it.
7   Installation is the reverse of the removal procedure. Apply thread locking compound to the panel retaining screws, install then and tighten them, a little at a time, in the sequence shown, until they are secure **(see illustrations)**.

19 Window regulator – removal and installation

1   Remove the door trim panel and water deflector (see Section 14).
2   Remove the door outer panel (Section 18).
3   Remove the door window glass (Section 17).
4   Remove the regulator cam bolts, drill out the retaining rivets, then detach the regulator.
5   Pull the regulator through the service hole to remove it.
6   Installation is the reverse of removal.

20 Bumpers – removal and installation

Front bumper

Warning: *On models equipped with a Supplemental Inflatable Restraint (SIR) system, always disable the airbag(s) before working in the vicinity of the steering wheel, instrument panel or SIR system components to avoid the possibility of accidental deployment of the airbag(s), which could cause personal injury (see Chapter 12).*
Caution: *On models equipped with a theft-deterrent radio, make sure you have the correct activation code, or the theft deterrent system is turned off, before disconnecting the battery.*

1   Apply the parking brake, block the rear wheels, lift the front of the vehicle and support it securely on jackstands.
2   Disconnect the negative battery cable from the battery and disconnect any wiring that would interfere with bumper removal.
3   On air conditioned models, remove the screws and detach air duct from the bumper cover.
4   Remove the front fascia bolts from inside the front wheel housings.
5   Remove the plastic fascia retainers.
6   Slide the fascia forward, disconnect the side marker lamp connectors and lift off the fascia.
7   Remove the grille from behind the fascia, if equipped.
8   Remove the headlamp assemblies. Remove the fog lamps if they interfere.
9   Remove the front bumper (impact bar) fasteners and remove it with the aid of an assistant.
10   Installation is the reverse of removal.

Rear bumper

11   In the rear compartment, remove the plastic retainers and detach bumper cover.
12   Remove the clips, nuts and bolts and detach the rear bumper assembly from the vehicle.
13   Installation is the reverse of removal.

21 Outside mirror – removal and installation

1   Remove the screw and pull off the mirror trim panel from the inside of the door.
2   Remove the retaining nuts, unplug the electrical connector (if equipped) and detach the mirror.
3   Installation is the reverse of removal.

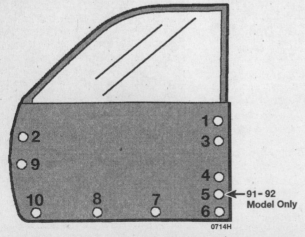

18.7a Typical front door outer panel retaining screw locations – when installing, tighten the screws in the sequence shown

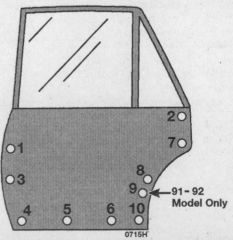

18.7b Typical rear door outer panel retaining screw locations – tighten the screws in the sequence shown when installing

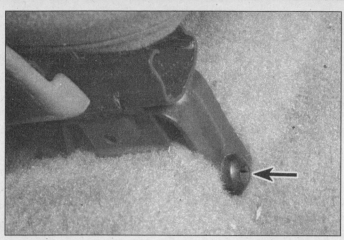

22.1 The front seats are held in place by Torx-head bolts (arrow)

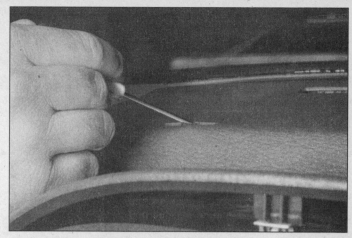

23.2a Pry out the bolt covers . . .

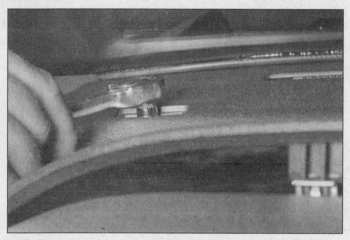

23.2b . . . then use a socket and ratchet to remove the bolts

23.4 Use a screwdriver to detach the clips at the rear edge of the instrument panel pad

22 Seats – removal and installation

Front seats

Refer to illustration 22.1

1   Remove the four retaining bolts, unplug any electrical connectors and lift the seats from the vehicle (see illustration).
2   Installation is the reverse of removal.

Rear seats

Cushions

3   On wagon and sedan models, press on the two retainer tabs at the bottom of the rear cushion, then rotate the front of the edge up and remove the cushion from the vehicle.
4   On coupe models, remove the rear center console, then detach and remove the cushions.

Seat back

5   On sedan and wagon models, remove the bolster by pulling on the top to detach the fastener, then lifting the bolster off the lower guide pin.
6   On all models, release the seat back latch, tilt the back forward and detach the carpet.
7   Slide the seat back toward the center of the vehicle to disengage it from the pivot pin, then remove the assembly.
8   Installation is the reverse of removal.

23 Instrument panel pad – removal and installation

Refer to illustrations 23.2a, 23.2b and 23.4

Warning: *On models equipped with a Supplemental Inflatable Restraint (SIR) system, always disable the airbag(s) before working in the vicinity of the steering wheel, instrument panel or SIR system components to avoid the possibility of accidental deployment of the airbag(s), which could cause personal injury (see Chapter 12).*

Caution: *On models equipped with a theft-deterrent radio, make sure you have the correct activation code (or the theft deterrent system is turned off) before disconnecting the battery.*

1   Disconnect the cable from the negative battery terminal.
2   Using a small screwdriver, pry the bolt covers from the top of the instrument panel pad and remove the bolts (see illustrations). On 1994 and earlier models, a bolt is located at each end of the pad. On 1995 and later models, two bolts are located at the center of the pad.
3   On 1995 and later models, carefully pry the retaining clips loose from each trim panel extension at each end of the instrument panel pad and remove the extensions.
4   Pry the clips loose along the rear edge of the pad (see illustration). Lift the rear edge of the pad up and pull it to the rear to disengage the clips along the base of the windshield.
5   On 1995 and later models, remove the trim panel insulator, if necessary.
6   Installation is the reverse of removal.

24 Instrument cluster bezel – removal and installation

Warning: *On models equipped with a Supplemental Inflatable Restraint (SIR) system, always disable the airbag(s) before working in*

11

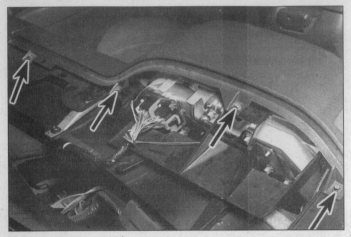

24.4 Remove the four bezel retaining screws (arrows)

25.2 Detach the clips and lift center panel off

26.3 After removing the screws, separate the cover halves

the vicinity of the steering wheel, instrument panel or SIR system components to avoid the possibility of accidental deployment of the airbag(s), which could cause personal injury (see Chapter 12).
Caution: On models equipped with a theft-deterrent radio, make sure you have the correct activation code (or the theft deterrent system is turned off) before disconnecting the battery.

1994 and earlier models

Refer to illustration 24.4
1   Disconnect the cable from the negative battery terminal.
2   Remove the instrument panel pad (see Section 23).
3   Remove the center trim panel (see Section 25).
4   Remove the four screws along the top of the bezel **(see illustration)**. Open the glove box, grasp the bezel and pull it back to disengage the retaining clips.
5   Installation is the reverse of removal.

1995 and later models

6   Disconnect the cable from the negative battery terminal.
7   Remove the knee protector (see Section 27).
8   Remove the steering column mounting bolts, lower the column and allow the steering wheel to rest on the seat.
9   Remove the screws at the top of the bezel and pull the bezel rearward to disengage the retaining clips. Disconnect the electrical connectors and remove the bezel.
10   Installation is the reverse of removal.

---

25 Center trim panel - removal and installation

Warning: On models equipped with a Supplemental Inflatable Restraint (SIR) system, always disable the airbag(s) before working in the vicinity of the steering wheel, instrument panel or SIR system components to avoid the

possibility of accidental deployment of the airbag(s), which could cause personal injury (see Chapter 12).
Caution: On models equipped with a theft-deterrent radio, make sure you have the correct activation code (or the theft deterrent system is turned off) before disconnecting the battery.

1994 and earlier models

Refer to illustration 25.2
1   Disconnect the cable from the negative battery terminal.
2   Starting at the bottom and working up, carefully pry the center trim panel retaining clips loose **(see illustration)**.
3   Pull the trim panel back and disconnect the electrical connectors, if equipped.
4   Installation is the reverse of removal.

1995 and later models

5   Disconnect the cable from the negative battery terminal.
6   Remove the retainers by pushing the center pin in and pulling the retainer body out.
7   Pull the center trim panel back and disconnect the electrical connectors, if equipped.
8   Installation is the reverse of removal.

---

26 Steering column cover – removal and installation

Refer to illustration 26.3
Warning: On models equipped with a Supplemental Inflatable Restraint (SIR) system, always disable the airbag(s) before working in the vicinity of the steering wheel, instrument panel or SIR system components to avoid the possibility of accidental deployment of the airbag(s), which could cause personal injury (see Chapter 12).
Caution: On models equipped with a theft-deterrent radio, make sure you have the correct activation code (or the theft deterrent system is turned off) before disconnecting the battery.
1   Disconnect the negative cable from the battery.
2   Remove the cluster bezel (see Section 24) and knee protector (see Section 25).
3   Lower the tilt column lever (if equipped) for access, remove the steering cover retaining screws and remove the upper cover **(see illustration)**.
4   Remove the ignition lock cylinder (see Chapter 12) and remove the lower cover.
5   Installation is the reverse of removal.

29.8a Remove the screws attaching the rear . . .   29.8b . . . and front of the center console

27 Knee protector – removal and installation

Warning: *On models equipped with a Supplemental Inflatable Restraint (SIR) system, always disable the airbag(s) before working in the vicinity of the steering wheel, instrument panel or SIR system components to avoid the possibility of accidental deployment of the airbag(s), which could cause personal injury (see Chapter 12).*
Caution: *On models equipped with a theft-deterrent radio, make sure you have the correct activation code (or the theft deterrent system is turned off) before disconnecting the battery.*
1   Disconnect the negative cable from the battery.
2   Remove the instrument cluster bezel (see Section 4).
3   Remove the retaining screws, detach the knee bolster and lower it from the instrument panel.
4   Installation is the reverse of the removal procedure.

28 Glove box – removal and installation

1   Open the glovebox door.
2   Remove the retaining screws and lift the glovebox out of the opening.
3   Installation is the reverse of removal.

29 Console – removal and installation

Refer to illustrations 29.8a, 29.8b and 29.9
Warning: *On models equipped with a Supplemental Inflatable Restraint (SIR) system, always disable the airbag(s) before working in the vicinity of the steering wheel, instrument panel or SIR system components to avoid the*

possibility of accidental deployment of the airbag(s), which could cause personal injury (see Chapter 12).
Caution: *On models equipped with a theft-deterrent radio, make sure you have the correct activation code (or the theft deterrent system is turned off) before disconnecting the battery.*
1   Disconnect the cable from the negative battery terminal.
2   Remove the left and right extension panels by pulling the front edge out to release the Velcro fasteners. Continue rotating the panels out to release the hinges from the console.
3   Remove the cupholder/ashtray and disconnect the illumination bulb, if equipped.
4   If equipped with power windows, pry the switch module forward, lift the rear edge up and disconnect the electrical connector. On 1995 and later models, disconnect the electrical connector to the cigar lighter and remove the illumination bulb from the console.
5   Apply the parking brake. On 1994 and earlier models, pry the parking brake filler panel from the console. On 1995 and later models, remove the screw retaining the parking brake cover to the lever and slide the

cover off.
6   If equipped with a manual transaxle, remove the shift knob. Remove the ash tray and the cup holder on later vehicles.
7   Remove the armrest/storage compartment, if equipped. If not equipped with an armrest, remove the rear console mounting screw cover.
8   Remove the screws retaining the console at the rear and at the front edge **(see illustrations).**
9   Lift the console to the rear and up over the park brake lever and shift lever **(see illustration).** Depress the shift lock button on the automatic transaxle lever to clear the console.
10   Installation is the reverse of removal.

30 Package tray – removal and installation

Refer to illustrations 30.3, 30.4a, 30.4b and 30.5
1   Pull the seat back down for access to the seat bolster.
2   Pull the bolster forward, then detach by lifting it up and out.

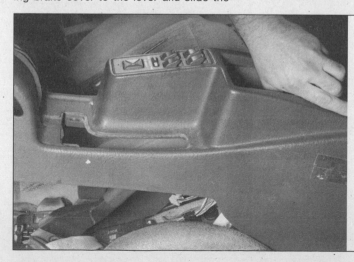

29.9 Lift the console straight up and over the parking brake and shift handles to remove it

11

30.3 Remove the screws, then detach the door opening trim piece

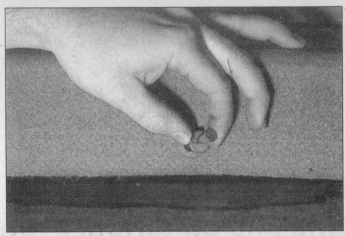

30.4a After prying out the centers of the plastic retainers, the retainers can be simply lifted out

3   Remove the screws, detach the clips and lower the door trim piece **(see illustration).**
4   Pull out the centers of the plastic retainers, remove them, then lift the package tray out **(see illustrations).**
5   Installation is the reverse of removal. After placing the tray in position, insert the retainers, then press the centers in to lock them in place **(see illustration).**

31 Seat belt check

1   Check the seat belts, buckles, latch plates and guide loops for any obvious damage or signs of wear.
2   Make sure the seat belt reminder light comes on when the key is turned on.
3   The seat belts are designed to lock up during a sudden stop or impact, yet allow free movement during normal driving. The retractors should hold the belt against your chest while driving and rewind the belt when the buckle is unlatched.
4   If any of the above checks reveal problems with the seat-belt system, replace parts as necessary.

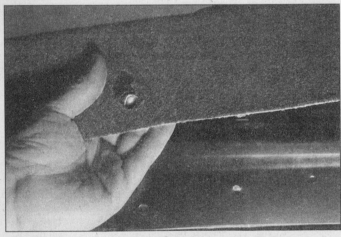

30.4b After the retainers have been removed, lift the tray out

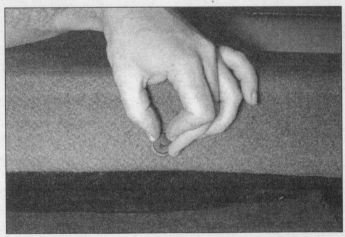

30.5 Press the retainer centers in to lock them in place

Chapter 12
Chassis electrical system

Contents

1 General information

The electrical system is a 12-volt, negative ground type. Power for the lights and all electrical accessories is supplied by a lead/acid-type battery which is charged by the alternator.

This Chapter covers repair and service procedures for the various electrical components not associated with the engine. Information on the battery, alternator, distributor and starter motor can be found in Chapter 5.

It should be noted that when portions of the electrical system are serviced, the cable should be disconnected from the negative battery terminal to prevent electrical shorts and/or fires.

2 Electrical troubleshooting - general information

A typical electrical circuit consists of an electrical component, any switches, relays, motors, fuses, fusible links or circuit breakers related to that component and the wiring and electrical connectors that link the component to both the battery and the chassis. To help you pinpoint an electrical circuit problem, wiring diagrams are included at the end of this book.

Before tackling any troublesome electrical circuit, first study the appropriate wiring diagrams to get a complete understanding of what makes up that individual circuit. Trouble spots, for instance, can often be narrowed down by noting if other components related to the circuit are operating properly. If several components or circuits fail at one time, chances are the problem is in a fuse or ground connection, because several circuits are often routed through the same fuse and ground connections.

Electrical problems usually stem from simple causes, such as loose or corroded connections, a blown fuse, a melted fusible link or a bad relay. Visually inspect the condition of all fuses, wires and connections in a problem circuit before troubleshooting it.

If testing instruments are going to be utilized, use the diagrams to plan ahead of time where you will make the necessary connections in order to accurately pinpoint the trouble spot.

The basic tools needed for electrical troubleshooting include a circuit tester or voltmeter (a 12-volt bulb with a set of test leads can also be used), a continuity tester, which includes a bulb, battery and set of test leads, and a jumper wire, preferably with a circuit breaker incorporated, which can be used to bypass electrical components. Before attempting to locate a problem with test instruments, use the wiring diagram(s) to decide where to make the connections.

Voltage checks

Voltage checks should be performed if a circuit is not functioning properly. Connect one lead of a circuit tester to either the negative battery terminal or a known good ground. Connect the other lead to a electrical connector in the circuit being tested, preferably nearest to the battery or fuse. If the bulb of the tester lights, voltage is present, which means that the part of the circuit between the electrical connector and the battery is problem free. Continue checking the rest of the circuit in the same fashion. When you reach a point at which no voltage is present, the problem lies between that point and the last test point with voltage. Most of the time the problem can be traced to a loose connection.

12

3.1a The main fuse block is located on the passenger's side of the center console, under a cover

3.1b The engine compartment fuse block is located next to the battery

Note: *Keep in mind that some circuits receive voltage only when the ignition key is in the Accessory or Run position.*

Finding a short

One method of finding shorts in a circuit is to remove the fuse and connect a test light or voltmeter in its place to the fuse terminals. There should be no voltage present in the circuit. Move the wiring harness from side to side while watching the test light. If the bulb goes on, there is a short to ground somewhere in that area, probably where the insulation has rubbed through. The same test can be performed on each component in the circuit, even a switch.

Ground check

Perform a ground test to check whether a component is properly grounded. Disconnect the battery and connect one lead of a self-powered test light, known as a continuity tester, to a known good ground. Connect the other lead to the wire or ground connection being tested. If the bulb goes on, the ground is good. If the bulb does not go on, the ground is not good.

Continuity check

A continuity check is done to determine if there are any breaks in a circuit - if it is passing electricity properly. With the circuit off (no power in the circuit), a self-powered continuity tester can be used to check the circuit. Connect the test leads to both ends of the circuit (or to the "power" end and a good ground), and if the test light comes on the circuit is passing current properly. If the light doesn't come on, there is a break somewhere in the circuit. The same procedure can be used to test a switch, by connecting the continuity tester to the power in and power out sides of the switch. With the switch turned On, the test light should come on.

Finding an open circuit

When diagnosing for possible open cir-

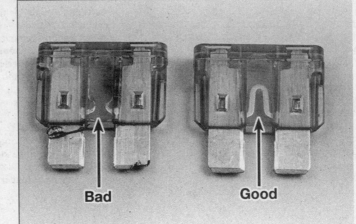

3.3 When a fuse blows, the element between the terminals melts - the fuse on the left is blown, the fuse on the right is good

Bad   Good

cuits, it is often difficult to locate them by sight because oxidation or terminal misalignment are hidden by the electrical connectors. Merely wiggling an electrical connector on a sensor or in the wiring harness may correct the open circuit condition. Remember this when an open circuit is indicated when troubleshooting a circuit. Intermittent problems may also be caused by oxidized or loose connections.

Electrical troubleshooting is simple if you keep in mind that all electrical circuits are basically electricity running from the battery, through the wires, switches, relays, fuses and fusible links to each electrical component (light bulb, motor, etc.) and to ground, from which it is passed back to the battery.

3   Fuses - general information

Refer to illustrations 3.1a, 3.1b and 3.3

The electrical circuits of the vehicle are protected by a combination of fuses, circuit breakers and fusible links. The fuse blocks are located under the right side of the center console, and in the engine compartment next to the battery **(see illustrations)**.

Each of the fuses is designed to protect a specific circuit, and the various circuits are identified on the fuse panel itself.

Miniaturized fuses are employed in the fuse block. These compact fuses, with blade terminal design, allow fingertip removal and replacement. If an electrical component fails, always check the fuse first. A blown fuse is easily identified through the clear plastic body. Visually inspect the element for evidence of damage **(see illustration)**. If a continuity check is called for, the blade terminal tips are exposed in the fuse body.

Be sure to replace blown fuses with the correct type. Fuses of different ratings are physically interchangeable, but only fuses of the proper rating should be used. Replacing a fuse with one of a higher or lower value than specified is not recommended. Each electrical circuit needs a specific amount of protection. The amperage value of each fuse is molded into the fuse body.

If the replacement fuse immediately fails, don't replace it again until the cause of the problem is isolated and corrected. In most cases, this will be a short circuit in the wiring caused by a broken or deteriorated wire.

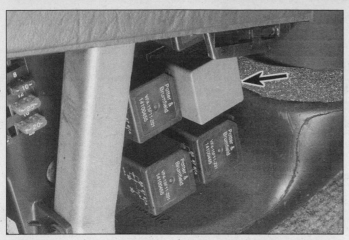

7.1 The turn signal hazard flasher unit (arrow) is located at the front of the fuse block on early models

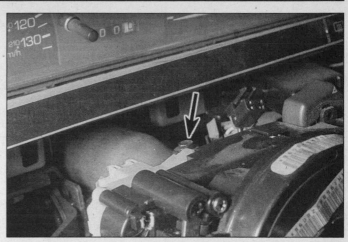

8.5a Remove the bolt at top (arrow) . . .

4 Fusible links - general information

Refer to illustration 4.2

Some circuits are protected by fusible links. The links are used in circuits which are not ordinarily fused, such as the ignition circuit.

On these models the fusible links are similar to fuses in that they can be visually checked to determine if they are melted, and are in fact called maxifuses. They are located in the underhood fuse block **(see illustration 3.1b)**.

To replace a maxifuse/fusible link, first disconnect the negative cable from the battery. Unplug the burned-out link and replace it with a new one (available from your dealer or auto parts store). Always determine the cause for the overload which melted the fusible link before installing a new one.

5 Circuit breakers - general information

Circuit breakers protect components such as power windows, power door locks and headlights which can experience intermittent overloads.

Because on these models the circuit breaker resets itself automatically, an electrical overload in a circuit breaker protected system will cause the circuit to fail momentarily, then come back on. If the circuit does not come back on, check it immediately. Once the condition is corrected, the circuit breaker will resume its normal function.

6 Relays - general information

Refer to illustration 6.2

Several electrical accessories in the vehicle use relays to transmit the electrical signal to the component. If the relay is defective, that component will not operate properly.

The various relays are mounted in several locations throughout the vehicle, although many key relays are located in the fuse/relay blocks located under the instrument panel and the engine compartment **(see illustrations 3.1a and 3.1b)**.

If a faulty relay is suspected, it can be removed and tested by a dealer or other qualified shop. Defective relays must be replaced as a unit.

Many of the relays on the vehicles covered by this manual are identical. If a faulty relay is suspected, simply unplug it and swap it with one (of the same part number) from another location. If the problem circuit now works properly, the cause was indeed a faulty relay.

7 Turn signal/hazard flashers - check and replacement

Refer to illustration 7.1

1 The turn signal/hazard flasher, a square, relay-size module, located behind the left kick panel on the fuse block **(see illustration)**, flashes the turn signals and hazard flashers.

2 When the flasher unit is functioning properly, an audible click can be heard during its operation. If the turn signals fail on one side or the other and the flasher unit does not make its characteristic clicking sound, a faulty turn signal bulb is indicated.

3 If both turn signals fail to blink, the problem may be due to a blown fuse, a faulty flasher unit, a broken switch or a loose or open connection. If a quick check of the fuse box indicates that the turn signal fuse has blown, check the wiring for a short before installing a new fuse.

4 To replace the flasher, remove the center console side panel, grasp the flasher unit and detach it from the fuse block.

5 Make sure the replacement unit is identical to the original. Compare the old one to the new one before installing it.

6 Installation is the reverse of removal.

8 Combination switch - removal and installation

Refer to illustrations 8.5a and 8.5b

Warning: *On models equipped with a Supplemental Inflatable Restraint (SIR) system, always disable the airbag(s) before working in the vicinity of the steering wheel, instrument panel or SIR system components to avoid the possibility of accidental deployment of the airbag(s), which could cause personal injury (see Chapter 12).*

Caution: *On models equipped with a theft-deterrent radio, make sure you have the correct activation code (or the theft deterrent system is turned off) before disconnecting the battery.*

1 Disconnect the cable from the negative battery terminal. Disable the airbag system, if equipped (see Chapter 12).

2 Position the steering wheel so the front wheels are pointing straight ahead and center the steering wheel. Remove the steering wheel (see Chapter 10).

3 Remove the knee protector and the steering column covers (see Chapter 11).

4 If equipped with an airbag, remove the airbag coil retaining screws, disconnect the horn and cruise control (if equipped) wiring harness connectors and remove the airbag coil, feeding the wiring harness through the combination switch. **Caution:** *DO NOT rotate the center hub of the airbag coil with the coil removed from the steering column, this will cause the coil to become uncentered. Damage to the airbag coil will result if it is installed off-center. Tape the center hub to the coil body to prevent it from moving.*

5 Disconnect the electrical connectors from the combination switch. Remove the mounting bolts and remove the switch **(see illustrations)**. **Note:** *On 1994 and earlier models, the switch is retained to the ignition switch module by two bolts, one at the top*

12

8.5b . . . and bottom (arrow), then remove the combination switch

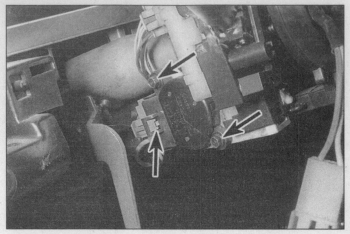

9.3 Unplug the connector, remove the bolts and detach the ignition switch

and one at the bottom. On 1995 and later models the switch is mounted with three bolts accessible from the rear (two at the top and one at the bottom).

6   Installation is the reverse of removal. If the airbag coil has moved off-center, perform the following steps to center the airbag coil:

a) *Rotate the center hub clockwise to the end of its travel. Rotate the hub gently and do not force it against its stop.*

b) *Rotate the hub counterclockwise, counting the number of turns required to reach its stop (approximately five turns).*

c) *Rotate the hub clockwise one-half the number of turns required in the above step (approximately 2.5 turns).*

9   Ignition switch - removal and installation

Refer to illustration 9.3

Warning: *On models equipped with a Supplemental Inflatable Restraint (SIR) system, always disable the airbag(s) before working in the vicinity of the steering wheel, instrument panel or SIR system components to avoid the possibility of accidental deployment of the airbag(s), which could cause personal injury (see Chapter 12).*

Caution: *On models equipped with a theft-deterrent radio, make sure you have the correct activation code, or the theft deterrent system is turned off, before disconnecting the battery.*

1   Disconnect the negative cable at the battery.

2   Remove the steering wheel if it interferes with access to the steering column covers on your vehicle (see Chapter 10). Remove the steering column covers (see Chapter 11).

3   Unplug the wires from the switch, remove the bolts and lift the switch off the steering column **(see illustration)**.

4   Installation is the reverse of removal.

10   Ignition lock cylinder - removal and installation

Refer to illustration 10.5

Warning: *On models equipped with a Supplemental Inflatable Restraint (SIR) system, always disable the airbag(s) before working in the vicinity of the steering wheel, instrument panel or SIR system components to avoid the possibility of accidental deployment of the airbag(s), which could cause personal injury (see Chapter 12).*

Caution: *On models equipped with a theft-deterrent radio, make sure you have the correct activation code, or the theft deterrent system is turned off, before disconnecting the battery.*

1   Disconnect the negative cable at the battery.

2   Remove the steering wheel if it interferes with access to the steering column covers on your vehicle (see Chapter 10).

3   Remove the knee bolster and the steer-

10.5 To remove the ignition lock cylinder, place the key in the "ACC" position, push in on the release tab with a screwdriver and pull the cylinder straight out

ing column cover screws (see Chapter 11).

4   Separate the steering column covers and pull the upper cover up for access to the lock cylinder release tab.

5   With the key in the ACC position, push the release tab on the left side of the column toward the front of the vehicle. Pull the lock cylinder straight out and remove it from the steering column **(see illustration)**.

6   Installation is the reverse of removal.

11   Rear window defogger - check and repair

1   The rear window defogger consists of a number of horizontal elements baked onto the glass surface.

2   Small breaks in the element can be repaired without removing the rear window.

Check

Refer to illustrations 11.4, 11.5 and 11.7

3   Turn the ignition switch and defogger system switches to the ON position.

4   When measuring voltage during the next two tests, wrap a piece of aluminum foil around the tip of the voltmeter negative probe and press the foil against the heating element with your finger **(see illustration)**.

5   Check the voltage at the center of each heating element **(see illustration)**. If the voltage is 6-volts, the element is okay (there is no break). If the voltage is 12-volts, the element is broken between the center of the element and the positive end. If the voltage is 0-volts the element is broken between the center of the element and ground.

6   Connect the negative lead to a good body ground. The reading should stay the same.

7   To find the break, place the voltmeter positive lead against the defogger positive terminal. Place the voltmeter negative lead with the foil strip against the heating element at the positive terminal end and slide it toward the negative terminal end. The point

11.4 When measuring the voltage at the rear window defogger grid, wrap a piece of aluminum foil around the negative probe of the voltmeter and press the foil against the wire with your finger

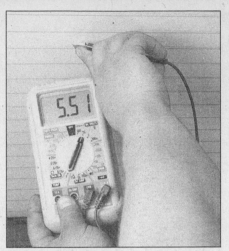

11.5 To determine if a heating element has broken, check the voltage at the center of each element - if the voltage is 6-volts, the element is unbroken - if the voltage is 12-volts, the element is broken between the center and the positive end - if there is no voltage, the element is broken between the center and ground

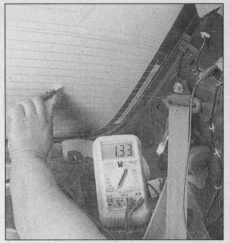

11.7 To find the break, place the voltmeter positive lead against the defogger positive terminal, place the voltmeter negative lead with the foil strip against the heating element at the positive terminal end and slide it toward the negative terminal end - the point at which the voltmeter reading changes abruptly is the point at which the element is broken

at which the voltmeter deflects from zero to several volts is the point at which the heating element is broken **(see illustration)**.

Repair

Refer to illustration 11.13

8   Repair the break in the element using a repair kit specifically recommended for this purpose, available at most parts stores. Included in this kit is plastic conductive epoxy.

9   Prior to repairing a break, turn off the system and allow it to cool off for a few minutes.

10   Lightly buff the element area with fine steel wool, then clean it thoroughly with rubbing alcohol.

11   Use masking tape to mask off the area being repaired.

12   Thoroughly mix the epoxy, following the instructions provided with the repair kit.

13   Apply the epoxy material to the slit in the masking tape, overlapping the undamaged area about 3/4-inch on either end **(see illustration)**.

14   Allow the repair to cure for 24 hours before removing the tape and using the system.

12   Radio and speakers - removal and installation

Warning: *On models equipped with a Supplemental Inflatable Restraint (SIR) system, always disable the airbag(s) before working in the vicinity of the steering wheel, instrument panel or SIR system components to avoid the possibility of accidental deployment of the airbag(s), which could cause personal injury*

(see Chapter 12).

Caution: *On models equipped with a theft-deterrent radio, make sure you have the correct activation code, or the theft deterrent system is turned off, before disconnecting the battery.*

Radio

Refer to illustrations 12.3 and 12.4

2   Remove the center trim panel (see Chapter 11).

3   Remove the two mounting screws and pull the radio out **(see illustration)**. On 1995 and later models, push the spring clips in on both sides of the radio

4   Unplug the electrical connector and the antenna lead, then remove the radio from the

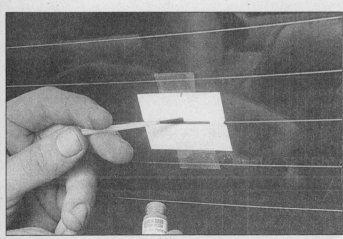

11.13 To use a defogger repair kit, apply masking tape to the inside of the window at the damaged area, then brush on the special conductive coating

12.3 Remove the two screws (arrows) and pull the radio out

12

12.4 Unplug the antenna and electrical connectors (arrows)

12.6 Use a small screwdriver to pry the speaker cover off

14.1 Detach the clips and unplug the headlight bulb electrical connector

14.3 Pull the old bulb out of the holder - when installing the new bulb, don't touch the surface (clean it with rubbing alcohol if you do)

instrument panel **(see illustration)**.
5   Installation is the reverse of removal.

Speakers

Refer to illustration 12.6
6   For access to the front door speakers pry off the covers **(see illustration)**. On rear speakers, remove the rear package tray.
7   Remove the speaker retaining screws, pull the speaker out and unplug the antenna and electrical connectors.
8   Installation is the reverse of removal.

13   Radio antenna - removal and installation

1   Disconnect the negative cable at the battery.
2   Use a small wrench to unscrew the antenna mast.
3   Installation is the reverse of removal.

14   Headlight - replacement

Refer to illustrations 14.1 and 14.3
Warning: *Halogen gas-filled bulbs are under*

pressure and may shatter if the surface is scratched or the bulb is dropped. Wear eye protection and handle the bulbs carefully, grasping only the base whenever possible. Do not touch the surface of the bulb with your fingers because the oil from your skin could cause it to overheat and fail prematurely. If you do touch the bulb surface, clean it with rubbing alcohol.

Fixed headlight models

1   Open the hood and unplug the electrical connector **(see illustration)**.
2   Rotate the bulb holder counterclockwise and withdraw it from the headlight housing.
3   Remove the bulb from the holder by pulling it straight out **(see illustration)**.
4   Without touching the glass with your bare fingers, insert the new bulb, install the bulb holder assembly into the headlight housing and plug in the electrical connector.
5   Test headlight operation, then close the hood.

Hidden headlight models

6   Turn on the headlights to raise them, then disconnect the cable from the negative terminal of the battery.
7   Remove the headlight bezel screws.

Rotate the headlight down manually to provide clearance at the lower inside corner of the bezel, the rotate it back up and lift the bezel off.
8   Remove the four headlight retainer screws. If you don't have an offset screwdriver, which will be necessary for removing the inside upper screw, disconnect the spring from the headlight carrier and detach the carrier from the adjusters.
9   Pull the headlight out and unplug the electrical connector.
10   Installation is the reverse of removal. Check and, if necessary, adjust the headlight(s) (see Section 15).

15   Headlights - adjustment

Refer to illustrations 15.1a, 15.1b, 15.1c and 15.3
Note: *It is important that the headlights are aimed correctly. If adjusted incorrectly they could blind the driver of an oncoming vehicle and cause a serious accident or seriously reduce your ability to see the road. The headlights should be checked for proper aim every 12 months and (on coupe models) any time a new headlight is installed or front end body*

15.1a Adjust the headlight vertical position by turning this Torx-head screw (arrow)

15.1b Insert a screwdriver behind the headlight and . . .

15.1c . . . turn the knob with the screwdriver to adjust the horizontal position

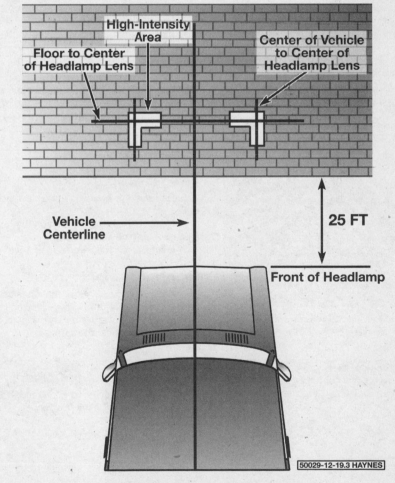

15.3 Headlight adjustment details

work is performed. It should be emphasized that the following procedure is only an interim step which will provide temporary adjustment until the headlights can be adjusted by a properly equipped shop.

1   Sedan and station wagon models have an adjusting screw on the top controlling the up-and-down movement and an adjusting knob on the side controlling left-and-right movement **(see illustrations).**

2   Coupe models also have two adjustment screws, one to the side controlling left-and-right movement and one below the light for up-and-down movement.

3   There are several methods of adjusting the headlights. The simplest method requires a blank wall 25 feet in front of the vehicle and a level floor.

4   Position masking tape vertically on the wall in reference to the vehicle centerline and the centerlines of both headlights.

5   Position a horizontal tape line in reference to the centerline of all the headlights. **Note:** *It may be easier to position the tape on the wall with the vehicle parked only a few inches away.*

6   Adjustment should be made with the vehicle sitting level, the gas tank half-full and no unusually heavy load in the vehicle.

7   Starting with the low beam adjustment, position the high intensity zone so it is two inches below the horizontal line and two inches to the right of the headlight vertical line. Twist the adjustment screws until the desired level has been achieved.

8   With the high beams on, the high intensity zone should be vertically centered with the exact center just below the horizontal line. **Note:** *It may not be possible to position the headlight aim exactly for both high and low*

12

16.3 Remove the bolts and lift the headlight housing off

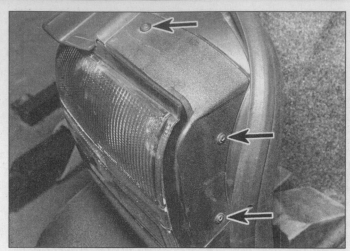

17.1a Remove the tail light housing attaching screws (arrows) . . .

beams. *If a compromise must be made, keep in mind that the low beams are the most used and have the greatest effect on driver safety.*
9   Have the headlights adjusted by a dealer service department at the earliest opportunity.

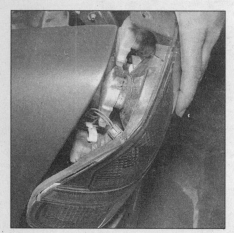

17.1b . . . and detach the housing for access to the bulbs

16 Headlight housing - removal and installation

Refer to illustration 16.3
Caution: *On models equipped with a theft-deterrent radio, make sure you have the correct activation code, or the theft deterrent system is turned off, before disconnecting the battery.*
1   Disconnect the negative battery cable from the battery.

Fixed headlight models

2   Remove the headlight and turn signal bulbs.
3   Remove the bolts and detach the headlight housing **(see illustration)**.
4   Installation is the reverse of removal.

Hidden headlight models

5   Detach the front half of the wheelhouse liner, then remove the nuts and detach the cover plate. Remove the bolts and lower the parking lamp **(see illustration)**.

6   Remove the screws and bolts, unplug the electrical connectors and detach the housing **(see illustration)**.
7   Installation is the reverse of removal.

17 Bulb replacement

Refer to illustrations 17.1a, 17.1b, 17.1c, 17.1d, 17.3a, 17.3b, 17.3c, 17.3d, 17.3e and 17.4
1   The lenses of many lights are held in place by screws. To gain access to the bulbs in these assemblies, simply remove the lenses or housings **(see illustrations)**.
2   The lenses or covers of some light assemblies are held in place by clips. You can remove them by unsnapping them or by prying them off with a small screwdriver.
3   Some bulbs can be removed by pushing them in and turning them counterclockwise while others can simply be pulled straight out of the socket **(see illustrations)**.
4   The instrument cluster bulbs are accessible after removing the cluster **(see illustration)**.

17.1c Remove the screws and detach the license plate light cover, then . . .

17.1d . . . loosen the screws and detach the lens assembly

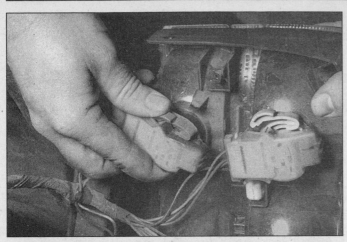

17.3a Press the tab on the tail light bulb holder and rotate counterclockwise to remove

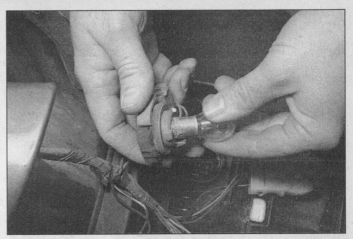

17.3b Push in and rotate the tail light bulb counterclockwise to release it from the holder

18   Wiper motor - removal and installation

Caution: *On models equipped with a theft-deterrent radio, make sure you have the correct activation code, or the theft deterrent system is turned off, before disconnecting the battery.*

1   Disconnect the cable from the negative terminal of the battery.

Windshield wiper motor

2   Flip up the wiper cover, remove the nuts and lift off the wiper arms, then remove the screws and detach the cowl vent grille.

3   Inside the vehicle, remove the instrument panel pad (see Chapter 11).

4   Detach the defroster nozzles for access to the wiper motor nozzle.

5   Remove the wiper module nuts and bolts, lift the module up and unplug the electrical connector, then remove the assembly from the vehicle. On later models it will be necessary to position the arm at 12 o'clock in order to ease removal.

6   Remove the nut from the crank arm and

disconnect the arm from the motor shaft.

7   Remove the bolts and detach the motor from the module.

8   When installing a new wiper motor, plug in the motor and turn it on momentarily so that it will cycle to the Park position when turned off.

9   Install the motor on the module. On early models, position the arm at 9 o'clock and install it on the motor using thread locking compound on the nut. On later models, the arm must be positioned at 12 o'clock as it was when it was removed. Tighten the mounting bolts securely.

10   Installation is the reverse of removal.

Rear window wiper motor

11   Lift up the cap, remove the nut and remove the rear wiper arm.

12   Remove the wiper shaft nut and washers.

13   Open the liftgate.

14   Remove the screws and detach the wedge blocks from the ends of the liftgate.

15   Remove the lower liftgate trim panel by pressing in on the centers of the plastic fasteners until they click, then detaching them

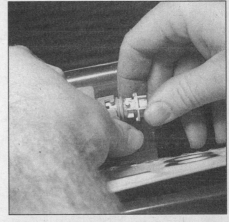

17.3c Rotate the license plate bulb holder counterclockwise, pull it out of the lens assembly, then . . .

and the panel.

16   Pry out the plastic cover in the wiper pivot housing, then insert a screwdriver and pry up to dislodge the trim panel clips.

17   Unplug the electrical connector, then

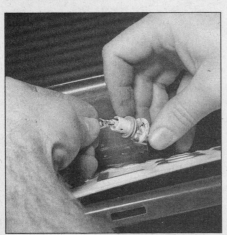

17.3d . . . pull the bulb straight out of the holder

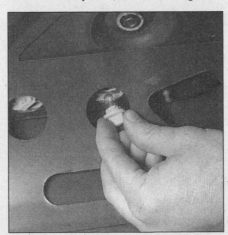

17.3e Pull the high-mounted brake light bulb (located in the trunk) straight out to remove it

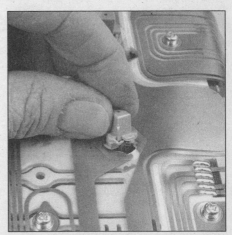

17.4 Most instrument cluster bulbs can be replaced after removing the cluster

19.3 Remove the screws (arrows) at each side of the instrument cluster. On 1995 and later models, two screws are located on the back of the cluster

20.9 Location of the airbag fuse on 1994 and earlier models (arrow)

remove the two bolts and detach the wiper motor assembly.

18   Installation is the reverse of removal.

19   Instrument cluster - removal and installation

Refer to illustration 19.3

Warning: *On models equipped with a Supplemental Inflatable Restraint (SIR) system, always disable the airbag(s) before working in the vicinity of the steering wheel, instrument panel or SIR system components to avoid the possibility of accidental deployment of the airbag(s), which could cause personal injury (see Chapter 12).*

Caution: *On models equipped with a theft-deterrent radio, make sure you have the correct activation code, or the theft deterrent system is turned off, before disconnecting the battery.*

1   Disconnect the cable from the negative battery terminal.

2   Remove the instrument cluster trim panel (see Chapter 11).

3   Remove the retaining screws and detach the cluster from the instrument panel **(see illustration)**.

4   Installation is the reverse of the removal procedure.

20   Airbag - general information

Later models are equipped with a Supplemental Inflatable Restraint (SIR) system, more commonly known as an airbag system. This system is designed to protect the driver (and passenger on 1995 and later models) from serious injury in the event of a head-on or frontal collision.

On 1994 and earlier models, the system consists of an airbag module located in the center of the steering wheel, two crash sensors located at the front of the vehicle (mounted to the radiator support at each side), an arming sensor located under the instrument panel pad and the diagnostic module located under the instrument panel pad.

On 1995 and later models, the system consists of a drivers airbag module located in the center of the steering wheel, a passengers airbag module located in the right side of the instrument panel, above the glove box and the diagnostic module located under the center console.

Airbag modules

The airbag modules contain a housing incorporating the cushion (airbag) and inflator unit. The inflator unit is mounted on the back of the housing over a hole through which gas is expelled, inflating the bag almost instantaneously when an electrical signal is sent from the system diagnostic module. On the drivers airbag, a coil assembly (mounted on the steering column under the steering wheel) carries the electrical signal to the inflator module. The coil assembly is able to transmit current regardless of steering wheel position.

Sensors

The discriminating sensors are basically pressure sensitive switches that complete an electrical circuit during an impact of sufficient G force. The arming sensor closes at deceleration rates lower than that of the crash sensors, reading the system for possible deployment.

1994 and earlier models use three sensors, two crash sensors and an arming sensor. On 1995 and later models, the sensors are incorporated into the diagnostic module.

Diagnostic module

The diagnostic module supplies current to the airbag modules in the event of a frontal collision. An energy reserve is available even if battery power is cut-off. The unit contains a self-diagnostic function that checks the system each time the ignition key is turned on or the vehicle is started. If the system is operat-

ing properly the AIRBAG warning light will flash several times, then go out. If there is a fault in the system, the light remain on and the diagnostic module will store a fault code. If the light is on steady while driving, the airbags system may fail to operate when needed. If the AIRBAG warning light stays on or comes on while driving, take the vehicle to a dealership for service as soon as possible.

Servicing components near the SIR system

Nevertheless, there are times when you need to work in the vicinity of components and wiring harnesses for the SIR system. SIR system wiring is easy to identify; they're all covered by a bright yellow conduit. Do not unplug the connectors for the SIR system wiring, except to disable the system. And do not use electrical test equipment on the SIR system wiring. *ALWAYS DISABLE THE SIR SYSTEM BEFORE WORKING NEAR THE SIR SYSTEM COMPONENTS OR RELATED WIRING.*

Disabling the system

Refer to illustration 20.9

To disable the airbag(s), perform the following steps:

a) *Position the front wheels pointing straight ahead, turn the ignition to the Lock position and remove the key.*

b) *Remove the AIR BAG fuse from the instrument panel fuse box and disconnect the negative cable from the battery.*

c) *Remove the lower steering column filler panel and disconnect the YELLOW two-way connector at the base of the steering column.*

d) *If equipped with a passenger airbag, remove the upper instrument panel pad (refer to Chapter 11). Remove the insulator beneath the pad. Disconnect the YELLOW two-way connector at the wiring harness leading to the airbag module.*

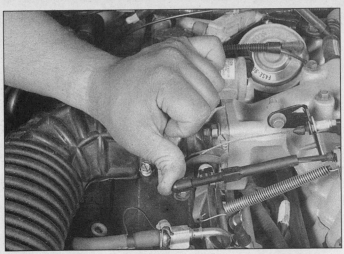

21.5 Make sure the cruise control and accelerator linkage mounted on the throttle body are not damaged and that they operate smoothly together when the throttle is opened

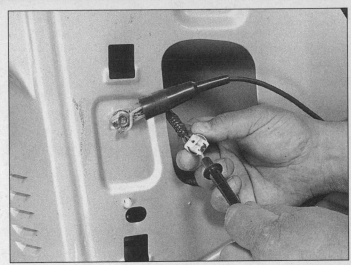

22.9 Check for voltage at the lock solenoid while the lock switch is operated

e) *Some systems have a steering centering device built into the steering column (just behind the air bag module). If disassembling steering system components, do not allow the steering wheel/shaft to rotate from its locked position.*

Enabling the system

To enable the airbag(s), perform the following steps:

a) *Position the front wheels pointing straight ahead, turn the ignition to the Lock position and remove the key.*
b) *Connect the YELLOW two-way connector to the driver and passenger (if equipped) air bag module wiring harness.*
c) *Install the AIRBAG fuse in the fuse box.*
d) *Turn the ignition On and verify that the AIRBAG warning light flashes seven times and goes out.*

21 Cruise control system - description and check

Refer to illustration 21.5

1   The cruise control system maintains vehicle speed with a servo motor located in the engine compartment on the driver's side fenderwell, which is connected to the throttle linkage by a cable. The system consists of the servo motor, brake switch, control switches, speed sensors and relays. Some features of the system require special testers and diagnostic procedures which are beyond the scope of this manual. Listed below are some general procedures that may be used to locate common problems.

2   Locate and check the fuse (see Section 3).

3   The brake pedal position (BPP) switch (or stop lamp switch) deactivates the cruise control system. Have an assistant press the

brake pedal while you check the stop lamp operation.

4   If the brake lights do not operate properly, correct the problem and retest the cruise control.

5   Check the control cable between the cruise control servo/amplifier and the throttle linkage and replace as necessary **(see illustration)**.

6   The cruise control system uses a speed sensing device. The speed sensor is located in the transmission. To test the speed sensor, see Chapter 6,.

7   Test drive the vehicle to determine if the cruise control is now working. If it isn't, take it to a dealer service department or an automotive electrical specialist for further diagnosis.

22 Power door lock system - description and check

Power door lock system

Refer to illustration 22.9

1   The power door lock system operates the door lock actuators mounted in each door. The system consists of the switches, actuators, and associated wiring. Diagnosis can usually be limited to simple checks of the wiring connections and actuators for minor faults which can be easily repaired.

2   Power door lock systems are operated by bi-directional solenoids located in the doors. The lock switches have two operating positions: Lock and Unlock. On later models with keyless entry the switches activate a module which in turn connects voltage to the door lock solenoids. Depending on which way the switch is activated, it reverses polarity, allowing the two sides of the circuit to be used alternately as the feed (positive) and ground side. On earlier models with out keyless entry the switches directly activate the door lock motors.

3   If you are unable to locate the trouble using the following general steps, consult your dealer service department.

4   Always check the circuit protection first. On these models the battery voltage passes through the 20 amp circuit breaker located in the passenger compartment fuse block.

5   Operate the door lock switches in both directions (Lock and Unlock) with the engine off. Listen for the faint click of the door lock solenoid (motor) or relay operating.

6   If there's no click, check for voltage at the switches. If no voltage is present, check the wiring between the fuse block and the switches for shorts and opens.

7   If voltage is present but no click is heard, test the switch for continuity. Replace it if there's no continuity in both switch positions.

8   If the switch has continuity but the solenoid doesn't click, check the wiring between the switch and solenoid for continuity. Repair the wiring if there's not continuity.

9   If all but one lock solenoids operate, remove the trim panel from the affected door (see Chapter 11) and check for voltage at the solenoid while the lock switch is operated **(see illustration)**. One of the wires should have voltage in the Lock position; the other should have voltage in the unlock position.

10   If the inoperative solenoid is receiving voltage, replace the solenoid. **Note:** *It's common for wires to break in the portion of the harness between the body and door (opening and closing the door fatigues and eventually breaks the wires).*

Keyless entry system

Refer to illustrations 25.13 and 25.14

11   The keyless entry system consists of a remote control transmitter that sends a coded infrared signal to a receiver which then operates the door lock system.

12   Replace the transmitter batteries when the red LED light on the side of the case

12

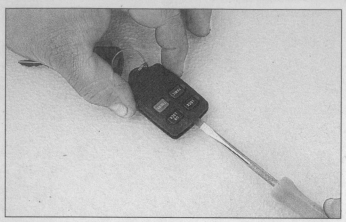

22.13 Use a small screwdriver to separate the transmitter halves

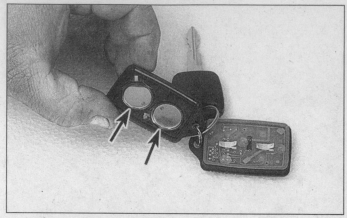

22.14 Replace the lithium batteries (arrows)

doesn't light when the button is pushed.

13   Use a small screwdriver to carefully separate the case halves **(see illustration)**.

14   Replace the lithium batteries **(see illustration)**.

15   Snap the case halves together.

23 Power window system - description and check

Refer to illustration 23.12

1   The power window system operates electric motors, mounted in the doors, which lower and raise the windows. The system consists of the control switches, the motors, regulators, glass mechanisms and associated wiring.

2   The power windows can be lowered and raised from the master control switch by the driver or by remote switches located at the individual windows. Each window has a separate motor which is reversible. The position of the control switch determines the polarity and therefore the direction of operation.

3   The circuit is protected by a fuse and a circuit breaker. Each motor is also equipped with an internal circuit breaker, this prevents one stuck window from disabling the whole system.

4   The power window system will only operate when the ignition switch is ON. In addition, many models have a window lockout switch at the master control switch which, when activated, disables the switches at the rear windows and, sometimes, the switch at the passenger's window also. Always check these items before troubleshooting a window problem.

5   These procedures are general in nature, so if you can't find the problem using them, take the vehicle to a dealer service department or other properly equipped repair facility.

6   If the power windows won't operate, always check the fuse and circuit breaker first.

7   If only the rear windows are inoperative,

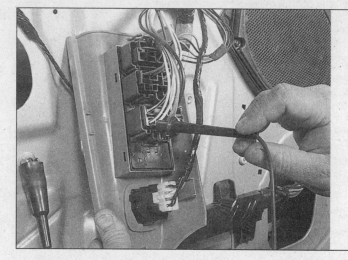

23.12 If no voltage is found at the motor with the switch depressed - check for voltage at the switch

or if the windows only operate from the master control switch, check the rear window lockout switch for continuity in the unlocked position. Replace it if it doesn't have continuity.

8   Check the wiring between the switches and fuse panel for continuity. Repair the wiring, if necessary.

9   If only one window is inoperative from the master control switch, try the other control switch at the window. **Note:** *This doesn't apply to the drivers door window.*

10   If the same window works from one switch, but not the other, check the switch for continuity.

11   If the switch tests OK, check for a short or open in the circuit between the affected switch and the window motor.

12   If one window is inoperative from both switches, remove the trim panel from the affected door and check for voltage at the switch **(see illustration)** and at the motor while the switch is operated.

13   If voltage is reaching the motor, disconnect the glass from the regulator (see Chapter 11). Move the window up and down by hand while checking for binding and damage. Also check for binding and damage to the regulator. If the regulator is not damaged and the window moves up and down smoothly,

replace the motor. If there's binding or damage, lubricate, repair or replace parts, as necessary.

14   If voltage isn't reaching the motor, check the wiring in the circuit for continuity between the switches and motors. You'll need to consult the wiring diagram for the vehicle. If the circuit is equipped with a relay, check that the relay is grounded properly and receiving voltage.

15   Test the windows after you are done to confirm proper repairs.

24 Wiring diagrams - general information

Prior to troubleshooting any circuits, check the fuses and circuit breakers (if equipped) to make sure they are in good condition. Make sure the battery is properly charged and has clean, tight cable connections (see Chapter 1).

When checking the wiring system, make sure that all electrical connectors are clean, with no broken or loose pins. When unplugging an electrical connector, do not pull on the wires, only on the connector housings themselves.

COMPONENT LOCATOR:

A/C COMP CLUTCH	E 24	HEADLT SW	A 36, B 36
A/C PRES SW	E 31	HIGH LEVEL STOP LTS	B 45
ABS SPLICE PACK	D 20	HORN SW	A 20, C 36
ABS SYSTEM	A-B 24-26	HVAC MODULE	C-E 24-25
AIRBAGS	B-E 26-27	I/P SPLICE PACK	D 21
ALDL CONN	A 15, A 19	IGNITION SW	E 7
ALTERNATOR	C 3	INST CLSTR	A-E 38-39
ASHTRAY LT	E 36	INST PANEL DIMMER SW	E 31
AUTO TRANS CONTROL	D-E 12-15, D-E 16-19	IPJB	B-D 28-35
BACK-UP LT SW	A 4	IPJB LAYOUT	A 33-34
BACK-UP/LICENSE LT ASSEMBLYS	C-D 44-45	LAP BELT RETRACTOR SW	A 43
BATTERY	A 3	LATCH PROXIMITY SW	A 40
BLOWER MOTOR	A 35	LEFT BODY GROUND SPLICE PACK	E 21
BLOWER RESISTOR	D 25-26	LEFT FRONT DOOR	
BTSI SOL	D 3	KEY CYLINDER SW	C 42
CARGO LT	D 46	LEFT FRONT DOOR SW	A 28
CARGO/MAP LTS	A 45	LEFT FRONT SOL	C 26
CIG LTR & ILLUM LT	E 36	LEFT REAR DOOR SW	A 28
CLUTCH START SW	E 5	LEFT SEAT BELT TRACK	C 40
COOLANT LEVEL SW	E 11	LOW BRAKE FLUID SW	E 31
COOLING FAN MOTOR	E 11	NEUTRAL START/BACK-UP LT SW	B 3
CRUISE CLUTCH SW	B 22	PARK BRAKE SW	A 28
CRUISE CONTROL	A-B 20-22	PARK SENSE SW	D 3
DEFOGGER GRID	A 44	PASSIVE RESTRAINT SYSTEM	A-C 40-43
DERM	D-E 27	PCM (PFI)	A-E 16
DIR/HAZARD SW	D 36	PCM (TBI)	A-E 16
DIS MODULE	A 22-23	POWER DOOR LOCKS	A-D 41-43
DOOR LOCK SWS	D 41	POWER MIRROR	E 44-45
ENGINE SPLICE PACK	D 20	POWER SPLICE PACK	E 5
FORWARD LAMP SPLICE PACK	E 20	POWER STEERING PRES SW	B 15, B 19
FUEL INJECTORS	C 13, C-D 18	POWER WINDOWS	C-D 40, D-E 42
FUEL PUMP	A 31	RADIO (PARTIAL)	E 36
GROUND A	A 2	REAR BODY GROUND SPLICE PACK	C 23, D 23
GROUND B	A 2	REAR DEFOG SW	A 44
GROUND C	D 14, C 17	RIGHT BODY GROUND SPLICE PACK	E 23
GROUND D	D 22	RIGHT FRONT DOOR SW	A 28
GROUND E	D 22	RIGHT FRONT SOL	C 26
GROUND F	D 20	RIGHT REAR DOOR SW	A 28
GROUND G	D 20	RIGHT SEAT BELT TRACK	B 40
GROUND H	D 20	SEAT BELT TRACKS	B-C 40
GROUND I	E 20	SHIFT MODE SELECTOR	E 14, E 18
GROUND J	E 20	STARTER	A 3
GROUND K	E 21	STOP LT SW	A 8
GROUND L	E 23	SUNROOF	A-C 44
GROUND M	D 23	TRANSAXLE SOLS	E 15, E 19
GROUND N	C 23	TRANSMISSION SELECT SW	D 15, D 19
GROUND O	A 35	UHJB	B-D 4-11
GROUND P	C 42	UHJB LAYOUT	D 5-7
GROUND Q	D 43	VEHICLE SPEED SENSOR	A 13, A 17
GROUND R	A 44	W/SHIELD WIPER	C 20-22
HEADLT DOOR CONTROL MOD	C 2	WASHER MOTOR	D 20

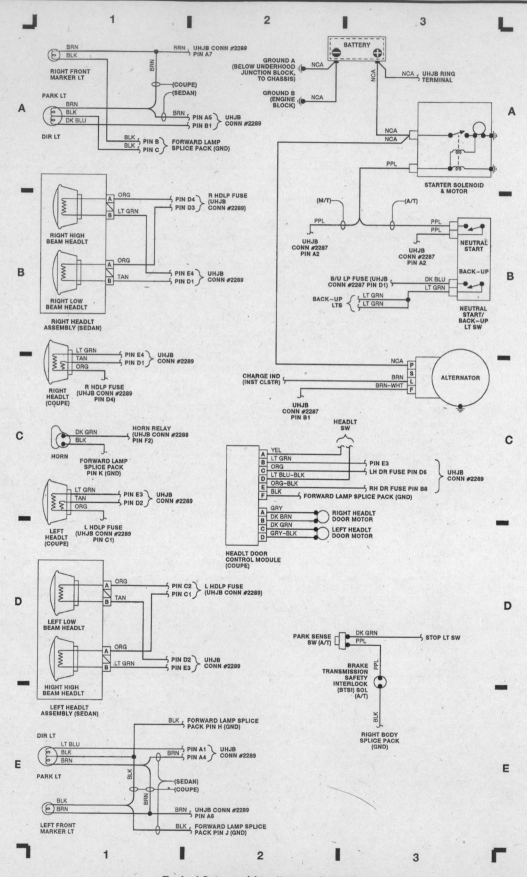

Typical Saturn wiring diagram (1 of 12)

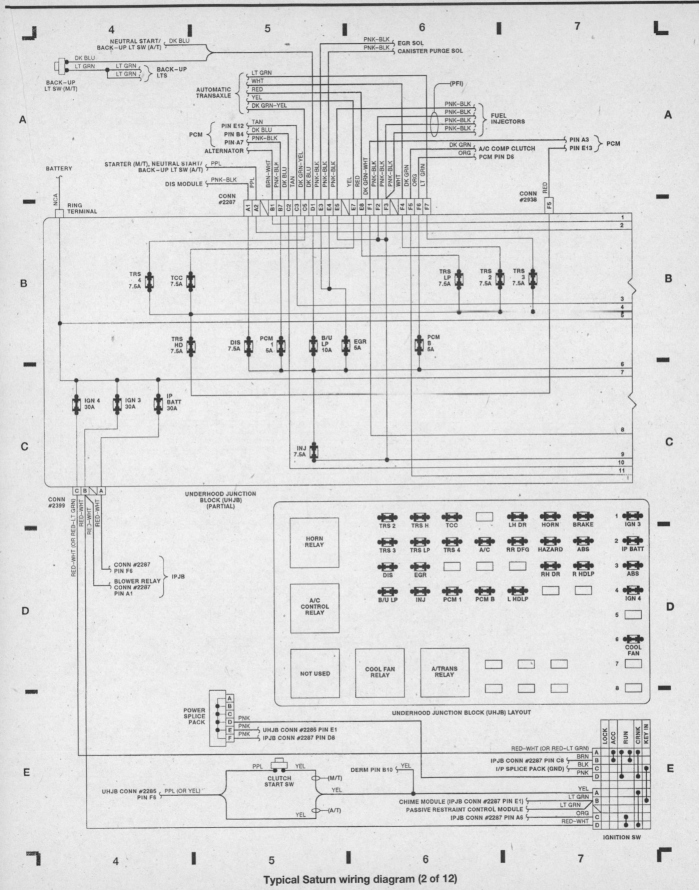

Typical Saturn wiring diagram (2 of 12)

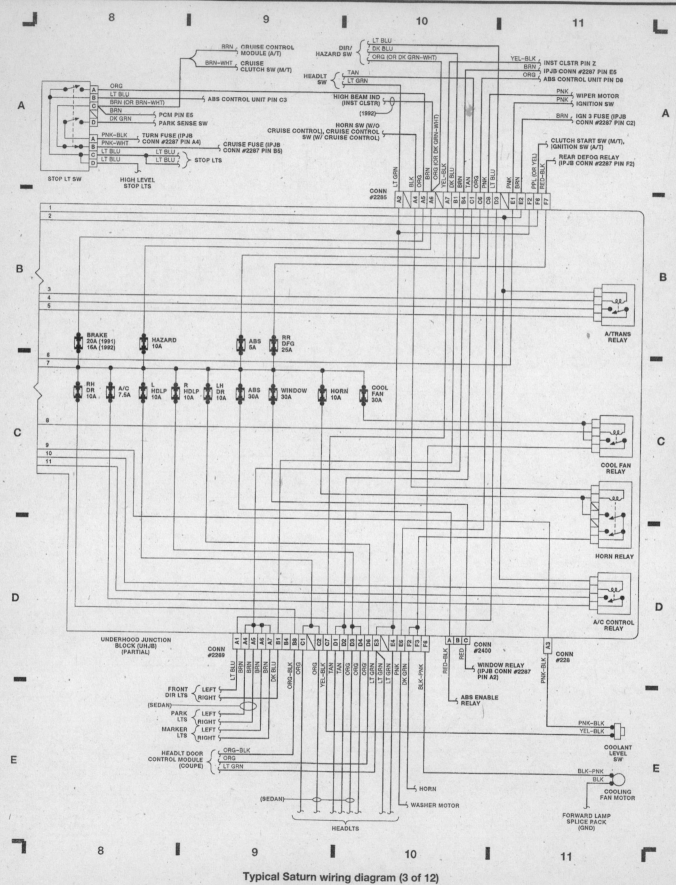

Typical Saturn wiring diagram (3 of 12)

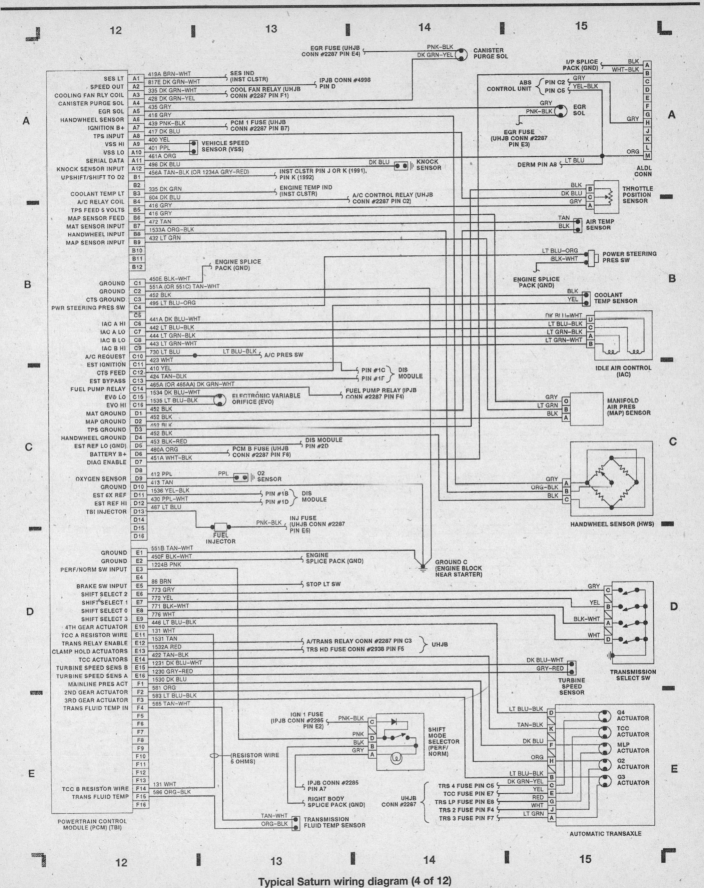

Typical Saturn wiring diagram (4 of 12)

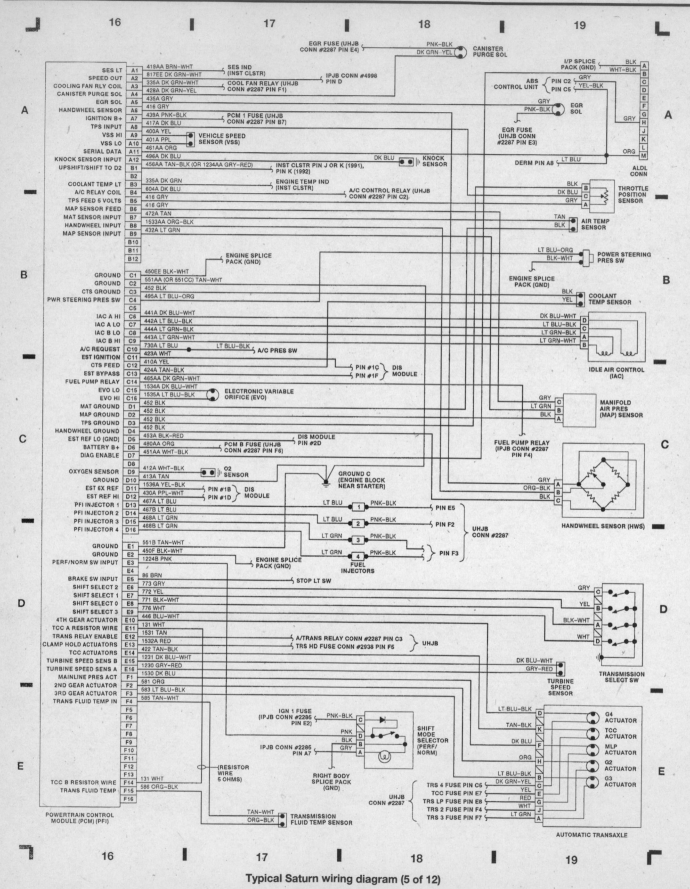

Typical Saturn wiring diagram (5 of 12)

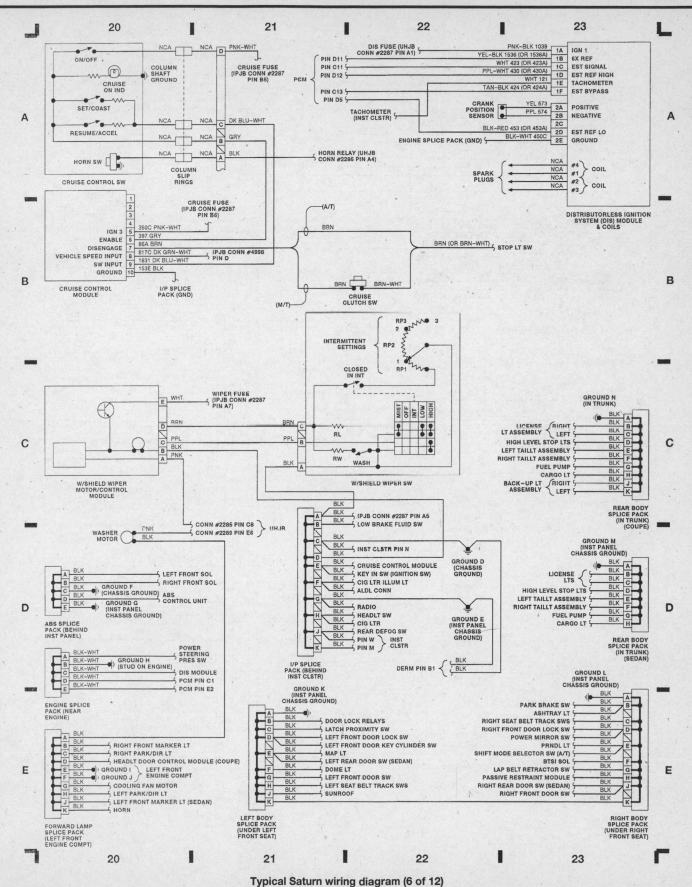

Typical Saturn wiring diagram (6 of 12)

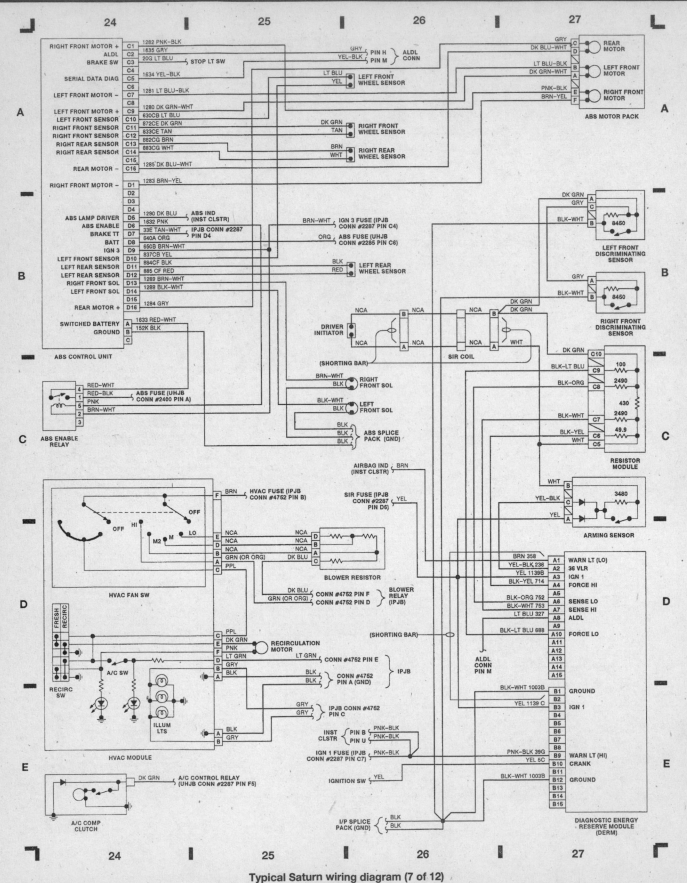

Typical Saturn wiring diagram (7 of 12)

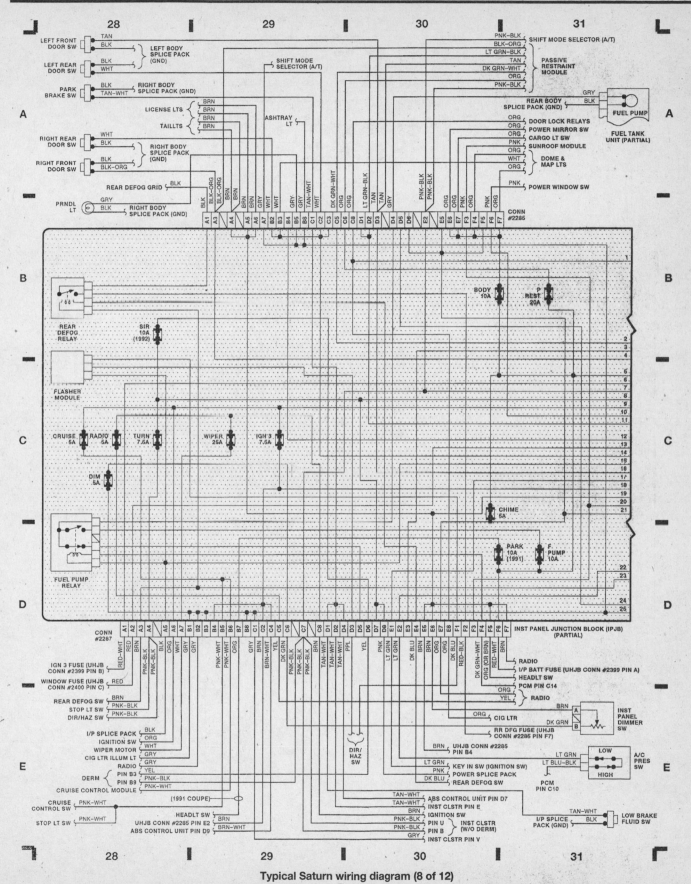

Typical Saturn wiring diagram (8 of 12)

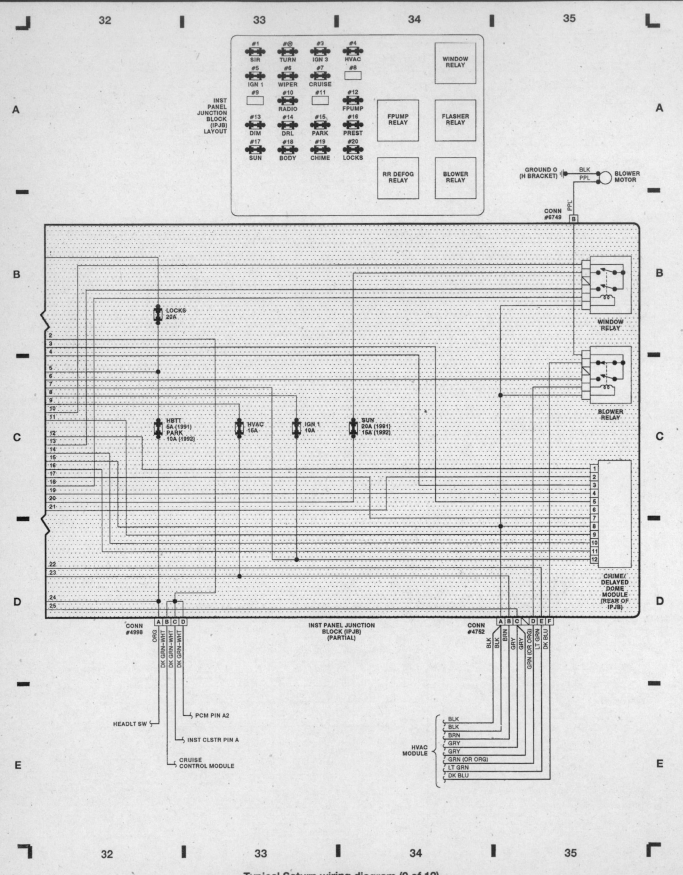

Typical Saturn wiring diagram (9 of 12)

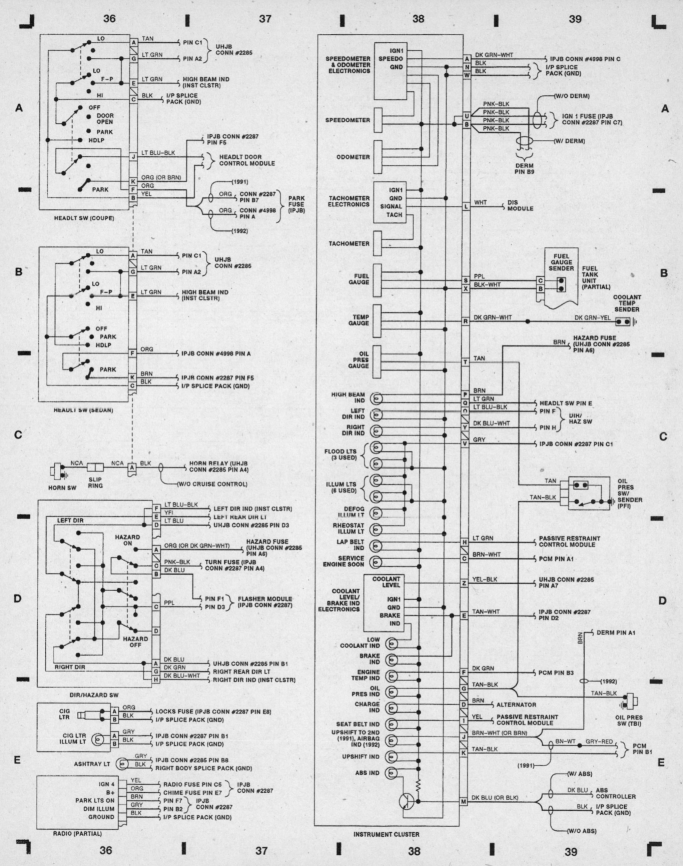

Typical Saturn wiring diagram (10 of 12)

12

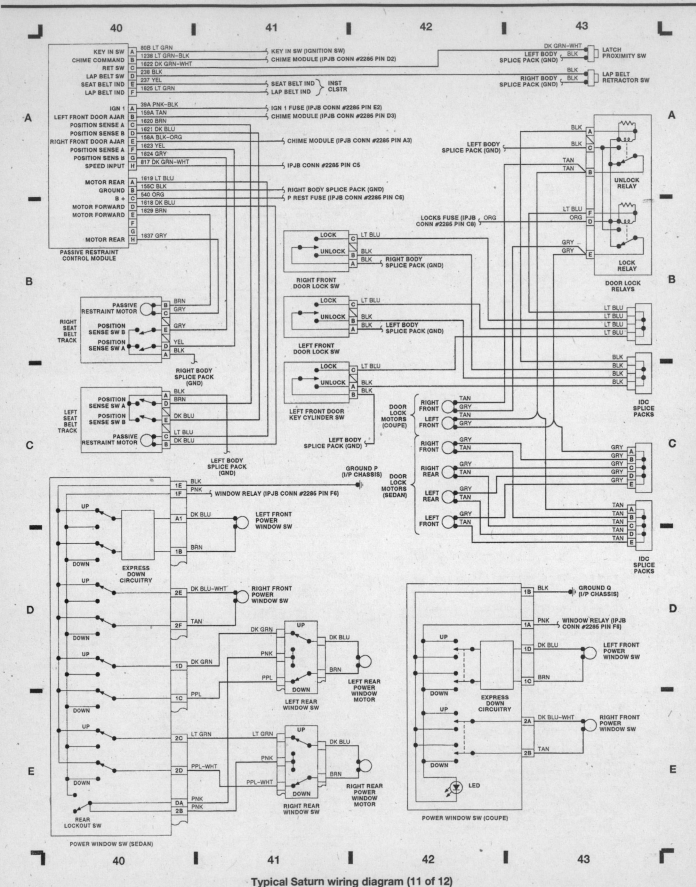

Typical Saturn wiring diagram (11 of 12)

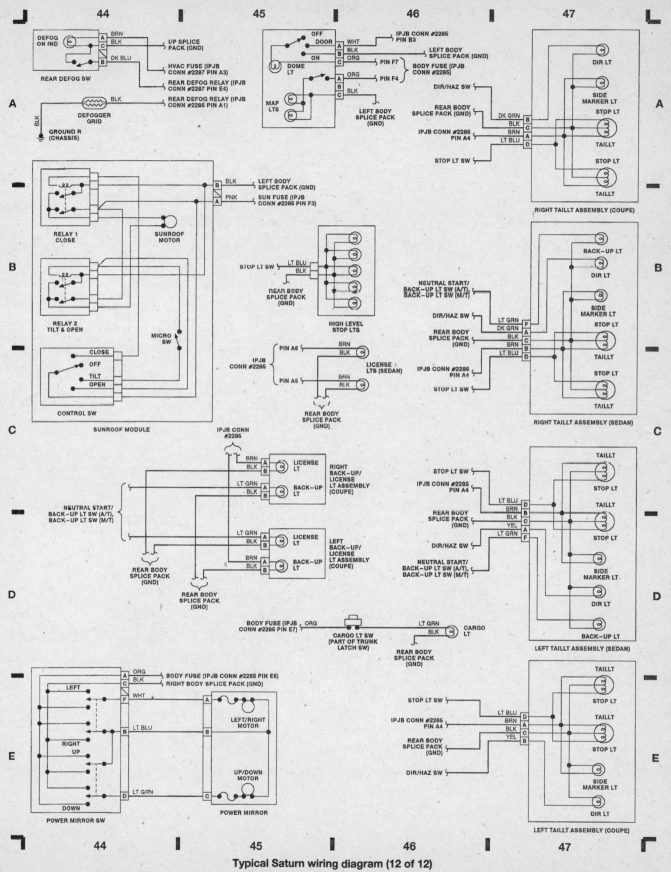

12

Notes

Index

Haynes Automotive Manuals

NOTE: New manuals are added to this list on a periodic basis. If you do not see a listing for your vehicle, consult your local Haynes dealer for the latest product information.

ACURA
*12020 Integra '86 thru '89 & Legend '86 thru '90

AMC
Jeep CJ - see JEEP (50020)
14020 Mid-size models, Concord, Hornet, Gremlin & Spirit '70 thru '83
14025 (Renault) Alliance & Encore '83 thru '87

AUDI
15020 4000 all models '80 thru '87
15025 5000 all models '77 thru '83
15026 5000 all models '84 thru '88

AUSTIN-HEALEY
Sprite - see MG Midget (66015)

BMW
*18020 3/5 Series not including diesel or all-wheel drive models '82 thru '92
*18021 3 Series except 325iX models '92 thru '97
18025 320i all 4 cyl models '75 thru '83
18035 528i & 530i all models '75 thru '80
18050 1500 thru 2002 except Turbo '59 thru '77

BUICK
Century (front wheel drive) - see GM (829)
*19020 Buick, Oldsmobile & Pontiac Full-size (Front wheel drive) all models '85 thru '98
Buick Electra, LeSabre and Park Avenue; Oldsmobile Delta 88 Royale, Ninety Eight and Regency; Pontiac Bonneville
19025 Buick Oldsmobile & Pontiac Full-size (Rear wheel drive)
Buick Estate '70 thru '90, Electra '70 thru '84, LeSabre '70 thru '85, Limited '74 thru '79
Oldsmobile Custom Cruiser '70 thru '90, Delta 88 '70 thru '85, Ninety-eight '70 thru '84
Pontiac Bonneville '70 thru '81, Catalina '70 thru '81, Grandville '70 thru '75, Parisienne '83 thru '86
19030 Mid-size Regal & Century all rear-drive models with V6, V8 and Turbo '74 thru '87
Regal - see GENERAL MOTORS (38010)
Riviera - see GENERAL MOTORS (38030)
Roadmaster - see CHEVROLET (24046)
Skyhawk - see GENERAL MOTORS (38015)
Skylark '80 thru '85 - see GM (38020)
Skylark '86 on - see GM (38025)
Somerset - see GENERAL MOTORS (38025)

CADILLAC
*21030 Cadillac Rear Wheel Drive all gasoline models '70 thru '93
Cimarron - see GENERAL MOTORS (38015)
Eldorado - see GENERAL MOTORS (38030)
Seville '80 thru '85 - see GM (38030)

CHEVROLET
*24010 Astro & GMC Safari Mini-vans '85 thru '93
24015 Camaro V8 all models '70 thru '81
24016 Camaro all models '82 thru '92
Cavalier - see GENERAL MOTORS (38015)
Celebrity - see GENERAL MOTORS (38005)
24017 Camaro & Firebird '93 thru '97
24020 Chevelle, Malibu & El Camino '69 thru '87
24024 Chevette & Pontiac T1000 '76 thru '87
Citation - see GENERAL MOTORS (38020)
*24032 Corsica/Beretta all models '87 thru '96
24040 Corvette all V8 models '68 thru '82
*24041 Corvette all models '84 thru '96
10305 Chevrolet Engine Overhaul Manual
24045 Full-size Sedans Caprice, Impala, Biscayne, Bel Air & Wagons '69 thru '90
24046 Impala SS & Caprice and Buick Roadmaster '91 thru '96
Lumina - see GENERAL MOTORS (38010)

24048 Lumina & Monte Carlo '95 thru '98
Lumina APV - see GM (38035)
24050 Luv Pick-up all 2WD & 4WD '72 thru '82
*24055 Monte Carlo all models '70 thru '88
Monte Carlo '95 thru '98 - see LUMINA (24048)
24059 Nova all V8 models '69 thru '79
*24060 Nova and Geo Prizm '85 thru '92
24064 Pick-ups '67 thru '87 - Chevrolet & GMC, all V8 & in-line 6 cyl, 2WD & 4WD '67 thru '87; Suburbans, Blazers & Jimmys '67 thru '91
*24065 Pick-ups '88 thru '98 - Chevrolet & GMC, all full-size pick-ups, '88 thru '98; Blazer & Jimmy '92 thru '94; Suburban '92 thru '98; Tahoe & Yukon '98
24070 S-10 & S-15 Pick-ups '82 thru '93, Blazer & Jimmy '83 thru '94,
*24071 S-10 & S-15 Pick-ups '94 thru '96 Blazer & Jimmy '95 thru '96
*24075 Sprint & Geo Metro '85 thru '94
*24080 Vans - Chevrolet & GMC, V8 & in-line 6 cylinder models '68 thru '96

CHRYSLER
25015 Chrysler Cirrus, Dodge Stratus, Plymouth Breeze '95 thru '98
25025 Chrysler Concorde, New Yorker & LHS, Dodge Intrepid, Eagle Vision, '93 thru '97
10310 Chrysler Engine Overhaul Manual
*25020 Full-size Front-Wheel Drive '88 thru '93
K-Cars - see DODGE Aries (30008)
Laser - see DODGE Daytona (30030)
*25030 Chrysler & Plymouth Mid-size front wheel drive '82 thru '95
Rear-wheel Drive - see Dodge (30050)

DATSUN
28005 200SX all models '80 thru '83
28007 B-210 all models '73 thru '78
28009 210 all models '79 thru '82
28012 240Z, 260Z & 280Z Coupe '70 thru '78
28014 280ZX Coupe & 2+2 '79 thru '83
300ZX - see NISSAN (72010)
28016 310 all models '78 thru '82
28018 510 & PL521 Pick-up '68 thru '73
28020 510 all models '78 thru '81
28022 620 Series Pick-up all models '73 thru '79
720 Series Pick-up - see NISSAN (72030)
28025 810/Maxima all gasoline models, '77 thru '84

DODGE
400 & 600 - see CHRYSLER (25030)
*30008 Aries & Plymouth Reliant '81 thru '89
30010 Caravan & Plymouth Voyager Mini-Vans all models '84 thru '95
*30011 Caravan & Plymouth Voyager Mini-Vans all models '96 thru '98
30012 Challenger/Plymouth Saporro '78 thru '83
30016 Colt & Plymouth Champ (front wheel drive) all models '78 thru '87
*30020 Dakota Pick-ups all models '87 thru '96
30025 Dart, Demon, Plymouth Barracuda, Duster & Valiant 6 cyl models '67 thru '76
*30030 Daytona & Chrysler Laser '84 thru '89
Intrepid - see CHRYSLER (25025)
*30034 Neon all models '95 thru '97
*30035 Omni & Plymouth Horizon '78 thru '90
*30040 Pick-ups all full-size models '74 thru '93
*30041 Pick-ups all full-size models '94 thru '96
*30045 Ram 50/D50 Pick-ups & Raider and Plymouth Arrow Pick-ups '79 thru '93
30050 Dodge/Plymouth/Chrysler rear wheel drive '71 thru '89
*30055 Shadow & Plymouth Sundance '87 thru '94
*30060 Spirit & Plymouth Acclaim '89 thru '95
*30065 Vans - Dodge & Plymouth '71 thru '96

EAGLE
Talon - see Mitsubishi Eclipse (68030)
Vision - see CHRYSLER (25025)

FIAT
34010 124 Sport Coupe & Spider '68 thru '78
34025 X1/9 all models '74 thru '80

FORD
10355 Ford Automatic Transmission Overhaul
*36004 Aerostar Mini-vans all models '86 thru '96
*36006 Contour & Mercury Mystique '95 thru '98
36008 Courier Pick-up all models '72 thru '82
36012 Crown Victoria & Mercury Grand Marquis '88 thru '96
10320 Ford Engine Overhaul Manual
36016 Escort/Mercury Lynx all models '81 thru '90
*36020 Escort/Mercury Tracer '91 thru '96
*36024 Explorer & Mazda Navajo '91 thru '95
36028 Fairmont & Mercury Zephyr '78 thru '83
36030 Festiva & Aspire '88 thru '97
36032 Fiesta all models '77 thru '80
36036 Ford & Mercury Full-size, Ford LTD & Mercury Marquis ('75 thru '82); Ford Custom 500, Country Squire, Crown Victoria & Mercury Colony Park ('75 thru '87); Ford LTD Crown Victoria & Mercury Gran Marquis ('83 thru '87)
36040 Granada & Mercury Monarch '75 thru '80
36044 Ford & Mercury Mid-size, Ford Thunderbird & Mercury Cougar ('75 thru '82); Ford LTD & Mercury Marquis ('83 thru '86); Ford Torino, Gran Torino, Elite, Ranchero pick-up, LTD II, Mercury Montego, Comet, XR-7 & Lincoln Versailles ('75 thru '86)
36048 Mustang V8 all models '64-1/2 thru '73
36049 Mustang II 4 cyl, V6 & V8 '74 thru '78
36050 Mustang & Mercury Capri all models Mustang, '79 thru '93; Capri, '79 thru '86
*36051 Mustang all models '94 thru '97
36054 Pick-ups & Bronco '73 thru '79
36058 Pick-ups & Bronco '80 thru '96
36059 Pick-ups, Expedition & Mercury Navigator '97 thru '98
36062 Pinto & Mercury Bobcat '75 thru '80
36066 Probe all models '89 thru '92
36070 Ranger/Bronco II gasoline models '83 thru '92
*36071 Ranger '93 thru '97 & Mazda Pick-ups '94 thru '97
36074 Taurus & Mercury Sable '86 thru '95
*36075 Taurus & Mercury Sable '96 thru '98
*36078 Tempo & Mercury Topaz '84 thru '94
36082 Thunderbird/Mercury Cougar '83 thru '88
*36086 Thunderbird/Mercury Cougar '89 and '97
36090 Vans all V8 Econoline models '69 thru '91
*36094 Vans full size '92 thru '95
*36097 Windstar Mini-van '95-'98

GENERAL MOTORS
*10360 GM Automatic Transmission Overhaul
*38005 Buick Century, Chevrolet Celebrity, Oldsmobile Cutlass Ciera & Pontiac 6000 all models '82 thru '96
*38010 Buick Regal, Chevrolet Lumina, Oldsmobile Cutlass Supreme & Pontiac Grand Prix front-wheel drive models '88 thru '95
*38015 Buick Skyhawk, Cadillac Cimarron, Chevrolet Cavalier, Oldsmobile Firenza & Pontiac J-2000 & Sunbird '82 thru '94
*38016 Chevrolet Cavalier & Pontiac Sunfire '95 thru '98
38020 Buick Skylark, Chevrolet Citation, Olds Omega, Pontiac Phoenix '80 thru '85
38025 Buick Skylark & Somerset, Oldsmobile Achieva & Calais and Pontiac Grand Am all models '85 thru '95
38030 Cadillac Eldorado '71 thru '85, Seville '80 thru '85, Oldsmobile Toronado '71 thru '85 & Buick Riviera '79 thru '85
*38035 Chevrolet Lumina APV, Olds Silhouette & Pontiac Trans Sport all models '90 thru '95
General Motors Full-size Rear-wheel Drive - see BUICK (19025)

(Continued on other side)

Listings shown with an asterisk (*) indicate model coverage as of this printing. These titles will be periodically updated to include later model years - consult your Haynes dealer for more information.

Haynes North America, Inc., 861 Lawrence Drive, Newbury Park, CA 91320-1514 • (805) 498-6703

NOTE: New manuals are added to this list on a periodic basis. If you do not see a listing for your vehicle, consult your local Haynes dealer for the latest product information.

GEO
Metro - *see CHEVROLET Sprint (24075)*
Prizm - '85 thru '92 see CHEVY (24060), '93 thru '96 see TOYOTA Corolla (92036)
*40030 **Storm** all models '90 thru '93
Tracker - *see SUZUKI Samurai (90010)*

GMC
Safari - *see CHEVROLET ASTRO (24010)*
Vans & Pick-ups - *see CHEVROLET*

HONDA
42010 **Accord CVCC** all models '76 thru '83
42011 **Accord** all models '84 thru '89
42012 **Accord** all models '90 thru '93
42013 **Accord** all models '94 thru '95
42020 **Civic 1200** all models '73 thru '79
42021 **Civic 1300 & 1500 CVCC** '80 thru '83
42022 **Civic 1500 CVCC** all models '75 thru '79
42023 **Civic** all models '84 thru '91
*42024 **Civic & del Sol** '92 thru '95
*42040 **Prelude CVCC** all models '79 thru '89

HYUNDAI
*43015 **Excel** all models '86 thru '94

ISUZU
Hombre - *see CHEVROLET S-10 (24071)*
*47017 **Rodeo** '91 thru '97; **Amigo** '89 thru '94; **Honda Passport** '95 thru '97
*47020 **Trooper & Pick-up**, all gasoline models Pick-up, '81 thru '93; Trooper, '84 thru '91

JAGUAR
*49010 **XJ6** all 6 cyl models '68 thru '86
*49011 **XJ6** all models '88 thru '94
*49015 **XJ12 & XJS** all 12 cyl models '72 thru '85

JEEP
*50010 **Cherokee, Comanche & Wagoneer Limited** all models '84 thru '96
50020 **CJ** all models '49 thru '86
*50025 **Grand Cherokee** all models '93 thru '98
50029 **Grand Wagoneer & Pick-up** '72 thru '91 Grand Wagoneer '84 thru '91, Cherokee & Wagoneer '72 thru '83, Pick-up '72 thru '88
*50030 **Wrangler** all models '87 thru '95

LINCOLN
Navigator - *see FORD Pick-up (36059)*
59010 **Rear Wheel Drive** all models '70 thru '96

MAZDA
61010 **GLC Hatchback (rear wheel drive)** '77 thru '83
61011 **GLC (front wheel drive)** '81 thru '85
*61015 **323 & Protegé** '90 thru '97
*61016 **MX-5 Miata** '90 thru '97
*61020 **MPV** all models '89 thru '94
Navajo - *see Ford Explorer (36024)*
61030 **Pick-ups** '72 thru '93
Pick-ups '94 thru '96 - *see Ford Ranger (36071)*
61035 **RX-7** all models '79 thru '85
*61036 **RX-7** all models '86 thru '91
61040 **626** (rear wheel drive) all models '79 thru '82
*61041 **626/MX-6 (front wheel drive)** '83 thru '91

MERCEDES-BENZ
63012 **123 Series Diesel** '76 thru '85
*63015 **190 Series** four-cyl gas models, '84 thru '88
63020 **230/250/280** 6 cyl sohc models '68 thru '72
63025 **280 123 Series** gasoline models '77 thru '81
63030 **350 & 450** all models '71 thru '80

MERCURY
See FORD Listing.

MG
66010 **MGB** Roadster & GT Coupe '62 thru '80
66015 **MG Midget, Austin Healey Sprite** '58 thru '80

MITSUBISHI
*68020 **Cordia, Tredia, Galant, Precis & Mirage** '83 thru '93
*68030 **Eclipse, Eagle Talon & Ply. Laser** '90 thru '94
*68040 **Pick-up** '83 thru '96 & **Montero** '83 thru '93

NISSAN
72010 **300ZX** all models including Turbo '84 thru '89
*72015 **Altima** all models '93 thru '97
*72020 **Maxima** all models '85 thru '91
*72030 **Pick-ups** '80 thru '96 **Pathfinder** '87 thru '95
72040 **Pulsar** all models '83 thru '86
*72050 **Sentra** all models '82 thru '94
*72051 **Sentra & 200SX** all models '95 thru '98
*72060 **Stanza** all models '82 thru '90

OLDSMOBILE
*73015 **Cutlass** V6 & V8 gas models '74 thru '88
For other OLDSMOBILE titles, see BUICK, CHEVROLET or GENERAL MOTORS listing.

PLYMOUTH
For PLYMOUTH titles, see DODGE listing.

PONTIAC
79008 **Fiero** all models '84 thru '88
79018 **Firebird** V8 models except Turbo '70 thru '81
79019 **Firebird** all models '82 thru '92
For other PONTIAC titles, see BUICK, CHEVROLET or GENERAL MOTORS listing.

PORSCHE
*80020 **911** except Turbo & Carrera 4 '65 thru '89
80025 **914** all 4 cyl models '69 thru '76
80030 **924** all models including Turbo '76 thru '82
*80035 **944** all models including Turbo '83 thru '89

RENAULT
Alliance & Encore - *see AMC (14020)*

SAAB
*84010 **900** all models including Turbo '79 thru '88

SATURN
87010 **Saturn** all models '91 thru '96

SUBARU
89002 **1100, 1300, 1400 & 1600** '71 thru '79
*89003 **1600 & 1800** 2WD & 4WD '80 thru '94

SUZUKI
*90010 **Samurai/Sidekick & Geo Tracker** '86 thru '96

TOYOTA
92005 **Camry** all models '83 thru '91
92006 **Camry** all models '92 thru '96
92015 **Celica Rear Wheel Drive** '71 thru '85
*92020 **Celica Front Wheel Drive** '86 thru '93
92025 **Celica Supra** all models '79 thru '92
92030 **Corolla** all models '75 thru '79
92032 **Corolla** all rear wheel drive models '80 thru '87
92035 **Corolla** all front wheel drive models '84 thru '92
*92036 **Corolla & Geo Prizm** '93 thru '97
92040 **Corolla Tercel** all models '80 thru '82
92045 **Corona** all models '74 thru '82
92050 **Cressida** all models '78 thru '82
92055 **Land Cruiser** FJ40, 43, 45, 55 '68 thru '82
92056 **Land Cruiser** FJ60, 62, 80, FZJ80 '80 thru '96
*92065 **MR2** all models '85 thru '87
92070 **Pick-up** all models '69 thru '78
*92075 **Pick-up** all models '79 thru '95
*92076 **Tacoma** '95 thru '98, **4Runner** '96 thru '98, & **T100** '93 thru '98
*92080 **Previa** all models '91 thru '95
92085 **Tercel** all models '87 thru '94

TRIUMPH
94007 **Spitfire** all models '62 thru '81
94010 **TR7** all models '75 thru '81

VW
96008 **Beetle & Karmann Ghia** '54 thru '79
96012 **Dasher** all gasoline models '74 thru '81
*96016 **Rabbit, Jetta, Scirocco, & Pick-up** gas models '74 thru '91 & Convertible '80 thru '92
96017 **Golf & Jetta** all models '93 thru '97
96020 **Rabbit, Jetta & Pick-up** diesel '77 thru '84
96030 **Transporter 1600** all models '68 thru '79
96035 **Transporter 1700, 1800 & 2000** '72 thru '79
96040 **Type 3 1500 & 1600** all models '63 thru '73
96045 **Vanagon** all air-cooled models '80 thru '83

VOLVO
97010 **120, 130 Series & 1800 Sports** '61 thru '73
97015 **140 Series** all models '66 thru '74
*97020 **240 Series** all models '76 thru '93
97025 **260 Series** all models '75 thru '82
*97040 **740 & 760 Series** all models '82 thru '88

TECHBOOK MANUALS
10205 **Automotive Computer Codes**
10210 **Automotive Emissions Control Manual**
10215 **Fuel Injection Manual, 1978 thru 1985**
10220 **Fuel Injection Manual, 1986 thru 1996**
10225 **Holley Carburetor Manual**
10230 **Rochester Carburetor Manual**
10240 **Weber/Zenith/Stromberg/SU Carburetors**
10305 **Chevrolet Engine Overhaul Manual**
10310 **Chrysler Engine Overhaul Manual**
10320 **Ford Engine Overhaul Manual**
10330 **GM and Ford Diesel Engine Repair Manual**
10340 **Small Engine Repair Manual**
10345 **Suspension, Steering & Driveline Manual**
10355 **Ford Automatic Transmission Overhaul**
10360 **GM Automatic Transmission Overhaul**
10405 **Automotive Body Repair & Painting**
10410 **Automotive Brake Manual**
10415 **Automotive Detailing Manual**
10420 **Automotive Electrical Manual**
10425 **Automotive Heating & Air Conditioning**
10430 **Automotive Reference Manual & Dictionary**
10435 **Automotive Tools Manual**
10440 **Used Car Buying Guide**
10445 **Welding Manual**
10450 **ATV Basics**

SPANISH MANUALS
98903 **Reparación de Carrocería & Pintura**
98905 **Códigos Automotrices de la Computadora**
98910 **Frenos Automotriz**
98915 **Inyección de Combustible 1986 al 1994**
99040 **Chevrolet & GMC Camionetas** '67 al '87 Incluye Suburban, Blazer & Jimmy '67 al '91
99041 **Chevrolet & GMC Camionetas** '88 al '95 Incluye Suburban '92 al '95, Blazer & Jimmy '92 al '94, Tahoe y Yukon '95
99042 **Chevrolet & GMC Camionetas Cerradas** '68 al '95
99055 **Dodge Caravan & Plymouth Voyager** '84 al '95
99075 **Ford Camionetas y Bronco** '80 al '94
99077 **Ford Camionetas Cerradas** '69 al '91
99083 **Ford Modelos de Tamaño Grande** '75 al '87
99088 **Ford Modelos de Tamaño Mediano** '75 al '86
99091 **Ford Taurus & Mercury Sable** '86 al '95
99095 **GM Modelos de Tamaño Grande** '70 al '90
99100 **GM Modelos de Tamaño Mediano** '70 al '88
99110 **Nissan Camionetas** '80 al '96, **Pathfinder** '87 al '95
99118 **Nissan Sentra** '82 al '94
99125 **Toyota Camionetas y 4Runner** '79 al '95

Listings shown with an asterisk (*) indicate model coverage as of this printing. These titles will be periodically updated to include later model years - consult your Haynes dealer for more information.

 Over 100 Haynes motorcycle manuals also available

5-98

Haynes North America, Inc., 861 Lawrence Drive, Newbury Park, CA 91320-1514 • (805) 498-6703